Gold in History, Geology, and Culture: Collected Essays

Gold in History, Geology, and Culture: Collected Essays

Published in Commemoration of the Bicentennial of
the First Discovery of Gold in the United States,
Cabarrus County, North Carolina, 1799

Edited by Richard F. Knapp and Robert M. Topkins

Division of Archives and History
North Carolina Department of Cultural Resources
Raleigh
2001

ISBN 0-86526-291-8

Publication of this volume was made possible by funding generously provided by the North Caroliniana Society, the North Carolina Mining Commission, and the Gold History Corporation.

Contents

Maps and Illustrations

Foreword

The twenty-two essays included in this volume originated as papers intended to have been read by their respective authors at a symposium titled "Gold in Carolina and America: A Bicentennial Perspective," which was to have taken place at the University of North Carolina at Charlotte, September 16-18, 1999. In the interest of safety, the symposium was canceled at the last minute because of the threat posed by the impending arrival of Hurricane Floyd (which indeed struck eastern portions of North Carolina on September 16, resulting in disastrous flooding and collateral damage). The abortive conclave, in the planning stage for many months, could not be rescheduled without interfering with various activities and commitments made by the respective participants, most of whom are currently engaged in ongoing scientific and/or academic endeavors. With the essays already in hand, Dr. H. G. Jones, honorary chairman of the American Gold Discovery Bicentennial Committee and former curator of the North Caroliniana Collection, University of North Carolina Library, Chapel Hill, and I recognized the merit of continuing with the planned publication of the proceedings of the symposium. I thereupon offered the services of the Division of Archives and History to oversee the project. Among the twenty-four speakers originally scheduled to present papers, twenty-two assented to that offer, and the following collection of essays is the result.

I wish to thank the North Caroliniana Society and the North Carolina Literary and Historical Association for their generous subsidies in support of the symposium and this publication. I also wish to thank Dr. Richard F. Knapp of the division's Historic Sites Section and Robert M. Topkins and Lisa D. Bailey of the Historical Publications Section for their skilled preparation and editing of this volume. It was worth the wait.

Jeffrey J. Crow, *Director*
Division of Archives and History

Acknowledgments

An anthology involves the work of a multitude of people, each of whom incurs debts to others. As editors we can merely name the people and institutions of most help to our own efforts. We do so with the knowledge that our listing is obviously incomplete in that it fails to acknowledge the efforts of numerous individuals who undoubtedly assisted the respective contributors of essays included in this volume. To all of those unsung helpers, who have made the work of authors and editors far easier, we offer our thanks and appreciation.

Our greatest debts are to Dr. H. G. Jones, author of the introduction to this volume and honorary chairman of the ad hoc American Gold Discovery Bicentennial Committee, and to Dr. Jeffrey J. Crow, chairman of that committee and director of the North Carolina Division of Archives and History. Dr. Jones, who has given his entire life in search and support of the breadth of North Carolina's four hundred years of recorded history, has at the same time spent nearly thirty years fostering interest in preservation of the state's gold-mining heritage. That latter endeavor ranges from his leadership in the 1970s in creating near the actual place at which gold was first discovered in 1799 a major state-operated historic site (Reed Gold Mine) to his recommendation for establishment of (and his very active honorary chairmanship of) the bicentennial committee. During all the intervening years, he has continued to support North Carolina mining history in numerous ways. Dr. Crow has, as leader of the committee and the division, been most helpful to the editors as we have gone about our business and (along with Dr. Jones) extremely patient as the process went on longer than anticipated.

For funding publication of these essays, we owe a particular debt to the North Caroliniana Society, which (at the suggestion of Dr. Jones) made available a special grant ($5,000) to publish the work. Another generous gift ($2,950) for the same purpose came from the North Carolina Mining Commission. The Gold History Corporation, a nonprofit support group that benefits the operations of Reed Gold Mine State Historic Site, backed the project through its sponsorship of the overall gold bicentennial of 1999.

We owe an incalculable debt to the authors of the essays, for without their stalwart efforts there would be no anthology. Each of them labored valiantly to produce an essay easily understandable by lay readers in spite of the fact that many such writings focused on arcane and esoteric topics in the realm of geology and the earth sciences. Their efforts in that regard made our job as editors—and, clearly, laymen—far less stressful. Several contributors went far beyond the call of duty to bring this book to print. Dr. William Bischoff, whose far-ranging essay concludes the volume, included dozens of superb color illustrations along with his submission. Although funding and other restraints precluded using all of them, he selflessly devoted countless hours (and well over one hundred worldwide letters, e-mail messages, faxes, telephone calls, and other communications) to obtaining, verifying, and securing permission to reproduce the images. Dr. Dennis LaPoint, a geologist, initially recruited many of the authors from his knowledge of their work and reputations and frequently assisted the editors in our attempts to comprehend and explain technical matters. Dr. Ray Smith made the whole job of assembling the essays easier with his consistent enthusiasm and support, which also was a crucial factor in the success of the spring 2000 miniconference that substituted for the cancelled bicentennial symposium.

Numerous individuals in North Carolina did much to make it possible for us to complete this venture and through their advice and support made our task easier. Dr. Jeff Reid of the North Carolina Geological Survey provided ready advice on geological matters when Dennis LaPoint was in South America doing fieldwork. Raleigh designer Sharon Dean, as she does with many published products, made our efforts visually attractive for a wide circle of readers. Within the Division of Archives and History, we are grateful to our colleagues Lisa D. Bailey, Rob Boyette, Matt Burton, Justin Chambers, Steve Massengill, Jim McPherson, Joe Mobley, Trudy Rayfield, Michael Southern, Elizabeth Sumner, and Jim Willard for a variety of recommendations, support, technical services, and encouragement during the process of assembling this volume. We alone are responsible for any errors or misstatements of the authors' original intent.

Richard F. Knapp
Robert M. Topkins

Introduction

Commemorations provide opportunities for recalling the past, and the confluence—during the years from 1996 through 1999—of three public anniversaries of gold rushes north of Mexico produced significant recognition of those important milestones in North American history. The commemorations proceeded in reverse chronological order—from the most recent gold rush to the earliest.

The centennial of the Klondike gold rush was observed in May 1997 during a scholarly conference held at an unlikely place, the University of Edinburgh in Scotland. The location had at least a tenuous connection to the Klondike, however, because brothers Colin Coates (director of the Centre of Canadian Studies at Edinburgh) and Ken Coates (then professor at Waikato University south of the equator) were natives of the Yukon and had not seen each other for years. So, with grants available to college professors, why not a reunion in sight of Edinburgh Castle? Even without a grant, I was there to spread the word that the history of gold in North America started in neither California or the Klondike. Yukon College published the conference papers in the *Northern Review* (No. 19, winter 1998).

The Edinburgh conference reminded the world that the Klondike gold rush originated in Canada, not Alaska, so it was natural that the largest state wanted to place an American imprint on the feverish activity that spread across its borders following the discovery at Rabbit Creek in 1896. Accordingly, a much larger and more diverse meeting, the International Symposium on Mining, was held at Fairbanks in September 1997. I was there to rain on the Alaska parade, again recounting events a century earlier in North Carolina. Festival Fairbanks published the edited papers in the 1998 publication *International Symposium on Mining Proceedings*.

The story of California's frantic gold rush, of course, claimed widespread attention on its sesquicentennial in 1998 through numerous programs and especially through a large exhibition mounted at the Oakland Museum of California History with assistance from the British Museum. A traveling portion of the exhibition attracted widespread attention. I took our Tar Heel story to California, first at an elderhostel at Sierra College in the center of gold country, then to the Sacramento meeting of the American Association for State and Local History, which gave equal time to the three states.

North Carolina, in preparing for a climactic bicentennial commemoration in 1999 of the *first* documented discovery of gold north of Mexico, sought to place the subject in its world context. North Carolinians recognized that historians had overlooked this state's role in the history of gold for two reasons: first, the invention of photography was forty years away, so virtually no visuals existed to call public attention to the stirring activities in the Piedmont following the discovery on John Reed's farm in 1799; and second, despite being the first, North Carolina's gold rush was minuscule when compared with those following the California and Klondike strikes. So in commemorating the bicentennial, with only written records to support its claim to fame, North Carolina simply sought to place itself in the progression of gold history, a link in the chain of developments from ancient times to the present—thus consciously living up to its motto, *esse quam videri*—to be rather than to seem.

Happily, the site of that first discovery had been preserved and thus provided the perfect setting for the planned bicentennial. In 1972 the public-spirited owners, members of the Kelly family of Ohio, donated to the state the historic mining portion of John Reed's farm and sold the remaining acreage at a bargain price. This eight-hundred-plus-acre site, within twenty miles of the burgeoning metropolis of Charlotte, came into state ownership little changed from its conditions when John Reed died in 1845. A small grant from the National Park Service (which had already declared the site a National Historic Landmark) was augmented by more generous funds appropriated by the state legislature at the urging of Gov. Robert W. Scott, and a master plan was produced to serve as a guide to the development of Reed Gold Mine as a state historic site. Some of the historical underground workings were opened for public exploration, a visitor center/museum was constructed (its shape suggested by a building at the Giant Yellowknife Gold Mine in Canada's Northwest Territories), and a stamp mill was erected and equipped with historic operating machinery from nearby defunct gold mines. Since its official opening in 1977, Reed Gold Mine State Historic Site has attracted nearly two million visitors.

Plans for an international symposium, titled "Gold in Carolina and America: A Bicentennial Perspective," were developed by an ad hoc American Gold Discovery Bicentennial Committee consisting of H. G. Jones, honorary chairman; Jeffrey J. Crow, chairman; and members Jack Claiborne, John B. Dysart, David R. Goldfield, Richard F. Knapp, Dennis J. LaPoint, Robert L. Remsburg III, and Jo Ann Williford. In this ambitious undertaking, the Division of Archives and History received human or financial support from the Gold History Corporation, the North Caroliniana Society, the North Carolina Literary and Historical Association, the University of North Carolina at Charlotte, Bank of America, and other donors. Even so, most of the work was performed by the home office staff of the Division of Archives and History and its on-site staff at Reed Gold Mine.

On September 14, 1999, nearly two dozen specialists in various aspects of the history of gold were preparing to fly or drive to Charlotte, North Carolina, to speak at the symposium. Their plans were interrupted by urgent telephone calls and e-mail messages from officials of the North Carolina Division of Archives and History with a traumatic announcement: the National Weather Service was predicting that Hurricane Floyd, then heading toward land from the Atlantic Ocean, would pass over the Queen City within two days. Prudence dictated a cancellation of the symposium. Although the hurricane changed course and might not have interfered inordinately with the planned event, it curved easterly and delivered history's most devastating blow to eastern North Carolina. The unprecedented flooding would have prevented even some of the members of the staff of the Division of Archives and History from reaching Charlotte. The decision, therefore, could not be second-guessed. The disappointment of those who had worked so hard to make arrangements for the symposium equaled the inconvenience caused to the speakers and the one hundred fifty or so preregistrants. Fortunately, funding allowed the reimbursement of nonrefundable expenses incurred by the speakers. And, ironically, for hundreds of Carolinians fleeing the approaching hurricane, the cancellation was a blessing, for the hotel space reserved by speakers and participants providentially became available to frantic travelers seeking shelter.

Because of unrecoverable losses resulting from the cancellation, funds were not available to reschedule the symposium. However, the indomitable Raymond L. Smith, who had agreed to serve as keynoter for the ill-fated event, suggested that some of the scheduled speakers might be willing to come at their own expense if a miniconference could be arranged. The result: about a hundred gold history enthusiasts gathered at Reed Gold Mine on March 31, 2000, for a scaled-down symposium at which Smith, Richard Knapp, Elizabeth Hines, Peter Maciulaitis, John Dysart, and Dennis LaPoint made presentations. Joan Antonson of Alaska, who had been scheduled to speak in September, also was present. After a barbecue lunch and a viewing of the exhibits in the museum, site manager John Dysart led participants on a guided tour of the mine, including the operating stamp mill and concentration table. Something indeed had been salvaged, thanks to the generosity of those who traveled at their own expense.

The speakers scheduled for the original symposium had agreed to the publication of their papers in a volume of the proceedings. Even so, demoralization might have short-circuited those plans except for the determination that the hurricane not waste the enormous effort put into the research and writing expended by those slated to speak. Happily, virtually all of them responded to the request for written papers, which the present volume now makes available through the Division of Archives and History's publications program. This is in fact the third publication resulting from the bicentennial, for the excellent *Gold Mining in North Carolina: A Bicentennial History*, by Richard F. Knapp and Brent D. Glass, and an improved and illustrated edition of Knapp's *Golden Promise in the Piedmont: The Story of John Reed's Mine* appeared during the bicentennial year. The bicentennial included as well other new programs, among them two state highway historical markers (one in downtown Charlotte), exhibitions at various museums around the state, two video productions, and a drama produced at the mine and a number of schools in the region. These activities, plus an effective interdisciplinary program in the public schools, heavy visitation at the physical site of the Reed discovery and mining activity, and widespread publicity, are calling attention as never before to North Carolina's role in the history of gold. Many fellow Americans now know about another gold rush—the first one, the one that occurred in North Carolina. The fact that only two (California and Alaska) of the three gold-rush states have been recognized by the United States Postal Service—which issued two special postage stamps—speaks more of political influence than of history.

As with all publications of oral presentations, the papers in this volume vary radically in approach, style, and content. Some are interesting historical summaries; others translate highly scientific data into the layman's language; still others take a conversational approach; and one was a planned after-lunch talk. Together, they demonstrate the diversity of interest surrounding that most alluring of metals, the subject of three major anniversaries during the past four years. The marvel is that Richard F. Knapp and Robert M. Topkins were able to harmonize their format and make them available in this compilation of essays for the convenience of a broad audience.

The value of commemorations such as the gold-rush anniversaries was considered at the 1999 meeting of the American Association for State and Local History in Baltimore. Joan Antonson (Alaska), Thomas Frye (California), and Jeffrey Crow (North Carolina) gave affirmative answers to the session's title question, "Was There Gold in the Gold Anniversaries?" All three concluded that, despite exaggerations and crass commercialism customary with commemorations, headway was made in creating interest in and promoting knowledge of the role of gold in the life of state and nation. *Gold in History, Geology, and Culture: Collected Essays* provides another example of the benefits produced by these commemorations.

H. G. Jones
July 2000

Editorial Method

Most of the authors of the essays that follow are professional geologists or other members of the scientific community. Many of them, accustomed to presenting the results of their experience and research to fellow specialists, have relied upon standard scientific forms of expression and annotation. To the greatest extent possible, the editors have preserved the authors' original style while making minimal modifications to text and endnotes to help lay readers comprehend sometimes highly technical writing. Moreover, the editors have standardized insofar as possible the form of citation employed in the endnotes that accompany most of the essays, although in a few cases the endnotes remain "scientific" in form, as in "(Johnson, 1991)." Finally, some authors have chosen to conclude their essays with a listing of sources instead of endnotes. In the interest of preserving this diversity, the editors have seen fit to allow the various forms of citation and attribution to coexist.

The Lure of Gold Rushes around the World

R. L. Smith

Raymond L. Smith, president emeritus of Michigan Technological University, holds a B.S. in mining engineering from the University of Alaska and both an M.S. and Ph.D. from the University of Pennsylvania. He served in the U.S. Army during World War II, was an assayer in Alaska, and has held numerous positions at the laboratories of the Franklin Institute in Philadelphia. The following paper was to have been the symposium's keynote address.

INTRODUCTION

The term "gold rush" is generally taken to mean the mass movement of people to reported gold fields, on a voluntary basis, over a relatively short span of time. Under that definition, the first gold rush in the Western Hemisphere was triggered in 1693 by a placer strike in Brazil. Historians not specifically concerned with metals sometime refer to gold rushes differently. For example, Hubert Herring wrote: "Spanish America was born of a gold rush, just as were, in later years, Australia, California, South Africa, and the Klondike. . . . Spain's greatest age was . . . the Century of Gold. . . . [Its] Captains moved from conquest to conquest until the flag of Spain was raised over the greatest Empire the world had yet known."[1] The Spanish, in their never ending quest for the yellow metal, had been risking their lives and enslaving the Indians for nearly two hundred years prior to the 1693 discovery. Columbus and those who followed lusted for gold and found some but were generally disappointed. Early occupation of the Caribbean was a failed venture. Within a quarter century the Spanish had decimated the population. The prolific historian Hugh Thomas said: "The Spanish Caribbean in 1518 seemed a ruined place. The Indians washed for gold and died young while the Spanish fed livestock and seduced women. The Crown seemed uninterested except where profit could be obtained to finance the royal venture in Italy."[2]

Spain's "gold rush" in the Americas started when, in 1517, Francisco Hernández de Córdoba set out with three ships and the blessings of Diego Velázquez, the governor of Cuba, to search for slaves and explore the islands to the west. In Yucatán, near what is now the seaport town of Champoton, Córdoba saw pyramids, engaged in battles with the Maya, picked up some gold, and was mortally wounded by an arrow. Those few gold ornaments changed the history of the Americas. It has been said that if the Indians had possessed the slightest prophecy, they would have thrown their baubles into the sea.[3] The discovery of the trinkets led to the appointment of Cortés by Velázquez and ultimately, the conquest of Mexico. Cortés represented the spirit of the time, yet history paints a different canvas. Historian Nigel Davies commented that Cortés was "filled with a fervor to convert the heathens to Christ and to save them from Hell, but intent upon extracting from them on their way to salvation, every ounce of gold they possessed."[4] Pizarro, in his conquest of Peru, was less of a dissembler than Cortés. He told a priest that he was not concerned about the souls of the Indians, saying "I have come to take away their gold."[5]

The Spanish abandoned prospecting and mining in the Caribbean when they discovered that it was easier to simply appropriate and melt the golden ornaments and votive offerings of the Aztecs, Maya, and Inca. With that source gone, the Spanish once again turned to mining. They had been sidetracked by the gold of Mesoamerica and South America—gold that they did not have to mine and extract from other minerals. Instead, they used force, guile, and torture rather than the miner's pan, pick, and shovel. In subsequent years the Spanish located and mined gold but never by the sweat of their own brows. Columbus and his followers changed the history of the Americas in their quest for gold, but there were no true gold rushes in the Americas prior to that of the Portuguese in Brazil.

BRAZIL

The early incursion of Europeans into Brazil was contemporary with the penetration of the Spanish in the Caribbean and is a story of competition among the Portuguese, Spanish, French, Dutch, and, to some extent, the British. It is a tale of enslavement of the natives, their decimation by disease, and the exploitation of their natural resources for the benefit of Europeans—but that was the way things were at the end of the Middle Ages. The church contended for control with the state, individuals

Map supplied by the author.

in power were messianic in religious beliefs, execration of one's enemies was endemic, and the Inquisition was in full sway with its auto-da-fé a popular form of public spectacle.

The opening of Brazil started when a Portuguese fleet anchored at Porto Seguro in April 1500, seventeen years before Córdoba touched on the Yucatán. It had been blown off course during a trading trip to India. The Portuguese had not rejected Columbus because of shortsightedness; rather, it was because at the time of his proposition they already had a fairly clear vision of the way to the spices of the Indies, and it was not west across the Atlantic. Columbus beat the Portuguese to the New World, but they beat him to the place he was looking for. The natives helped load one of their ships with brazilwood, a source of reddish dyes, and that article drew other European traders to the area. Thus, it was not the lure of gold that opened up Brazil, it was trees; and the first gold rush had to wait for nearly two more centuries. Besides, the Portuguese were busy extracting gold from the west coast of Africa as an item of trade—as one example, in 1505, more than sixteen hundred pounds (170,000 doubloons @ 4.4 grams each) from the Cape of Three Points area on the west coast of Ghana.[6]

For the first part of the sixteenth century, the Portuguese slighted Brazil because they were occupied with trade in silk, spices, and jewels from India by way of their newly discovered sea route around Africa. Finally, in 1532, in order to protect Portugal's interests against other Europeans, Martim de Souza founded a settlement in Brazil at São Vicente. It was located on the coast southeast of present-day São Paulo and backed by a strong Portuguese fleet. Even though Spain ruled Portugal between 1580 and 1640, it paid scant attention to Brazil since the primitive natives there, unlike their counterparts in Inca or Aztec nations, did not possess handcrafted gold objects ready to be melted. By 1580 the population of Brazil included about twenty thousand Portuguese, fourteen thousand Negro slaves, and many thousands of pacified Indians.[7]

Brazilwood, cotton, chocolate, tobacco, and sugarcane were the major items of export to the European markets. With the utilization of slave labor, the sugarcane industry flourished from São Vicente along the coast to the north, with Pernambuco (the Olinda and Recife area) eventually becoming the major center. The Portuguese had established a post there in 1521, but the French took it over in 1530, only to lose it two years later to de Souza. The French persisted in attempts at colonization until the early 1700s, when they were finally expelled from Brazil. The Dutch were more persistent. They took over the Brazilian coastal town of Recife in 1630 and then gained control of the Angolan coast, thus cutting off the primary source of African slaves for the Portuguese of Brazil. That action by the Dutch damaged the Portuguese sugarcane industry. The Dutch abandoned their presence in Brazil in 1654 and outmaneuvered the Portuguese by focusing the sugarcane industry in the Caribbean. The foundation for another venture was now in place as the Portuguese shifted emphasis to the search for gold in Brazil.[8]

Brazil's First Gold Rush

There had been minor gold strikes near São Paulo between 1560 and 1597. The strikes were not large, but they intrigued the crown enough so that in 1674 it officially blessed the search for gold. During the 1670s, strikes were made southwest of São Paulo at Paranaguá and Coritiba. In 1693 two expeditions found gold in the upper basin of the Rio São Francisco. One group was led by a priest and the other by the Paulista Antonio Rodriques Arzão, who died shortly after his discovery. Arzão's brother-in-law, Bartolomeu Bueno de Siqueira, made other discoveries nearby and was farming and mining in the area.

Bandeirantes (Portuguese, Amerindians, and mestizos who traveled together under the bandeira, or flag of a leader) were explorers, frontiersmen, and ruthless hunters of slaves. In 1695, during the course of a slave hunt in the Serra do Espenhaco range, Paulista *bandeirantes* stumbled across Arzão's mining location on the Rio das Velhas, a tributary of the Rio São Francisco near Sabará, a town founded in 1674. They took samples south to the town of Taubaté, and the news of a strike was soon out. By 1697 a full-fledged gold rush was in progress. Portuguese clamoring for passage to Brazil had great difficulty finding a ship with room on it. Farming was neglected, and a severe famine ensued in 1701, causing many of the mines to shut down.[9] The Taubaté area became the state of Minas Gerais (General Mines). Rex A. Hudson wrote: "Word of the discoveries set off an unprecedented rush, the likes of which would not be seen again until the California gold rush of 1849. The Paulistas soon found themselves competing for control with adventurers from the Northeast, Portugal and from elsewhere. By 1709-1710 the Vale São Francisco had become a lawless region filled with the dregs of the Portuguese world."[10]

The rush spread south to Ouro Prêto, which is about one hundred miles due north of Rio de Janeiro. Between 1708 and 1710 the Paulistas and interlopers (the Paulistas called them *emboabas*) battled over who was to control the area. The crown interceded, and the Paulistas' control of the mines diminished severely, inasmuch as they were outnumbered by the newcomers. Ouro Prêto became the unofficial capital of Minas Gerais. Mariana, just a few miles east of Ouro Prêto, eventually emerged as the major mining center. São João del Rio, about one hundred miles southwest of Mariana, was founded in 1696, with gold being discovered nearby in 1719. That area became a hub of gold mining. Itabira, a new district northeast of Sabará, was opened in 1720. The disenchanted Paulistas prospected to the west and made important gold discoveries at Cuiabá in 1718, Goiás (west of Brasilia) in 1725, and on the Rio Sarare of the Matto Grosso field in 1734.[11] The gold miners were supported by heavy imports of slaves from Africa. Minas Gerais grew from very few people in 1690 to 30,000 in 1709 and 320,000 (of whom 22 percent were white) by 1782.[12] In 1821 the population was 514,000.[13] M. Schoen and W. Herzberg described the physical conditions of the boom well: "[W]ild boom towns arose in the mountain valleys; Sabará, Mariana, São João del Rio and the great Ouro Prêto (Rich Town of Black Gold). Wealthy merchants built opulent mansions and churches. Crime, gambling, drinking and prostitution ruled the streets."[14]

The steep terrain around the old mining centers was not suitable for a metropolitan area, so in the 1890s the capital of Minas Gerais—Belo Horizonte—was established just west of Sabará.[15] Gold mining caused a shift of political and economic power from the capital of Bahia in the east to Rio de Janeiro in the south. By the middle of the 1700s the boom was in decline, with marginal enduring economic benefit to Brazil. Much of the profit had gone to Portuguese traders who funneled it into European commerce, which helped fuel the Industrial Revolution. In that period, however, Portugal extended the boundaries of Brazil far west of those set by the 1494 Treaty of Tordesillas, which had divided the Western Hemisphere between Spain and Portugal. More important, however, is the fact that, except for minor pockets of Indians, miscegenation and acculturation of three centuries have endowed the people of Brazil with a solid religious base, a unified language, a true appreciation of art and music, and a sense of humor.

Brazil's Last Gold Rush

Skipping ahead in time in order to continue with Brazil, let us examine the last and greatest gold rush in this century—the Serra Pelada. Like most gold rushes, that one was preceded by gold mining throughout Brazil. For example, during the 1800s Indians and especially *quilombos* (runaway slaves) had mined extensively in the western part of the state of Maranhão along the Gurupí, Maracassumé, and Turiaçú Rivers. Several attempts in the 1850s by outside interests to mine a deposit in Maranhão failed. The Urubú Indians did not take kindly to strangers, and it was not until the early 1930s that a very minor rush took place in that area. In 1956 gold was discovered near Santarém in the Tapajós area of Pará, and raft mining rapidly evolved with pumps and diving equipment. The small-scale miner (*garimpeiro*) expanded his mining equipment beyond the pan, rocker, and long tom and became a major producer of Brazilian gold.[16] Just prior to the big Serra Pelada strike, *garimpeiros* were scattered all over Minas Gerais, Pará, Mato Grosso, Roraima, and Amazonas, with a concentration in the far north. The seeds of the gold strike were sown when in the 1960s geologists evaluated the Precambrian shield of Brazil. The geology of the area fits quite nicely with that of Africa, when the continents were one. According to Cleary et al., in 1968 a military helicopter "developed engine trouble and landed on one of the hills in the Carajas range. . . . [T]he geologists on board discovered to their astonishment . . . [that] they were standing on a hill composed almost entirely of high-grade iron ore."[17]

The government-owned company, CRVD (Companhia do Vale do Rio Doce), with the help of United States Steel, examined the area and eventually developed a world-class iron ore deposit. Carl Hogberg of U.S. Steel described an inspection of the district with these words. "I visited the area, flying from Belem and landing on a provisional airstrip on top of one of the ore bodies, then spending several days reviewing and viewing the discovery from a helicopter and landing at several places where the features matched the Serra Pelada (Pelada meaning bald). Shortly after the discovery, our people worked with the government CRVD people preparing concession agreements. This was when the project was under my wing a couple of levels up. Involved were several billions of tons of [more than] 67% iron ore. I had no idea at the time of gold on the property."[18]

Minor gold prospects had been worked in the Marabá area for many years. In late 1979 nuggets were found on a farm about fifty-five miles southwest of Marabá, the area known as Serra Pelada. The farm was owned by Genésio Ferreira da Silva, who built an airstrip and initially assigned mining plots to the *garimpeiros*. By May 1980 there were thirty thousand *garimpeiros* there. That concentration of people, accompanied by the increase in support population in Marabá, alarmed the government, which did not want a repeat of a rather serious communist disturbance in the Marabá area that had occurred about ten years earlier. The military intervened, supported by the mining bureaucracy, which had never approved of the *garimpeiro* system of mining, inasmuch as the industry had not been able to control individual small-scale miners. CRVD was also favorably inclined because it could see the possibility of regaining its foothold on the property.

Because Brazil had never experienced a rush involving thousands of people who were not seasoned miners flooding into an area, army troops took control of Marabá in May 1980. The military set up a store, a post office, and a medical clinic. It allocated claims, enforced the rule that mining could be undertaken only by manual technology, and set a limit of forty thousand on the number of claims to be mined. It also prohibited the use of mercury, which killed fish and poisoned streams, and in addition banned guns and alcohol.

The Serra Pelada rush became highly political in nature because the miners and those who depended upon them in Marabá and nearby represented a solid voting bloc, and in the state elections of 1982 political control of the area turned from the landed oligarchy to the political party backed by the military. The large mineral powers such as CRVD, in alliance with the government mining bureaucracy, feared that the *garimpeiro* system of mining was becoming formally accepted by the government and lobbied to get Serra Pelada turned over to CRVD for mechanized mining. Their lobbying effort almost succeeded but was stymied by a bill approved by the Brazilian legislature allowing the *garimpeiros* to hold forth at Serra Pelada. President João Figueiredo vetoed the bill, and by mid-1984 the pot nearly boiled over, with the *garimpeiros* burning some of CRVD's buildings, staging marches, and holding government people hostages at Serra Pelada. The president backed down and CRVD was awarded sixty million dollars for its mining interest at Serra Pelada, the award to be gained from a gold tax.[19]

The miners excavated an open pit that was more than six hundred feet in depth and a half mile in diameter. Excitement escalated in 1983 when two nuggets were found, one at 79 pounds and the other at 137 pounds. At that time a miner typically carried a ton of ore a day out of the pit, climbing upward on rickety ladders. His tools were the shovel, pick, and sledgehammer.[20] While the rush lasted, it was one of the most remarkable sights of all times, with thousands of men laboring like slaves in a monstrous pit they had carved out by hand. Photographer Jorge Araújo graphically illustrated the working conditions at Serra Pelada.[21] Writer Norman Gall likened it to Dante's Inferno: "an enormous pit in the jungle that once was a mountain but has been transformed in recent years into a teeming hive of formigas (ants)—tens of thousands of half naked men carrying plastic sacks of earth and rock in endless chains up the continually improvised and deepening escarpments of the pit, muddied bodies climbing precarious wooden ladders perched like match sticks on each shelf of pocked diggings."[22]

Because the Serra Pelada phenomenon tapered off within the decade, one might conclude that the last big gold rush had ended; however, as Sutter's Mill was to our Old West, there was more to come. The *garimpeiro* preferred the excitement of searching for gold with a good chance of finding it to any other life open to him. He had the opportunity of working for himself, for someone else on a percentage basis, or as a laborer with a sure financial return. Although the labor of mining was hard and dangerous, most miners earned ten times what they had previously. The numbers are unreliable, but estimates are that between 250,000 and 400,000 *garimpeiros* are scattered in the jungles of Brazil searching for their El Dorado. As Serra Pelada wound down, the action increased on the alluvial beds of the Rio Madiera and Rio Tapajós, tributaries of the Amazon, and also to the north in the state of Roraima.

During that period, for small-scale mining on the stream beds, the technology of the pick, shovel, and pan had been largely supplanted by rubber rafts, scuba gear, water-and-gravel pumps, and portable stamp mills. The Brazilian law requiring *garimpeiros* to use manual methods was not enforced. Bush pilots flew into the jungle with prospectors, equipment, food, liquor, and prostitutes. In some areas the miners have decimated wild game, polluted the rivers, and spread disease among the indigenous people. In the early 1990s an attempt was made to help the Indians repossess their land from the *garimpeiros*, who outnumbered the Indians by six to one. The primary focus of the police effort was in the Roraima area north of Manaus. The major center of the area was Boa Vista, a city of 35,000 people in 1980 that as a result of the gold rush grew to 175,000 by 1989.[23] The police had limited success in removing the miners, inasmuch as they were faced by a powerful political contingent of miners, pilots, and businessmen who resented outside interference and considered their treatment of the Indians to be no different from the way the forty-niners interacted with Native Americans during the gold rush to California. The mining camps of northern Brazil are rough and ready. During 1988-1989 about four hundred miners were killed in gunfights and plane crashes. James Brooke wrote: "Here in Roraima, a territory that is scheduled to achieve full statehood next year [1990], claim jumping, gun fights, and plane crashes in the rain forest are the currency of the day."[24]

Brazil: Summary

Eventually large-scale gold mining may replace the individual miner just as it did in our West; but the *garimpeiro* culture has become a strong influence on mineral exploitation in Brazil and may remain in place for many years to come. Geologic exploration indicates that the potential for new discovery is good. There is an undeveloped gold belt emerging north of the Amazon River and a more defined one stretching east-west all across Amazonia.[25] There has been considerable exploration activity on the eastern end of the Guiana Shield, especially on the Gross Rosebell deposits. Two Canadian companies—Gold Star Resources and Cambior Ltd.—in cooperation with the Surinam government have delineated a strong potential for the recovery of two million ounces of gold.[26]

AUSTRALIA: New South Wales

The first gold rush in Australia took place shortly after that in California. Hundreds of other rushes were to follow as Australia took its place as one of the world's top gold producers. If there had not been a California rush, the first one in Australia would have been delayed indefinitely. The mining laws in the land down under were not favorable to the gold fossicker (prospector): gold on public lands belonged to the Crown; thus, there was very little incentive to search for it. Gold had been found in the early part of the 1800s, but such an event was not particularly newsworthy. The early finds by surveyor James McBrien in 1823, Count Paul de Strezlecki in 1839, the Reverend W. B. Clarke in 1841, mineralogist William Tipple Smith in 1849, and several others were known to the Crown but were not followed up.[27] Literature on the subject often notes that Smith's gold was from one of the original discoveries in New South Wales. Smith bought some gold specimens from a Mr. Trappit, who had bought them from a shepherd. In 1848 Smith sent samples to Sir Roderick Murchison, an English geologist who in turn notified the secretary of state and of the colonies, but the matter was dropped after Murchison told Smith that the

Map supplied by the author.

government did not wish to encourage mining in Australia. Early in 1849 Smith took the gold specimens to the colonial secretary and asked him to show them to the governor; he also pledged that in return for an award he would disclose the location of the find. The matter was never followed up.[28]

At that time the convict population was substantial, with the first census of 1828 indicating 20,930 free whites and 15,668 convicts.[29] Apparently it was feared that if there were rumors of gold, the impressed criminals would dash off and start towns ridden with crime and, furthermore, that the landholders would then lose their source of cheap labor. Representative of such a belief is the oft-quoted story that when the Reverend W. B. Clarke showed his gold specimen to Gov. Sir George Gipps, the governor said, "Put it away Mr. Clarke or we shall all have our throats cut." The eminent Australian historian Geoffrey Blainey pointed out that Clarke himself was the author of that statement and that Clarke saw gold as a moral menace.[30] Nevertheless, Clarke was not above laying claim to a reward for his part in the first discoveries. Simpson Davison summarized the Report of the Committee on the Claims to Original Discovery of Gold Fields in Victoria. The committee cited the work of the Committee for New South Wales, which gave Clarke sound credit as the original discoverer; however, the latter body gave Edward Hargraves the majority of the credit for actually focusing the public's attention on gold.[31]

The discoverer of gold in Australia in the early days was between a rock and a hard place. If he reported the gold, he would be honored as the first discoverer and supposedly would be given a reward. However, if he could keep the find secret long enough, he could dig sufficient amounts of the precious metal to far outweigh any reward he might receive. The problem was that the Bathurst area was covered with herdsmen, and if one happened upon the prospector he could report the gold and get the reward. Moreover, the case might be like that of William Tipple Smith, who reported the shepherd's find and still did not get a reward. In general, when one reads of who discovered gold for a particular strike, the name of the finder derives from committee findings such as the instance cited above. (This subject is discussed more fully in the section on Victoria.)

Predicting what would have happened would be difficult if Australian fossickers had not joined the California rush—three thousand of them between June 1849 and September 1850.[32] Edward Hargraves and Simpson Davison met each other aboard the ship *Elizabeth Archer* in June 1849. When they sailed into the harbor of San Francisco in September, the crew deserted. Flags of many nations flew from other deserted ships. (One of those ships was appropriated and hauled to San Quentin Point as a prison ship, thus laying the groundwork for the founding of San Quentin prison.) Such a scene was to repeat itself in Melbourne harbor four years later. Hargraves and Davison prospected in California together and, according to Davison's account, he told Hargraves of the old gold discoveries in Wellington near Bathurst, which is why Hargraves prospected in that area when he got back home.[33] Hargraves gives a different account of why he prospected where he did upon his return to Australia, saying that he recognized the similarity of the Sacramento gold area to places he knew near Bathurst in New South Wales and predicted that he would be the first to start an Australian rush. His California partner, Davison, reported that Hargraves had bragged in California that when he returned to Australia the queen of England would appoint him one of her gold commissioners.[34]

From his adventures in California, Hargraves learned that a gold rush meant far more than panning riches from a stream and that a more certain way to a fortune was by providing those things necessary to sustain a mining venture, especially when it meant there might be thousands of gold seekers involved. He could see the advantages of promoting a rush and, with the knowledge of alluvial mining he had gained in the California fields, was confident that he would be successful. Once back in Australia, Hargraves tried to inform the public and the government that Australia had economic gold deposits. Because the California rush had increased Australia's shipping trade and pulled the nation out of a severe depression, there was no need for an economic prop; therefore, he could generate little interest. Hargraves wanted to generate publicity about himself so was persistent in promoting the notion of gold in Australia. After finally persuading a backer to lend him enough money to buy a horse, he set out across the Blue Mountains and stopped at an inn in Guyong near Bathurst run by a Mrs. Lister. His objective had been Wellington, near the site of the shepherd's find; however, he noticed some mineral specimens on the mantelpiece of the inn and decided to prospect in the immediate area instead of going on to Wellington. His account of why he did that does not square with facts. Much has been made of this; however, the point is that on February 12, 1851, with John Lister as guide, Hargraves found traces of gold on Lewis Ponds Creek, a tributary of Summer Hill Creek. He wrote:

I told him [Lister] that we were now in the gold fields, and that gold was under his feet as he went to fetch the water for our dinner. He stared with incredulous amazement, and, on my telling him that I would now find some gold, watched my movement with most intense interest. My own excitement, probably, was far more intense than his. I took the pick and scratched the gravel off a schistose dyke, which ran across the creek at right angles with its side; and, with the trowel, I dug a panful of earth, which I washed in the water-hole. The first trial produced a little piece of gold. "Here it is!" I exclaimed, and then I washed five panfuls in succession, obtaining gold from all but one. . . . "This," I exclaimed to my guide, "is a memorable day in the history of New South Wales. I shall be a baronet, you will be knighted, and my old horse will be stuffed, put into a glass cage, and sent to the British Museum!" At that instant I felt like a great man.[35]

Hargraves said that upon returning to the inn he wrote a memorandum of the find to the colonial secretary and gave it to him later. With James Tom and John Lister, he prospected about eighty miles to the Macquarie River and was satisfied by actual trial that at least seventy miles of the area were auriferous. He next went to Wellington to visit a friend, Cruikshank, and found gold on his property. From there he went to nearby Mitchell's Creek and wrote:

It was not difficult to discover where the old shepherd procured his gold, though he had not touched the alluvial soil. I returned by nearly the same route [back to the inn], observing as I rode along a good deal of promising country, which has since proved to be very productive, when there is sufficient water to wash the soil. In the meantime, my two former guides, Lister and James Tom[,] had returned home[,] bringing with them some fine gold which they had obtained on the Turon River, which, from its character, held out prospects of an abundantly rich field.[36]

Hargraves then reported that he taught his guides how to build and use a cradle and that he then returned to Sydney to report the find to the colonial secretary and ask for a reward. In response to his request, he received a letter from the colonial secretary, Deas Thomson, stating that it was necessary for him to reveal the whereabouts of his gold before the Crown could grant his request. Hargraves was faced with a dilemma because he did not have a good prospect. Blainey pointed out that at that time Hargraves had not made a significant strike before leaving Guyong for Sydney but that with the cradle, Lister and Tom found four ounces of gold about two miles downstream from Hargraves's first find. They posted notice to Hargraves, who came back to Guyong, bought the gold, and sent some to the colonial secretary after convincing the discoverers that it was better to report the gold than to try to mine it. He named the strike Ophir and the richest bar the Fitzroy, after the governor.[37] The strike was announced on April 29, 1851, and by mid-June four thousand fossickers were encamped in the Bathurst area of New South Wales.[38] By then the government was very serious about stopping the exodus of labor from the country and was willing to promote anything that appeared to be a strike. Hargraves was a master promoter, and, as R. S. Anderson wrote in 1852, he had shouted, "Gold!" until a rush began.[39] Hargraves did not get to own his own mine but received a reward of £10,000 and fulfilled his long-held wish of becoming a commissioner of mines, just as he had predicted. In addition, he met Queen Victoria.

As the Ophir strike tapered off, new strikes were made all along the Turon River (beginning in June 1851), and the town of Sofala was established. Strikes were likewise made on tributaries of the Macquarie River, the Murrumbidgee River, the Peel River, the Bingara River, the Snowy River, and many other locations in New South Wales (about five hundred separate rushes were recorded in Australia in the last half of the 1800s.) Most of the locations were peaceful, but, as gold dwindled, tempers flared and anger was particularly directed at the Chinese. By 1858 there were two thousand Chinese on the fields around the Turon. Several vicious attacks were made on them in attempts to drive them from the gold fields. They were beaten, robbed, a few killed, and many had their pigtails torn out. In early 1851, at Lambing Flats on the Burrangong field, troops had to be brought in to restore order. The miners even attacked the police; however, order was finally restored. Gold found at Lambing Flats had generally played out by 1876.[40]

A nugget containing 630 pounds of gold (the Holterman Nugget) was found at Hill's End in October 1872, and another with 106 pounds of gold (the Kerr Nugget) was unearthed near Hargraves in July 1851. By 1900 the major mine operating in New South Wales was the Cobar.

Victoria

Hoping to stem the tide of men rushing to the diggings of New South Wales, a group loyal to the budding town of Melbourne formed a gold committee and posted a £200 reward for the first to find gold in the vicinity. Although most historians agree that gold had already been noted in Victoria and that James Esmonds was the discoverer, the 1853 hearings on the subject clearly state otherwise—that the official discoverer was Louis Michel. Modern-day lawyers would have had a field day over the whole affair. So that the reader can draw his or her own conclusions, a partial transcript from *The Report of the Committee on the Claims to Original Discovery of the Gold Fields of Victoria* follows. The committee pointed out that in March 1850

> Mr. W. Campbell discovered gold on the station of Mr. Cameron, of Clunes. . . . The circumstance was narrowly concealed at that time, from an apprehension that its announcement would prove injurious to M. Cameron's run. . . . A party formed by Mr. Louis John Michel, consisting of himself, Mr. William Haberlin, James Furnival, James Melville, James Herdon, and B. Gruenig, discovered the existence of gold in the quartz rocks of the Yarra ranges at Anderson's Creek in the latter part of June, and showed it on the spot to Dr. Webb Richmond, on behalf of the Gold Discovery Committee above mentioned. . . . About the same time James Esmonds . . . set out in the search along with Messrs. Pugh, Byrns, and Kelly and obtained gold in the quartz rocks of the Pyrenees, near Mr. Donald Cameron's station. This was exhibited by Mr. Esmonds at Geelong, to Mr. Clark and Mr. Patterson, on the 5th of July and the general fact of the discovery of gold at the Pyrenees was published in the "Geelong Advertiser" on the 7th. . . . Dr. Georg H. Bruhn, a German physician, in the month of January, 1851 [i.e., before Mr. Hargraves discovery at Summerhill] . . . upon arriving at Mr. Cameron's station was shown by that gentleman specimens of gold at what is now called Clunes diggings. . . . Dr. Bruhn forwarded specimens which were received by the Gold Discovery Committee on 30th Jun, 1851. [He had also told Esmonds of the find] . . . finally James Esmonds, a prospector from the California fields, made a discovery at Clunes in July 1851. . . . The precise locality of Mr. Esmond's discovery was not made known until the 22nd, while Mr.

Campbell and Mr. Michel developed the precise spot on the 5th, but it appears in evidence that previous to the former gentlemen making the exact spot known, Mr. Esmond's party was already at work on it. Mr. Michel and his party have therefore, in the opinion of your committee, clearly established their claim to be held as first discoverers of a gold field in the Colony of Victoria. . . . Mr. Thomas Hiscock, a resident at Buninyong, induced by the writings of the Rev. W. B. Clarke . . . discovered an auriferous deposit in the gully of Buninyong Ranges now bearing his name, on the 8th of August, and he commemorated the fact, with its precise locality, to the editor of the "Geelong Advertiser" on the 10th of that month. . . . Your committee scarcely consider it within the scope of their instruction to pursue further discoveries. But the . . . deposits of Mt. Alexander render it interesting to record that the honour of first finding gold there is claimed by Christopher Thomas Peters . . . on the 20th July, at Specimen Gully. . . . (Signed) A. F. A. Graves

The committee recommended an award of £1,000 each to Michel, Esmonds, Campbell, and Hiscock and £500 to Bruhn for their discoveries in Victoria. It also recommended £1,000 to Clarke and £5,000 to Hargraves because their work led to the Victoria fields. The legislature cut the total amount to £5,000, with about half of it going to Hargraves and the rest evenly divided but with Campbell left out entirely.[41]

The Hiscock find opened Ballarat; and then in December 1851 Henry Frenchman announced the Bendigo field. (Bendigo peaked out in 1872.)[42] Mount Alexander opened in the early spring of 1852, then a rush was made to Omeo in the far east of Victoria and to the Ovens River the following October. November brought in Korong, west of Bendigo. Strikes followed at Wedderburn, Heathcote, Beechworth, Maryborough, and other locations. Those discoveries triggered a rush and an enormous inflow of population that the Crown could not control. Mount Alexander had forty thousand diggers by March 1852.[43] Australian author Paul McGuire wrote: "[I]n the five years '52-'57, one hundred thousand Englishmen, fifty thousand Scots, sixty thousand Irish, four thousand Welshmen, three thousand Americans, twenty-five thousand Chinese, eight thousand Germans, and people from every country on earth poured through Melbourne."[44]

Fossickers thought they could find and keep gold before the government could stop them. The Crown compromised by letting the digger have a small plot of ground (eight square feet), for which he paid a £2 license fee every three months whether or not he found gold. The police collected the fee and made surprise visits to the diggings to see if anyone was cheating. This and other, real, injustices so riled up the fossickers that finally several hundred of them, under the leadership of Peter Lalor, staged a revolt and destroyed their licenses. Their petitions were ignored, so they put together a makeshift fort at Eureka near Ballarat and defied the law. In a fight that ensued on December 3, 1854, about two dozen miners and several police were killed. Lalor escaped with a missing arm.[45] The news was fast to spread, and the populace, believing that unnecessary force had been used, railed against the government. The license fee was reduced to £1 per year. The Eureka affair set into motion numerous changes in Australia's mining policies.

Other riots followed Eureka, but they were directed at Chinese miners because of prejudice and fear of competition. In 1858 there were 147,000 diggers in Victoria, with 34,000 of them Chinese.[46] The Australians borrowed from the 1850 California Foreign Miner's Tax, which had been aimed at Mexicans and Chinese. A £6 annual tax was levied on each Chinese; but the measure failed to satisfy the white miners. A mob raided two camps that housed several hundred Chinese. The rioters pillaged, stole gold, and pulled out the pigtails of many of the Chinese to display them like scalps. As with Eureka, the public and juries sided with the miners, in this case the white miners. While mining declined in New South Wales, it expanded in Victoria, so that by 1900 there were still five hundred to six hundred active mines.[47]

Queensland

Gold had been found in Queensland in the early 1850s, and in September 1858 a rush of about fifteen thousand fossickers flocked to the Canoona cattle run on the Fitzroy River north of Rockhampton. It turned out to be a false alarm, but over the next couple of years there were many minor finds in the Rockhampton area—for example, Clarence River, Peak Downs, Calliope River, Darling Downs, Crocodile Creek, and others. (J. W. Collinson provides an excellent chronological listing.)[48] In the fall of 1867 James Nash found gold near Maryborough about 120 miles north of Brisbane.[49] That discovery precipitated a major rush that opened up the Gympie field, and by 1868 the town of Gympie was a typically raucous Australian mining town. Strikes at Mount Wheeler, New Zealand Gully, and Ravenswood followed in 1868. In August 1872 George Clarke, Hugh Mosman, and a Mr. Fraser announced the Charter Towers field eighty miles southwest of Townsville; by 1900 it was second only to Kalgoorlie in Western Australia in gold production.

The fossickers moved north; and J. V. Mulligan, following up on William Hann's report of gold on the Palmer River, discovered gold on forty miles of the Palmer, and a strike was announced in August 1873. The gold rushes that followed on the Cape York Peninsula were between the Normandy River in the north and the Hodgkinson River in the south. Cooktown was established in late 1873 to serve the gold area about one hundred miles to the west. By 1877 there were 1,500 Europeans and 18,000 Chinese in the gold fields. The area was plagued by cannibals. Holterhouse wrote that although the Europeans were roasted for food, the cannibals preferred the rice-fed Chinese: "Chinese were manna from heaven. Hundreds of them . . . were eaten at leisure." "Rumor was [that] . . . Chinese were taken there [to Hell's Kitchen near the Palmer track] by the dozen and hung on trees by their pigtails until they were needed for killing and eating."[50] There were numerous other mines and rushes in Queensland: the Croydon field in 1885 and, about the same time, one of the most famous mines of all, Mount Morgan, just south of Rockhampton. It paid £1 million in dividends in a single year.[51]

Western Australia

Further discoveries were made in Western Australia, a land that was opened with the founding of Freemantle in May 1829 and Perth in June 1829. By September there were about thirteen hundred settlers in the area. Growth was slow until the first shipload of convicts arrived in 1850; the last shipload was landed in 1868. With the help of convict labor, Western Australia struggled along, and by 1890 there were about forty-eight thousand people in the district. In 1861 the government offered a reward for the first to find gold within 150 miles of Perth; later the area was extended and the reward raised. In 1862 Edward Hargraves, on a government assignment, explored the Darling Range, which in 1869 was once again prospected, but in both cases without good results. Better prospects were found in northwest Australia between 1882 and 1885 by Philip Saunders, Adams Johns, E. T. Hardman, and John Forrest. The reports by Hardman, a government geologist, stimulated Charles Hall to put together a prospecting party. Hall and Jack Slattery made a good strike about 270 miles east of Derby in the far north of Western Australia, and by 1886 a minor rush developed involving about two thousand people.[52] Hall Creek, with the strikes that followed in the north, became known as the Kimberley field. (In modern times it lived up to its name and became a major diamond producer.)

As the gold rush diminished, a fifteen-year-old youth named Jimmy Withnell found gold seventy miles east of Roebourne in 1887, sparking a minor rush that led to the development of the Pilbara gold field with strikes at places like Marble Bar and Nullagine. Areas opened up rapidly: Southern Cross in the Yilgarn field about two hundred miles east of Perth (1888); Nannine, in the Murchison field three hundred miles northeast of Geraldton (1890); and Cue, in the Cue field just southwest of Nannine (1890). In 1892 Arthur Bayley and William Ford made a major strike about a hundred miles east of Southern Cross and mined it until their secret leaked. Once announced on September 17, 1892, Coolgardie was ushered into the mining world as one of the all-time greats. It was just a few miles to the north that Patrick Hannan, Thomas Flanagan, and Daniel Shea discovered what became Kalgoorlie in June 1893. A major rush took place, and eventually the Kalgoorlie reef—a gold-bearing field seven miles long and one and one-half miles wide—was developed. A critical factor in developing the area was that water had to be brought in from the coast. Prospectors depended upon dry blowing and dry panning. The boom was stimulated in 1894 by the Golden Hole, which became the highly promoted but short-lived Londonderry Gold Mine. It was followed by new rushes—Norseman south of Coolgardie and White Feather, or Kanowna, a few miles northeast of Kalgoorlie. Other strikes followed, with mining by large companies expanding so that by 1903 Kalgoorlie alone produced 1,150,000 ounces of gold and became one of the world's leading gold districts. One of the major mining figures in Kalgoorlie in the early 1900s was Herbert Hoover, later an American president, and it was there that he won his first fortune.[53]

Northern Territory

Gold had been known in the Northern Territory as early as 1865; then, in 1872, John Westcott, with a party of nine men, seriously prospected the Darwin area. He noted gold at the Shackle River, Pine Creek, Pam Creek, Margaret River, and later what was to be known as the Port Darwin Camp. A telegraph line had been laid through the north in order to connect Australia with the rest of the world, and, as a result, word of new discoveries, such as that on Pam Creek, was quickly relayed to the public. Minor rushes moved through the Port Darwin Camp, and a few rich surface deposits were located. In 1880 there was a rush to the Margaret River area, with the Chinese dominating. According to E. M. Christie, in 1881 the population of the Northern Territories included 717 Europeans and 3,934 Chinese, and by 1886 there were 7,000 Chinese. Interest in gold dwindled because of the remoteness of the area, disease, climate, flies, and hostile aborigines. Other minerals were to become dominant in the Northern Territory.[54]

Australia: Summary

Britain's Transportation Act of 1784 paved the way for the first shipment of 736 convicts to arrive in Australia in January 1788; sixty-three years later the gold rush had ended the blight of transportation. By 1853 Australia had passed California in gold production. Although gold became a staple in Australia's economy, the initial rushes of New South Wales and Victoria ended within a decade; but in that period the population of the country tripled, and the mix of people changed from a substantial proportion of criminals to one of middle-class workers and businessmen. The economic building blocks for this emerging country were put into place. Melbourne, like San Francisco, became a bustling city. Because of gold, the economy changed dramatically, more roads were built, farming and ranching bloomed, and mining in general prospered. A gold rush that had to be literally forced into being changed Australia forever. Presently Australia remains one of the world's largest producers of gold, but many other metals and minerals make mining in general one of its major industries. In speaking of the impact of gold, author Robert Hughes, an authority on Australia's convict era, wrote: "[I]ts discovery plucked the last rags of terror that clung to the name of Australia."[55] Gold rushes have changed regions and countries, but no rush has ever changed the social and economic fabric of an entire continent. For good measure, the gold diggers bequeathed the term "mate" to the Australian language, and Australian gold brought the clipper ship, the fastest sailing vessel of its day, to the world.

AFRICA: The Coming of the Europeans

Metals have played a significant part in the history of Africa as reflected in the early centuries of Egyptian and Carthaginian civilizations. Later Henry the Navigator (1394–1460) was most responsible for building Portugal's naval prowess and opening the western coast of Africa to the Europeans. In 1471 the Portuguese opened up gold trade with Ghana, which became known as the land of gold. In the southern part of Africa, mining, minerals, and metals became major players on the world scene. South African political historian Geoffrey Wheatcroft wrote: "A poor and unknown country became rich, famous and powerful all because of the mines. Governments were subverted or overthrown and bolstered up because of gold. . . . [T]he country has seen its history changed by mining. . . . All that Africa has been and will become relates directly to this one industry."[56]

By the time the first Europeans touched on the coast of South Africa, the dominant coastal natives were the Khoikhoi (called Hottentots by the Dutch). In 1488 Bartholomeu Dias of Portugal was the first European to make contact with the Khoikhoi at Mossel Bay, just east of the Cape of Good Hope. He was followed by Vasco da Gama, also of Portugal, who rounded the Cape in 1497 and opened up trade with the East. In the ensuing years a number of Europeans visited the Cape area. From then on, the story of southern Africa became one of cattle, wine, ivory, diamonds, gold, politics, power, and conflict among British, Boers, Germans, Portuguese, and natives. It is a story of broken treaties, shifting boundaries, and unscrupulous deals, with a nexus in time with the Pilgrims at Plymouth Rock, for in that year (1620) the British established a resupply station just west of the southern tip of Africa but abandoned it. In 1647 a ship belonging to the Dutch East India Company ran aground at Table Bay, and the sailors from the vessel erected a temporary haven. A report to the company on that episode led to the appointment of Jan van Riebeeck to establish the Fort of Good Hope as a supply station for the company's ships. Van Riebeeck accomplished that task in 1652 and operated the station for ten years before turning the duty over to others.[57]

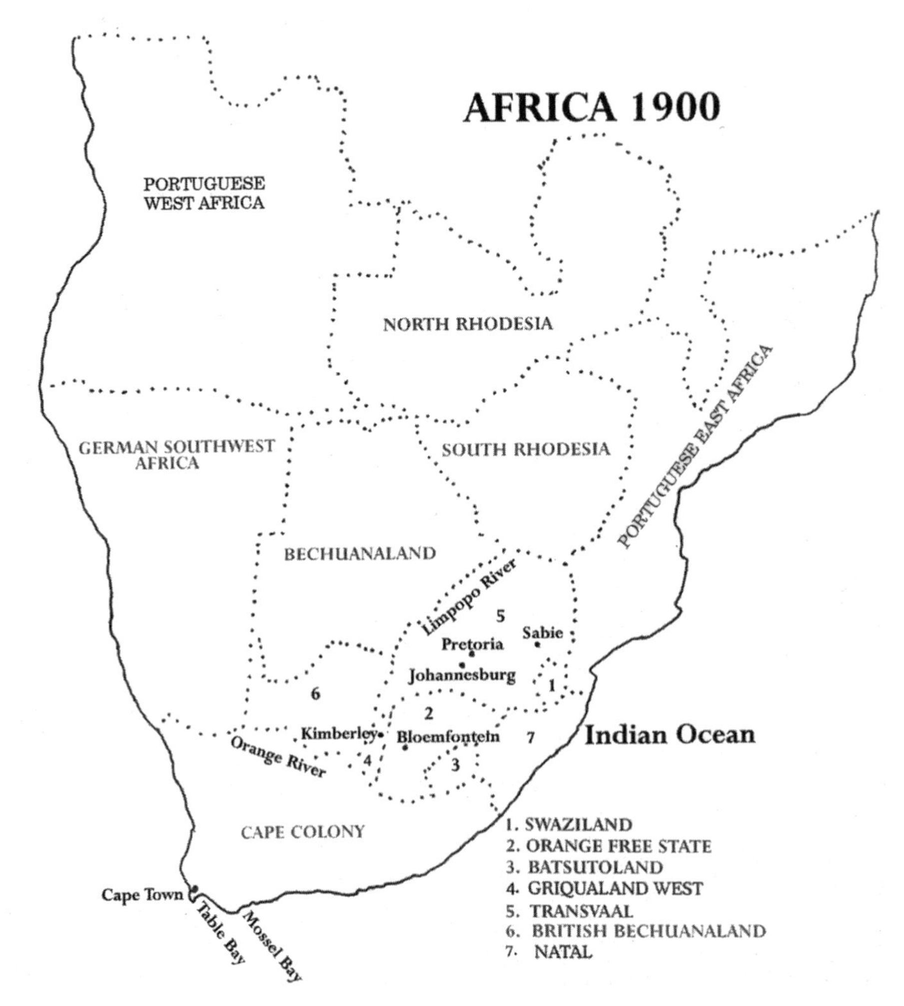

Map supplied by the author.

To save money, the company released a few men to trade and farm on their own and granted them about 28 acres of the Khoikhoi land. Many of those men became prosperous burghers who developed a flourishing wine and maize industry. Not every free white could obtain land grants, so a poor, landless white class known as the trekboers developed and adopted the pastoral life of the Khoikhois. As time passed the trekboers infringed on native lands and stole Khoikhoi cattle. Many of the Khoikhois who had lost their land and cattle became servants to the trekboers. Meanwhile, the prosperous burghers at the Cape acquired slaves from Portuguese slave ships from Angola. The trekboers added to the natives' woes by displacing them as traders with the Dutch East India Company settlement at the Cape. The Dutch contingent grew as the natives were forced away from the coast. That expansion created many years of armed conflict, with the natives at the short end of the stick because of lack of weapons and military skills to match those of the intruders.

North from the Cape to Gold

The Cape economy prospered until the British took over Cape Town forcibly in 1795, returned it in 1803, then took it back in 1806. Britain's vacillation was a fallout from the Napoleonic Wars. Britain received formal cessation of the Cape from the Netherlands in 1814. At that time the slave population equaled that of the Afrikaners (Dutch Calvinists, German Protestants, and French Huguenots). In 1834 the British emancipated the slaves of Cape Colony. To escape British domination and to acquire better pastoral land, many Afrikaners moved north in mass migration. They became known as Voortrekkers (many trekboers joined with them). The Voortrekkers now faced natives other than the Khoikhoi—natives such as the Zulu, the Ngwane, the Xhosa, and the Mfengu. The whites' mode of operation was to defeat the natives, take their lands and cattle, and steal their children.[58] The ensuing decades saw lands to the north and east of Cape Colony change hands as the Voortrekkers, the British, and various native tribes contested control. By the mid-1850s a quasi-peace had set in, with the Boers settled in two major areas—the Orange Free State, between the Orange and Vaal Rivers, and the Transvaal, between the Vaal and Limpopo Rivers—with ill-defined eastern and western boundaries. The Boers had moved out of Natal when the British took over in the 1840s.

The Boers did not want their pastoral lives disturbed, so they suppressed news of gold finds and discouraged prospecting. In 1852 a Welshman, John Davis, discovered gold near what is now Johannesburg and took a sample to the president of the South African Republic, M. W. Pretorius. At that time the Boers of the Transvaal operated under three republics. Pretorius headed one of them, Frans Joubert another, and Stephanus Schoeman the other. The entities combined in the 1860s. Pretorius gave Davis a small reward and kicked him out of the country. About a year later Pieter Marais, a veteran of the California and Australia gold strikes, found gold and applied for a permit to prospect. He was able to extract from the authorities an agreement stipulating that if the prospector disclosed any finds to those other than appropriate government officials, the penalty was death without appeal.[59]

The situation was far different in the western part of the Orange Free State, known as Griqualand West. The diamond strikes in the 1860s opened up that area, with the diamond business forming the basis of the social and economic fabric of Kimberley and providing the legal and technical skills critical for the development of the gold fields yet to come in Transvaal. During that period the diamond area was in contention among natives, the Transvaal, the Orange Free State, and the British. In 1871 the British assumed control of the diamond region and incorporated it into Cape Colony in 1880.[60] In 1877 they annexed the Transvaal. Events of that nature seeded the First Boer War of 1881, which the Boers won easily; as a result, the British gave them self-government.[61]

The Boers of Transvaal had come upon hard times. Their obstinacy concerning prospecting for gold had slowed discovery. But after depleting the treasury in winning the First Boer War, they took a more lenient attitude about prospecting. Rumors had spread of mineral wealth in the Transvaal along a line of hills that became known as the Witwatersrand (White Water's Ridge). Because there were no substantial gold finds in these early days and, moreover, the gold rush of Australia had diverted prospectors away from South Africa, the Boers were able to gain a few years of peace. Then, in 1881, an Englishman, Edward Button, made a strike near Sabie in the eastern end of the Transvaal. He followed this with strikes to the north in the Murchison Range. By the middle of the 1880s the gold mines of the Transvaal were not exciting enough to stimulate a good rush; nevertheless, the reef prospected earlier by Davis eventually became a bonanza. In March 1886 George Harrison prospected in the area in which Davis had found gold and made a substantial strike on an outcropping in the Witwatersrand. He sold his discoverer's claim for £10. It became Johannesburg.[62]

Witwatersrand

Mining diamonds was considerably different from placer mining of gold, so most of the small operators in the Kimberley sold out to venture capitalists who had made their fortunes early in the diamond field; thus, interest started shifting to the Transvaal. Prospectors finally unraveled the geology of the Rand reefs, but it was not until Kimberley diamond magnates such as J. B. Robinson, Alfred Beit, and a few others bought up large holdings on the Rand that a rush got started.[63] But, like the diamond fields, the Rand gold was difficult to mine, since the gold reefs extended for miles and penetrated thousands of feet underground, thus requiring large capital investment. Nevertheless, people swarmed to the Transvaal from all corners of the earth. In 1886 the government laid out the new township of Johannesburg and opened lots for auction. A stock exchange was opened in 1888. By 1889 Johannesburg had a population larger than Cape Town and double that of either Virginia City or the Klondike at their peak. There were so many Uitlanders (outsiders) in the Transvaal with English in the majority that President Kruger became deeply concerned they would dominate the population of the country. He kept strict control and refused the Uitlanders a vote by imposing long residency requirements. This outraged the Uitlanders, who now exceeded the Boer population. The boom continued, and the Rand soon surpassed the United States, Australia, and Russia in gold output. It was called the richest spot on earth. Like all mining boom towns, Johannesburg had its seamy side. In 1899 there were 97 brothels in town staffed by French, German, and Russian prostitutes.

Unrest continued, and in the mid-nineties the Uitlanders formed the Transvaal National Union as a prelude to strength. The wealthy mining industrialists—Cecil Rhodes, Alfred Beit, and Julius Wernher—conspired to annex the Transvaal. Joseph Chamberlain, secretary of state for the colonies, gave them unofficial backing.[64] The British Foreign Office wanted the Transvaal as a colony. The Germans wanted the Boers to remain in power but with Germany's influence expanded. The Portuguese wanted an east-west corridor across Africa. Unrest escalated as the Uitlanders conspired to take power. Mining magnates Cecil Rhodes and Alfred Beit were major players. Dr. L. S. Jameson, the right-hand man of Rhodes, was to organize an army of fifteen hundred men, assemble in bordering Bechuanaland, and, when given the word, head for Johannesburg to join the sympathetic Uitlanders and take control. About forty thousand Uitlanders signed the last petition to the government for their rights. Jameson did not get the go-ahead from the plotters in Johannesburg but started his raid anyway with fewer than five hundred men. It was December 29, 1895. Rhodes had wired him to cease operations, but the telegraph lines had been cut. The Jameson raid ended in absolute disaster at Doornskop. Some of the men were killed, and the rest were marched off to jail. Jameson, with some of the others, was sent to England for trial. They were fined and sentenced to fifteen months in prison. Some ringleaders were likewise fined but in addition were given two years of prison time in Johannesburg, to be followed by banishment. The five top plotters were sentenced to death, among them John Hays Hammond, a famous American mining engineer working for Rhodes, and Frank Rhodes, the brother of Cecil. The death sentence was commuted, and the prisoners were fined and released.[65] Cecil Rhodes stepped down as premier of Cape Colony, but his popularity and influence remained firm. A committee appointed by the British government to investigate the affair instead whitewashed it, prompting the public to characterize the body as the "Lying-in-State Committee."[66]

Africa: Conclusion

Chamberlain appointed Alfred Milner high commissioner to South Africa. Many Boers were disposed to maintaining friendly relations with Great Britain, but Milner encouraged the dissidents. In May 1899 he wired Chamberlain: "the case for intervention is overwhelming." Mines started closing, and refugees left for Cape Colony—twenty thousand by September 1899. The stock exchange closed. The second Boer War began on October 11, 1899, and ended on May 31, 1902, with a British win.[67] A major objective of the war was to gain control of the Rand mines.[68]

By 1905 gold production had recovered. In 1910 the British formed the Union of South Africa out of the Transvaal, the Orange Free State, Natal, British Bechuanaland, and Cape Province. Stability came about in the mid-twenties, with the Union producing half the world's gold. In subsequent years, geophysical prospecting opened huge new gold fields in the Transvaal and also in what had been the Orange Free State. No other region in the world has been so influenced by gold as has the Transvaal. It caused wars and altered the population mix, language, and political control. Unlike other gold rushes, the one in South Africa built an economic prop that has endured for a hundred years.

NEW ZEALAND

In 1840 the British, in an agreement with the native Maoris, declared sovereignty over New Zealand. The agreement paved the way for an aggressive European immigration, primarily by English from Australia and Britain. Officials looked with envy at the California and Australian gold rushes and, as a result, in 1852 offered a reward for a new discovery. Minor discoveries were made, with one at Massacre Bay on the northwest tip of the South Island that attracted a few hundred prospectors. In 1858 came passage of the Gold Fields Act, which defined the respective rights of the miners and the government, allowing for the orderly development of mining.

In June 1861 Gabriel Reed reported a strike in the Tuapeka district of South Island southwest of Otago (now Dunedin). This was New Zealand's first true rush in that it involved thousands of people. Prospectors poured through Dunedin at the rate of two thousand per week. The central government through its Gold Field Act, acting in concert with Otago's provincial government, maintained good control of the strike. Prospectors moved to other locations, but by July 1862 there were seven thousand miners still remaining in Gabriel's Gully.

In August 1862 a new strike was made on the Clutha River near the Dunstan Mountains, just northwest of Dunedin. By October there were 9,000 new arrivals, with additional strikes in the area at the Kawarau River, the Shotover, and Lake Wakatipu. By July 1863 the gold-field population in that area was about 24,000, but 3,000 were drained off to a new strike east of Dunstan near Mount Ida. This was

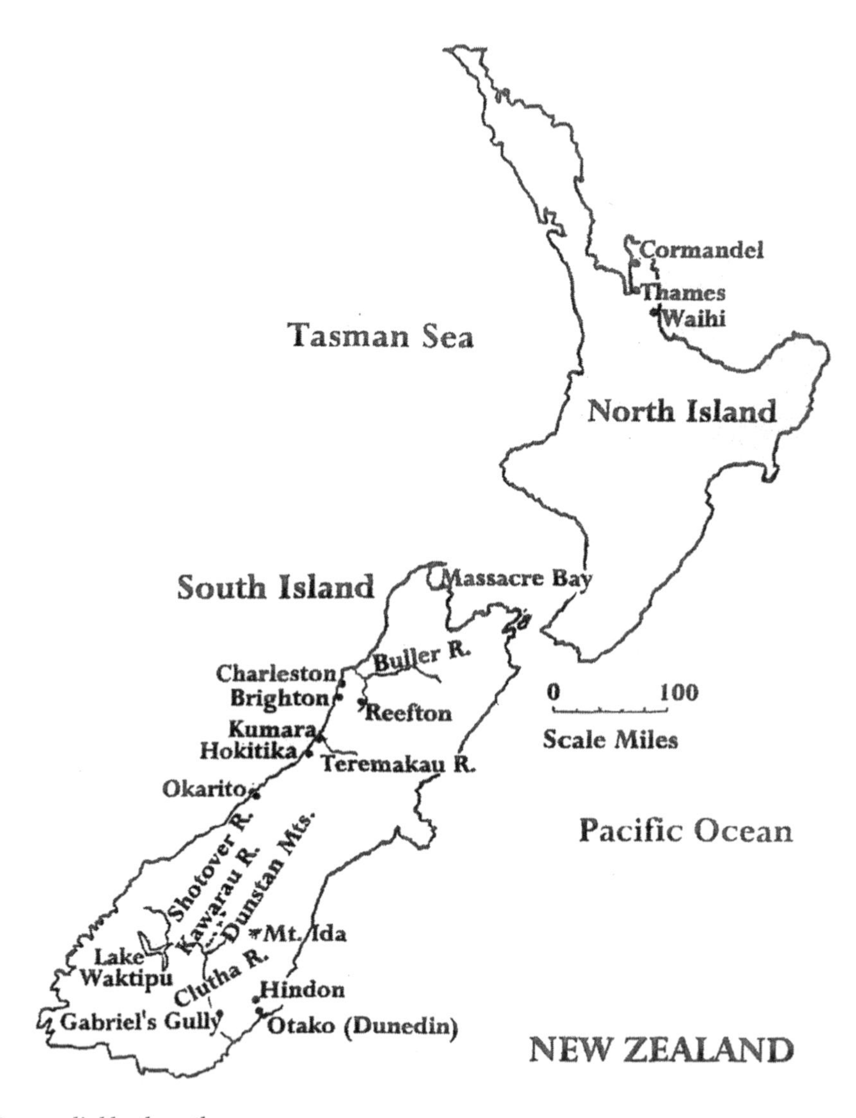

Map supplied by the author.

quickly followed by a rush to Hindon, only twenty miles north of Dunedin.[69] At that time Chinese from the Australian gold fields were joining the rushes. (However, they had sent scouts ahead to determine if pigtail hunting was a favorite sport.)[70]

A small rush to the Teremakau area in 1864 grew to a major one in 1865. Hokitika on the west coast of the South Island became a boom town. An interesting aspect of that west coast rush were the beach sands at Okarito, where three miles of beach had been staked fifty-five miles south of Hokitika. By July 1866 the population of the area had soared to between 25,000 and 30,000 people. To the north, the gold mining towns of Charleston and Brighton were founded with a combined population of 12,000. In May 1867 another rush took place at Buller, close to Charleston. A modest rush took prospectors to the Coromandel Peninsula on the North Island, and by 1868 the Thames goldfield on the peninsula had attracted about 11,500 miners. In 1870 many were drawn off to Queensland, Australia, and back to the South Island, where a rich prospect had been located on the Inangahua, a tributary of the Buller. The mining town of Reefton was established. In 1876 the best alluvial field in New Zealand was opened at Kumara, northeast of Hokitika.[71]

By the early 1900s New Zealand had produced £56 million worth of gold and was still producing £1 million per year, but gold mining had declined sharply. J. H. Curle examined New Zealand's mines at the turn of the century and noted that 91 English mines had been floated, 48 had failed, 2 were absorbed by working mines, 25 were operating but without good prospects, and 16 had good promise.[72] Gold mining was a stable industry for many years and was instrumental in establishing numerous communities in the South Island such as Dunedin. Nevertheless, it never supplanted the pastoral or agriculture industries as key factors in the economy.

SIBERIA: The Early Days

In the thirteenth century, Mongols overran and gained control of most of Siberia. Russian Cossacks under Ermak Timofeevich invaded western Siberia in the early 1580s, and by 1639 the Cossacks had reached the Pacific Ocean. Their brutality spurred the natives to turn to the Chinese for help, and in 1689 Russia lost control of the gold-rich southeastern area of Amur. It was not until 1858 that Russia regained solid governance of the Amur region. But at that time gold was not the main attraction—nothing could compete with the superb and abundant furs to be found there. It was also a place to begin a new life, so exiles and prisoners were a significant proportion of the population. Historian Bruce Lincoln pointed out that although Siberia was a dumping ground for riffraff and criminals, under the Law Code of 1649, it was "also a land of opportunity where men and women oppressed by their masters or broken by misfortune could build new lives. Thus long before the centennial of Ermak's victory, Russian soldiers . . . craftsmen . . . peasants . . . and a Russian way of life were all taking their places in Siberia. . . . Whole caravans, sometimes with as many as a hundred families . . . began to make their journey in two wheel carts, their belongings piled high and their livestock following. . . . [R]unaway serfs crossed the Urals singly or in small groups to build hidden villages in the Siberian wilderness."[73]

Siberian Discoveries

There are numerous references to Siberian gold rushes in the literature, but such events were not true "rushes" in the sense of a very large number of people invading an area spontaneously and willingly upon news of a gold strike. Although gold was a significant factor in opening up Siberia, and prisoners were sent there to mine it, the people who went there voluntarily did so for opportunity and escape. In 1800 three thousand prisoners mined gold in Czarist Siberia, but within a half century the number of free miners stood at sixty thousand.[74]

After 1933, hundreds of thousands mined and searched for gold under the Soviet regime. Substantial gold-mining communities were established, but it was through a government-controlled, steady growth—not at all like the stampedes of California, Australia, or the Klondike. The vastness of the area, with its endless tundra,

Map supplied by the author.

high mountains, cold weather, lack of transportation systems, and tight government control of the land and lives of people, precluded, to large extent, a typical gold rush. Yet, the gold was there and had been known of for many years before the semblance of a rush started. In the period before the Russian Revolution of 1917, gold expeditions into the numerous gold fields of Siberia were usually carried out by small, organized groups of men led by someone such as a merchant armed with appropriate government clearances. Issued at various times were edicts that made prospecting possible; but most of the miners were relegated to the job of a laborer. Nevertheless, mine work improved their lives, and a Siberian mining camp, except for the prison nearby, might resemble the brawling camps of Australia or the United States.

Peter the Great had issued an edict that made prospecting on crown lands possible, but it was more than a hundred years later, in 1824, that individuals were licensed to prospect. Prior to that date, discoveries were made in the mid-1700s at Berezov and Ekaterinburg in Siberia's far west. The latter became such an important mining resource that by 1829 about forty individual working mines existed there. Those discoveries in turn brought many laborers into the area and might be considered as Russia's first gold rush, albeit minor in terms of masses of people. In 1830 a Colonel Begger, working for the crown, made a placer discovery hundreds of miles to the east between the Ob and Tom Rivers. In 1832 a merchant, Riazanov, with his partner, Ataschev, made an important find on a tributary of the Chulym (east of Tomsk) and additional discoveries to the northeast on the Birjussa River. They prospected with great success between the Upper and Stony Tungaska Rivers, which opened up the Yenisei basin.[75] Discoveries were made on the Angara in 1836 and at Yakutsk (about 2,500 miles east of Ekaterinburg) in 1840.[76] Government auditors were sent to the distant mines to investigate cheating. Everyone was encouraged to watch each other and report any misdeeds. Accurate records of gold production were kept; they indicate that production peaked at 962,300 ounces in 1847 but picked up again in 1854 with new discoveries east of Lake Baikal.

Soon after, the richest fields in Siberia were discovered on the Vitim River, a tributary of the Lena River, and on tributaries of the Olekma and Mama Rivers. The city of Irkutsk had served that area since its founding year of 1652 and during its early gold-rush days was known as the Paris of Siberia. After the southeastern area of Siberia was recovered from China, prospecting was allowed in 1865, and discoveries were made in tributaries of the Amur and Amgun Rivers. Gold production in Siberia was slightly short of two million ounces by 1914.[77] In the Amur region, Blagoveshchensk, founded in 1856, was the nearest comparison to an Australian or western United States boomtown. It grew from a population of three thousand in 1880 to thirty thousand within twenty years as a result of a gold strike on the Zeya River. When construction started on the Trans-Siberian Railroad in the late 1890s, the population of Siberia stood at about five million. The road was officially opened in 1900, and by 1914 the population had risen to 12.8 million. Timber, wheat, coal, iron, and butter were major economic drawing cards, but to the investor, gold still held a high priority. In 1908 the major commercial enterprise in eastern Siberia was Lena Gold Fields Ltd., which operated about four hundred placers under atrocious working conditions. More than three thousand miners went on strike there in 1912. Under official orders to liquidate the protesters, soldiers shot more than five hundred strikers, killing half of them. Lincoln, in quoting Lenin, wrote that the Lena incident was to be the "pivotal event in transforming the masses' revolutionary mood into a new revolutionary upsurge."[78]

The Lena massacre was the culmination of years of oppression.The forced colonization of Siberia, administrative exile, and convict exportation were conducted with no regard for human beings. It was a history of Czarist bureaucracy at its worst, mixed with greed, depravity, and indifference to human suffering, as detailed by George Kennan, who in 1885-1886 traveled eight thousand miles through Siberia visiting prison camps. Kennan noted that it was not the mine work so much as the terrible prison conditions that decimated the convicts. In 1857 at the Kara gold mines, owned by the Tsar, more than a thousand convicts died from disease and overwork. In the Nérchinsk mining district in the village of Kavwíkuchigazunúrskaya, a mining technologist told Kennan that the prisons there were "the very worst in the Empire and that the convicts were ill-treated, beaten by everybody, . . . forced to work when sick and killed outright with explosives which the overseers were too ignorant or too careless to handle. . . ."[79]

After the Revolution

Friedrich Engels tried to explain the upsurge in capitalism in the last half of the 1800s to Karl Marx by saying that it was only an anomaly caused by California and Australian gold.[80] Karl Marx, Vladimir Lenin, and Leon Trotsky believed that gold would be virtually valueless under a collectivist government, and by 1922 gold mining had ground almost to a standstill. Lenin said that money would be done away with, and his fellow communists felt that "Gold they would use to pave the streets, gold would be the material of which lavatory pans were made."[81] However, Stalin was a maverick in that respect, inasmuch as he was a serious student of the gold rush that populated the western United States. He was completely familiar with American mining historians such as T. A. Rickard and wanted to stimulate the industry—not for the monetary value of gold but to populate the eastern reaches of the Russian empire, especially on the far southeastern border, where he feared encroachment by the Japanese. He initially faced strong opposition because he was not in full power; however, he prevailed in the late 1920s and formed the Gold Trust of Russia, with a superb engineer, Alexander Serebrovsky, in charge. In order to establish the trust, Stalin had to compromise and agree to the liquidation of prospectors, who were considered capitalists. Serebrovsky visited America and hired the Alaskan mining engineer John Littlepage to reactivate the Russian gold industry. Littlepage reported directly to Serebrovsky, who in turn reported to Stalin. As a result, gold mining became a most-favored activity, especially as dictator Stalin consolidated power.[82] Moreover, by that time most of the Bolshevik bosses knew gold was needed. A gold fever started in late 1929; it did not, however, involve a gold rush but rather a theft of private property, with thousands arrested and their property confiscated because they were suspected of hoarding gold. Those arrested were presumed to have had such an opportunity in their business as retailers or to have earned wages in a craft.[83]

By 1933 Littlepage had made significant progress in revitalizing the gold industry, mastered the language, and gained the complete confidence of Serebrovsky. He was made deputy chief engineer of the Gold Trust—a highly unusual move inasmuch as foreign engineers were usually consultants reporting to bureaucrats who knew little of what they were supervising. Prospecting by individuals was allowed—an incredible deviation of Communist policy when considering the millions of kulaks (displaced peasants) who were torn from their farms to work for the collectivist system. This action changed the whole nature of mining for gold. Littlepage wrote: "Notice was given that men and women of all Soviet races would be eligible to join in the search for gold, that this would be considered one of the most honorable occupations, and that the prospectors and lessees would be richly rewarded for their services to the state."[84]

Lack of communications, bureaucratic stumbling blocks, antipathy, and mistrust of workers made it difficult to put that notion into action rapidly, but it eventually worked. The Soviet government owned all of the land and through the Gold Trust controlled every aspect of prospecting and did not want gold rushes to go uncontrolled, as they had been in California and Alaska. The Gold Trust established a Prospector's Department and ended up with several hundred thousand people working under it. Many of the so-called miners were not prospectors but simply people assigned to known areas to work old mines; but there were sufficient incentives and a degree of freedom for true prospectors to search in undeveloped areas. A prospector had to be very careful because if he received a black mark against him from the Gold Trust, he was washed up. Nevertheless, prospecting by individuals was stimulated and strikes were made, but they did not trigger gold rushes in the traditional sense of tens of thousands of people taking off in one grand event.

Because of its policy of strict control of the people, the Soviet system was not designed to tolerate gold rushes.

Major strikes were made after the revolution near the old mining districts of Lake Baikal in 1928 and later, then in 1929 at Baley, about two hundred miles east of Chita. By 1933 that field was producing one-third of the entire yield of gold in eastern Siberia. Unfortunately, the Baley area is presently plagued by high levels of radioactivity from careless mining of monocite, a mineral that contains uranium and thorium 232. More than 95 percent of the children in the Chita region suffer minor physical deformities, and 54.8 percent of them have speech problems.[85]

The search for gold and the mining of it was a major factor in populating the eastern regions of the Soviet empire. Littlepage commented: "The opening up of Asiatic Russia, through the gold rush and dozen of similar devices for forcing colonization and exploitation, has been personally supervised and closely watched by Joseph Stalin." Yet, it was Stalin who finally had Alexander Serebrovsky—the man most responsible for making Russia competitive in gold production with the United States, Africa, and Australia—arrested and most probably shot.[86] There were no rushes to the gold-rich area of the Arctic to the northeast because that region was under strict government control. The cruelty of the Tsars in operating their mines and the diet of the Japanese POWs on the River Kwai paled in comparison to conditions at the Kolyma gold fields of Siberia's Arctic northeast after 1937.[87]

The Kolyma mines deserve attention when discussing gold rushes. If one were to include in the definition of a gold rush an event during which, in a period of twenty years, hundreds of thousands of people annually left their homes, families, and belongings at a moment's notice to spend either the rest of their lives or from five to twenty-five years in the gold fields, then Kolyma would qualify as one of the greatest gold rushes in history. But instead, Kolyma qualifies as the blackest incident in the history of gold mining, unmatched even by ancient mines in which slaves died by the thousands.

In far eastern Siberia, prisoners (mostly political) arrived at Vladivostok on the Trans-Siberian Railway. The end cars, bearing the corpses of those who had died in transit, were uncoupled, and the half-starved, dehydrated convicts, ridden with lice and many decimated by disease, were held to await shipment north to Nagayeva, a coastal port on the Sea of Okhotsky, and then moved a few miles inland to the transit camp of Magadan, which was built by slave labor. The mines were dispersed north of Magaden between the Indigirka River and the Chukotsky Peninsula but were centered primarily on the Kolyma River and its tributaries. It was a vast gold-bearing area that over the years included numerous workings known as the Kolyma mines. Historian Robert Conquest made detailed estimates and arrived at 125 penal camps in the Kolyma area, with 70 of them for mining gold.[88] Prior to 1937, work in Kolyma under Eduard Berzin, head of the Dalstroy (NKVD's Far Northern Construction Trust, which operated the concentration camps), was regarded as better than at other convict camps. Berzin, along with several thousand of his supposed collaborators, was shot in Stalin's first purge for being too soft on the miners. Berzin's object had been to produce gold, and for that he needed healthy workers. (R. C. Cole wrote me on August 17, 1999: "My own role was indirectly involved . . . when Roosevelt opened diplomatic relations with Russia and Stalin sent endless shiploads of gold ores to the Tacoma smelter to establish dollar credits. At one time there were as many as five freighters anchored off the smelter dock. . . . [H]elp was needed in the assay office . . . my career took a favorable turn.") There was a dramatic change when Berzin was replaced by Carp Alexandrovich Pavlov, and Kolyma soon became one of the most dreaded places in Siberia. Goals for gold production were set higher.[89] Pavlov's solution was to increase the number of counterrevolutionists working in the mines and simply work them to death. The only hope of surviving the concentration camps was to avoid general assignment work in the mines. That meant trying to get posted cutting wood or doing camp duty. Dissident writer Aleksandr Solzhenitsyn wrote that 80 percent of the prisoners worked at general assignment, and they all died.[90]

In the late 1930s about two hundred thousand convicts a year were sent to work in the Kolyma mines. So many died that there were seldom more than about two hundred thousand extracting gold at any given time. Three and a half million died in the Kolyma mines.[91] If the convict failed to fulfill his quota (for example, so many ounces of gold panned), his already starvation ration was reduced. As he starved and produced less, he was given smaller rations until he fell dead at his wheelbarrow or pan. Some overseers were too impatient to wait for the slacker to die. Eleanor Lipper, who survived eleven years in Siberia's camps, wrote that in one year twenty-six thousand prisoners were shot at Serpantinka for nonproduction and dozens of other counterrevolutionary activities such as protesting while being beaten.[92] Robert Conquest points out that that number exceeded "the total executions throughout the Russian Empire for the whole of the last century of Czarist rule."[93] Pavlov's policy reduced gold output to about half of the four hundred tons produced under Berzin, yet Stalin and his underlings could not bring themselves to ease up on the prisoners. Stalin historian Edvard Radzinski wrote: "In Kolyma, in the northeast corner of Asiatic Russia, a godforsaken land of marshland and permafrost, a wild beast called Garanin was let loose as commandant. He used to parade sick prisoners who were suspected of malingering, walk along the ranks and shoot them point blank, while camp guards followed with a change of pistols. The bodies were stacked by the camp gates, and parties of prisoners on their way to work were told they would get the same treatment if they tried slacking."[94] Miners of the world know of Comstock, Ballarat, and the Klondike, but few know of mines like the Eighth Apparatus of Shturmovoy, where the life expectancy of the miner was rumored to be one month.[95]

GENERAL SUMMARY

Gold rushes have changed history. Try to imagine the world today if they had never taken place. Australia simply did not allow a person to dig gold; nor did the early Boers. What would have happened had the United States taken the same path? It did not, so gold mining by individuals on their own claims changed history. Many parts of the world have felt the impact of gold rushes. The United States was unique in its approach to mining—particularly in California, where people from all parts of the globe got off a ship, found the gold fields, and staked a claim, relatively free from government interference, thus setting the stage for others to follow. Of course, if they had never experienced gold rushes, regions such as South Africa, Australia, Brazil, Canada, and the western United States would have developed and prospered just as gold-free countries have; but in most cases they would have been far different from what they are today, just as we, as individuals, are different because of some turn in our lives. A case in point worth pondering is that of the career of Sergie Pavlovich Korolev, a superb Russian engineer. He most likely would have died in the Kolyma gold mines as not even a footnote in history, but he was pulled out of the death mines and moved into his own niche of rocket technology. He was the Soviet Union's NASA, literally creating that nation's entire space program—the ICBM, Sputnik, the first spy satellite, the first man in space, and the first launch vehicle and spacecraft to reach the moon. If Kolyma gold had swallowed Korolev as it did countless thousands of others, space history would have been different. Reflect on the point that America's space program was a direct reaction to Sputnik.[96] Travel writer John Gunther may have overstated the case when he asserted that without minerals, South Africa would be a country of sheep herders and petty agriculturists; nevertheless, minerals still form the bulwark of its economy, and gold has been a leader for the past century.[97]

Who discovered gold in an area is unimportant. Major changes took place when there was enough gold in an area to create a mass movement of people sufficient to influence social factors such as language, religion, politics, miscegenation, and acculturation. For the United States and Australia, changes primarily involved power and politics; for South Africa, language and religion were added features; and for Brazil, it was all of the above. It is superfluous to argue the good and the bad of gold. The fact remains that gold rushes have dramatically affected the denouement of events in world history.

*In deference to other speakers, this paper does not include gold rushes in the United States or Canada. Many nonreferenced parts of the paper are derived from a keynote lecture given by the author at the International Symposium on Mining, held in Fairbanks, Alaska, in September 1997 and published in a continuing series of articles in the *Journal of Metals* (*JOM*) from April 1998 to the present. This paper revises the Fairbanks presentation and the *JOM* articles to cover gold in greater depth and to eliminate sections that dealt with metals other than gold. The sections on Siberia and New Zealand are new.

Notes

1. Hubert Herring, *A History of Latin America* (New York: Alfred A. Knopf, 1955), 74, 198.

2. Hugh Thomas, *Conquest* (New York: Simon and Schuster, 1993), 69.

3. Eric Rosenthal, *Gold! Gold! Gold!* (New York: MacMillan Company, 1970), 71.

4. Nigel Davies, *The Toltec Heritage* (Norman: University of Oklahoma Press, 1980), 18.

5. Albert Marrin, *Inca and Spaniard* (New York: Atheneum, 1989), 65.

6. Pierre Vilar, *A History of Gold and Money, 1450–1920* (London and New York: Verso, 1991), 52.

7. Herring, *A History of Latin America*, 222.

8. Rex A. Hudson, ed., *Brazil: A Country Study* (Washington, D.C.: Federal Research Division, Library of Congress, 1998), 4–22.

9. William Parker Morrell, *The Gold Rushes* (New York: MacMillan Company, 1941), 18–21.

10. Hudson, *Brazil*, 24.

11. Morrell, *The Gold Rushes*, 24–28.

12. Hudson, *Brazil*, 24, 25.

13. Morrell, *The Gold Rushes*, 37.

14. M. Schoen and W. Herzberg, *Brazil: Travel Survival Kit* (New York: Lonely Planet Publications, 1989), 17.

15. David Cleary, Dilwyn Jenkins, and Oliver Marsh, *Brazil* (London: Rough Guides Ltd., 1998), 133.

16. David Cleary, *Anatomy of the Amazon Gold Rush* (Iowa City: University of Iowa Press, 1990), 12, 30, 37, 38.

17. Cleary, Jenkins, and Marsh, *Brazil*, 334.

18. Carl Hogberg, retired vice-president of United States Steel, letter to author, March 26, 1997.

19. Cleary, *Anatomy of the Amazon Gold*, 164–185.

20. Ron Arias, "A Hunger For Gold," *People* 30 (August 1, 1988), 40.

21. Ricardo Kotscho, *Serra Pelada* (São Paulo, Brazil: Editor Brasilia, 1984), 45.

22. Norman Gall, "The Last Gold Rush," *Harper's Magazine*, December 1984, 172.

23. Anders Price, "Blood & Gold: Amazon Agony," *Earth Island Journal* 5 (summer 1990), 36.

24. James Brooke, "This Is El Dorado? Gold and Death in the Jungle," *New York Times*, November 30, 1989.

25. Norman Gall, "Gold Rush in Amazonia," *Forbes*, May 21, 1984, 172.

26. M. Wasel et al., "History of the Gross Rosebell Gold Project in Surinam," *Mining Engineering* 50 (October 1998), 29–38.

27. Vincent Buranelli, *Gold: An Illustrated History* (Maplewood, N.J.: Hammond, 1969), 68.

28. Charles Barrett, ed., *Gold in Australia* (London: Cassell and Company Ltd., 1951), 3.

29. Ross Terrill, *The Australians* (New York: Simon and Schuster, 1987), 66.

30. Geoffrey Blainey, *The Rush that Never Ended* (Melbourne, Australia: Melbourne University Press, 1993), 8.

31. Simpson Davison, *The Gold Rushes in Australia, Their Discovery and Development* (London: Longman, Green, Longman, and Roberts, 1861), 117–121.

32. Brian Fitzpatrick, *The Australian People, 1788–1945* (Melbourne: Melbourne University Press, 1946), 154.

33. Davison, *The Gold Rushes in Australia*, 37–38.

34. Barrett, *Gold in Australia*, 5.

35. Edward Hammond Hargraves, *Australia and Its Gold Fields* (London: H. Ingram and Company, 1855), 115, 116.

36. Hargraves, *Australia and Its Gold Fields*, 118, 119.

37. Blainey, *The Rush that Never Ended*, 17, 18.

38. Jay Monaghan, *Australians and the Gold Rush* (Berkeley: University of California Press, 1966), 171–178.

39. R. S. Anderson, *Australian Gold Fields: Their Discovery, Progress and Prospects* (1852; reprint, Sydney: G. Mackness, 1956), 12.

40. Barrett, *Gold in Australia*, 12–14.

41. Davison, *The Gold Rushes in Australia*, 118–121.

42. Peter Philip, "Reopening the Bendigo Goldfield in Victoria, Australia" *Mining Engineering* 51 (July 1999), 46.

43. Robyn Annear, *Nothing but Gold* (Melbourne, Australia: Text Publishing, 1999), 102, 175–176.

44. Paul McGuire, *Australia, Her Heritage, Her Future* (New York: Frederick Stokes Co., 1939), 248–249.

45. Roderick Cameron, *Australia: History and Horizons* (New York: Columbia University Press, 1971), 185–195.

46. Fitzpatrick, *The Australian People*, 164.

47. Nancy Keesing, ed., *Gold Fever* (Sydney and Melbourne, Australia: Angus and Robertson Ltd., 1967), 249–256.

48. Barrett, *Gold in Australia*, 41–55.

49. Hector Holthouse, *Gympie Gold* (Sydney, Australia: Angus and Robertson, 1973), 2.

50. Hector Holthouse, *River of Gold: The Wild Days of the Palmer River Gold Rush* (North Ryde, New South Wales: Angus and Robertson, 1967), 72, 93, 94, 196.

51. Ernest Scott, *A Short History of Australia* (Melbourne: Oxford University Press, 1916), 225.

52. Neil Jarvis, ed., *Western Australia: An Atlas of Human Endeavour* (Perth, Western Australia: Department of Lands and Surveys in Association with the Education Department of Western Australia, 1986), 44; Barrett, *Gold in Australia*, 60; Blainey, *The Rush that Never Ended*, 159–162.

53. Blainey, *The Rush that Never Ended*, 165–175, 197; Morrell, *The Gold Rushes*, 294–297, 305–308.

54. Barrett, *Gold in Australia*, 69, 70.

55. Robert Hughes, *The Fatal Shore* (New York: Alfred A. Knopf, 1987), 571.

56. Geoffrey Wheatcroft, *The Randlords* (New York: Atheneum, 1986), xv.

57. Dougie Oakes, ed., *Illustrated History of South Africa* (Capetown, London, New York, Sydney: Reader's Digest Association Ltd., 1994), 32, 34, 36.

58. Oakes, *Illustrated History of South Africa*, 146.

59. Rosenthal, *Gold! Gold! Gold!*, 14.

60. Wheatcroft, *The Randlords*, 33–37.

61. Rosenthal, *Gold! Gold! Gold!*, 108, 109.

62. Oakes, *Illustrated History of South Africa*, 200, 201.

63. Robert T. Rotberg, *Cecil Rhodes and the Pursuit of Power* (New York: Oxford University Press, 1988), 196, 213.

64. Leonard Thompson, *A History of South Africa* (New Haven: Yale University Press, 1990), 139.

65. Brian Roberts, *Cecil Rhodes, Flawed Colossus* (New York and London: W. W. Norton and Co., 1987), 216–220.

66. John Marlowe, *Cecil Rhodes: The Anatomy of Empire* (New York: Mason and Lipscomb, 1972), 242.

67. Thompson, *A History of South Africa*, 140.

68. Thomas Pakenham, *The Boer War* (New York: Random House, 1979), 451, 501, 606.

69. Morrel, *The Gold Rushes*, 261–270.

70. Robin May, *Gold Rushes* (New York: Hippocrene Books, 1978), 56.

71. Morrell, *The Gold Rushes*, 273–280.

72. J. H. Curle, *The Gold Mines of the World* (London: Waterlow and Sons Ltd., 1902), 198.

73. W. Bruce Lincoln, *The Conquest of a Continent: Siberia and the Russians* (New York: Random House, 1994), 88, 164, 188, 212.

74. Lincoln, *The Conquest of a Continent*, 186, 188.

75. Morrell, *The Gold Rushes*, 48, 49.

76. Lincoln, *The Conquest of a Continent*, 186.

77. Morrell, *The Gold Rushes*, 58, 60, 64, 69.

78. Lincoln, *The Conquest of a Continent*, 233, 243, 268, 271–274, 352.

79. George Kennan, *Siberia and the Exile System*, 2 vols. (New York: Century Co., 1891), 2:148, 282–283, 307.

80. Fitzpatrick, *The Australian People*, 156.

81. Edvard Radzinsky, *Stalin* (New York: Doubleday, 1996), 123.

82. John D. Littlepage and Demaree Bess, *In Search of Soviet Gold* (New York: Harcourt, Brace and Company, 1937), 23–30.

83. Aleksandr I. Solzhenitsyn, *The Gulag Archipelago* (New York: Harper and Row, 1973), 52.

84. Littlepage and Bess, *In Search of Soviet Gold*, 116, 120, 123, 129.

85. *www.igc.apc.org/ei/baikal/supp95lb.html*, October 29, 1998.

86. Littlepage and Bess, *In Search of Soviet Gold*, 278, 307.

87. Lincoln, *The Conquest of a Continent*, 339–341.

88. Robert Conquest, *Kolyma: The Arctic Death Camps* (New York: Viking Press, 1978), 215.

89. Vladimir Petrov, *Soviet Gold* (New York: Farrar, Straus and Co., 1949), 244.

90. Solzhenitsyn, *The Gulag Archipelago*, 564.

91. Conquest, *Kolyma: The Arctic Death Camps*, 227.

92. E. Lipper, *Eleven Years in Soviet Prison Camps* (Chicago, Henry Regnery Co., 1951), 107.

93. Conquest, *Kolyma: The Arctic Death Camps*, 229.

94. Radzinsky, *Stalin*, 413, 414.

95. Petrov, *Soviet Gold*, 384.

96. James Harwood, *Korolev* (New York: John Wiley and Sons, 1995), 2, 3, 209, 215.

97. John Gunther, *Inside Africa* (New York: Harper and Brothers, 1955), 545.

The Geology and Science of Gold

P. Geoffrey Feiss

Dr. Feiss is professor of geology and dean of the faculty of arts and sciences at the College of William and Mary, Williamsburg, Virginia. He holds degrees in geology from Princeton (B.A.) and Harvard (M.A., Ph.D.). He has taught at Harvard, Albion College, and primarily the University of North Carolina at Chapel Hill (where he was also senior associate dean) before going to William and Mary.

Introduction

The papers in this conference volume, which celebrates the wholly accidental discovery of gold in North Carolina two centuries ago, cover a wide range of topics, including the history of mining, social history, and the geology of gold, with specific reference to the deposits of the southeastern and other areas in the United States. The charge to this author from the conference organizers was to describe the geology and "science" of gold for a diverse audience that might have interests in any and all of the previously mentioned subjects. That seems a tall order. The literature on both subjects, especially the geology of gold, is vast and extensive. The author's quandary is: where to start and who to please?

I have decided to start at the beginning—of the universe, that is—and to seek a middling level of detail that is unlikely to please anyone as a result. Nonetheless, I plunge on. The paper that follows is organized along five themes:

(1) What is gold?
(2) Where does gold come from?
(3) Where is it found in the earth?
(4) How does gold behave?
(5) Where do we find it?

I expect that there will be readers of this volume more interested in a satisfying technical presentation of these subjects. I believe such readers will find R. P. Foster's *Gold Metallogeny and Exploration* (1991) to be a useful starting point. In addition, a visit to your local university library, where you can access the GeoRef searchable database, will readily provide a surfeit of references to satisfy the most curious of students of gold "science" and geology.

What Is Gold?

Gold is a metal, one of many metals. Our daily life is filled with metals, and most of us can classify a wide range of materials intuitively as metals. We do this based on the physical properties of those materials. Metals, in our daily experience:

- are lustrous and shiny (they reflect light)
- have high electrical and thermal conductivity
- are malleable (can be hammered) and ductile (can be drawn into wires)

On a less experiential basis, they also:

- display photoelectric and thermionic effects
- are dense, with high boiling points
- are electron donors and thus form positive ions

Chemists tell us that metals are a group of chemical elements characterized by metallic bonding. In spite of the tautological-sounding character of that definition, it is important. All the metals share a common electronic structure such that outer valence electron orbitals, typically the *s* orbitals, are unfilled. This means that they have low, positive valence states, often +1 or +2. But, more important, in their molecular, elemental state, the energy differences between these unfilled outer valence orbitals become very small and become even smaller as more atoms are bonded in the lattice structure. As a consequence, the electrons in these unfilled orbitals are free to move throughout the structure of the material, though only within tight energy bands. A not wholly fanciful analogy is to think of metals as a superstructure of metal cations (positively charged ions) surrounded by a sea of electrons. That analogy is useful if one recalls that this "sea of electrons" analog is applicable only to the small subset of electrons "resident" in the outer, unfilled valence shells and that these electrons occupy specific, quantized energy bands.

This "sea of electrons" accounts for most of the physical properties enumerated above. The mobility of electrons can explain conductivity, photoelectric and thermionic effects, and the opacity of metals. Malleability and ductility are a consequence of the fact that there is no anisotropy, no directional character to metallic bonds—they are as strong in one direction as another: the environment of a cation in the lattice is the same before as after deformation. Metallic bonding also explains the near ideal mixing of one metal with another of similar valence electron structure to form alloys such as stainless steel or gold-silver mixtures.

Gold, then, is a metal—one of many metals. It has atomic number 79 and an atomic weight of 196.97. Gold's electronic configuration is:

$$\mathrm{Xe}\ 4f^{14}\ 5d^{10}\ 6s^{1}$$

where Xe is the xenon core structure of the noble gases. Along with copper, gold is the only metal to have color in its elemental state, and that color is so well known that we use the name to refer to it. Gold's malleability and ductility is legendary. It can be patiently hammered into foils 0.000001 cm thick (300 atoms thick) and drawn into wires .005 cm in diameter. It is very dense, almost twice the density of lead. It is rare, thus expensive, and carries a magic property unique to chemical substances.

The Origin of Gold

Of course, readers of this volume share a common interest: gold. More specifically, many are interested in where gold can be found. And, to address that matter intelligently, it might be useful to know where gold came from in the first place.

Gold is rare. We knew that, but what I mean here is that gold is truly rare. It is not only rare on Earth; it is rare in the universe. Ninety-nine percent of the stuff of the universe is hydrogen and helium. If you were to count atoms in the universe, you would likely count a trillion, that's 10^{12} atoms of hydrogen for every atom of gold. To put that ratio in perspective, if you were the cosmic census taker and could count an average of one atom a second, you would encounter a gold atom about once every 31,710 years. Or, if you were patient and avaricious enough to use such a methodology to sort through the universe for gold, it would take about 752 x 10^{25} years to amass an ounce of gold. By the way, that is more than a quintillian (10^{15}) times the age of the universe.

Where did even this little bit of gold come from? The easy answer is "from the interior of stars." Our sun is a hydrogen-burning star whose internal temperatures of tens of millions of degrees Celsius allow hydrogen atoms to fuse into helium atoms and release enormous amounts of energy. Larger and larger stars "burn" heavier and heavier elements until the very largest, those at least twenty times the size of the sun, "burn" silicon and other mid-range elements to make iron. Iron has atomic number 26; gold is 79. So we are still, apparently, even in these silicon-burning stellar bodies, a long way from making gold. But in the last days of these giant stars, at temperatures approaching 3 to 6 billion degrees Celsius, gold along with all the other heavy elements of the fourth through seventh rows of the periodic table are produced.

How? As a silicon-burning star runs out of fuel and thus out of energy needed to sustain its form and volume, it begins to collapse. This implosion produces endothermic nuclear reactions, which rob the center of the star of energy and convert the star's core to a plasma of neutrons, electrically neutral particles. These neutrons begin to stream outward, bombarding the surrounding shell of iron nuclei like shot from a hundred impatient hunters as the first duck of the season comes in for a landing. The iron nuclei admit neutrons until they become unstable, emit an electron, and, lo and behold, become cobalt.

$$^{56}\mathrm{Fe} + 5n \rightarrow {}^{61}\mathrm{Fe} \rightarrow {}^{61}\mathrm{Co} + e^{-} + v$$

And then, the cobalt nucleus has the same experience. It is bombarded by neutrons and transforms to nickel and so on and on until the alchemist's dream is realized and gold is made. All of this is happening in a relatively short time in astronomical terms because no sooner have the heavier elements been produced in this collapsing, but now unstable, star than the star explodes—spewing its new elements into the universe.

This final explosion lasts days at most—we call it a supernova. Six of these supernova events have occurred in our galaxy alone in the past millennium. We have observed supernovas. In 1987 astronomers at the University of Toronto photographed Supernova 1987A in the Large Magellenic Cloud—a neighboring galaxy to our own Milky Way. Supernovas have been seen before—for example, the famous one in A.D. 1054 that was as bright as Venus and noted by observers in China, Europe, and the Americas. But the importance of 1987A is that one day prior to its discovery by optical means, a brief flash, thirteen seconds long, of neutrinos was detected in two separate locations in Japan and the United States, as predicted by the model of stellar nucleosynthesis described above.

In fact, not only do we have a sense of how often these events occur and that they follow our models of stellar evolution pretty well; we even have a sense of how a supernova disperses its contents. Nearly a thousand years after the A.D. 1054 supernova, a stellar explosion that occurred about 370 million A.U.s (astronomical units;

1 A.U. = 90 million miles, or the distance from the Earth to the sun) from our solar system, or right around the corner in our galaxy, we can study its aftermath—the Crab Nebula, a hazy mass of stellar debris scattered over a patch of space with a diameter about 400,000 times the distance from the Earth to the sun. There's gold in "them thar" interstellar clouds. And, thus, the gold we seek is literally the product of the dying gasps of ancient stars. Each atom was created on a billion-degree anvil of thermonuclear reactions in stars no longer in existence.

Where Is Gold Found on Earth?

Everywhere and nowhere. Gold atoms make up only one part in a trillion in the universe. The good news is that gold is one part in a billion (1 ppb) in the Earth. Thus, the collapse of the solar nebula that resulted in our solar system some 4,500 million years ago managed a thousandfold concentration of gold on Earth over gold in the universe. We still need at least another thousand- or ten-thousand-fold increase in abundance—from 1 ppb to 1-10 ppm (part per million)—to permit the economic recovery of gold.

But perhaps the *average* gold content of the Earth is misleading. Just as the average percentage of the U.S. population over sixty-five years of age is 12.6, we might not be surprised to learn that in Maine it might be 8 percent, while in Arizona it might be 18 percent and, in Sun City, Arizona, 98 percent. People who are mobile, even older people who might seem less mobile than the young, migrate in response to economic pressure and variations in temperature. Are there geologic and geochemical pressure and temperature variations that would allow gold to migrate? The answer is obviously yes, or we would never find a gold deposit in the first place. Before we find the gold equivalents of Sun City, what are the gold equivalents of Arizona and Maine—the geological environments in which gold is more or less likely to be found in greater abundance than average. Some generalizations:

- Gold is more abundant in the Earth's crust than in the mantle (2 ppb v. 1 ppb).
- Gold is more abundant in mafic igneous rocks than in felsic igneous rocks (2-4 ppb vs. 1.5 ppb).
- Gold is more abundant in rocks from orogenic belts than those from anorogenic belts.
- Gold is more abundant in old (Archean) rocks than in young ones.
- Gold is more abundant in coarse-grained detrital sediments than in fine-grained or chemical sediments.

All that said, these generalizations are of remarkably little value in finding gold deposits. That might seem contradictory until one realizes that these are averages, statistical norms. Gold deposits are anything but normal. Gold deposits are found when conditions deviate from the norm, when a sequence of rare and unusual geological events occurs to concentrate a rare and dispersed element into localized concentrations. Ore deposits, especially gold deposits, are the end product of a series of low-probability events—understandable, even predictable, but low-probability. It's not that we're looking for a needle in a haystack; we're looking for one needle in a million in a haystack.

Geology is like that. Averages can be deceiving; it is the singular events that are often important and often have the most lasting impact on Earth systems. Ask anyone who has experienced a hurricane or an earthquake about that.

How Does Gold Behave?

Remembering that gold is a metal, we know that it has the properties described in a previous section. We should add one other important property of metals for geological purposes: substances characterized by metallic bonding are, like covalently bonded compounds, relatively insoluble in dilute aqueous solutions—such as most surface waters on the Earth. Gold also belongs to a class of metals called noble metals. So too do silver and platinum. What does this mean? In people this means distinguished, exalted, magnificent—all of which gold is. But chemically it means chemically inert—but inert in the special context of difficult to oxidize.

Oxidation, at first pass, is the tendency of a substance to react with oxygen. Thus, iron in contact with the atmosphere readily oxidizes to form rust—in its simplest form, an iron oxide called hematite, Fe_2O_3. To a chemist, oxidation has a more generic meaning: the tendency of a substance to give up electrons and form positive ions, cations. In the presence of oxygen, a ready electron-receptor, oxidation results in the formation of oxide minerals such as hematite.

Gold and the other noble metals do not readily form oxides. They are insoluble in dilute solutions of water. Thus, gold is relatively inert in the Earth's oxygen-rich atmosphere and surface hydrosphere. This, of course, is a positive attribute of a metal whose warm color and easy workability make it attractive and desirable as jewelry and coinage. We want our body adornments to retain their color and mass. We also are eager to see our coins, our wealth, retain their mass through the vagaries of weather, burial, and sequestration.

But this very stability would seem to present a geochemical conundrum. If gold is rare, found only in the parts-per-billion range in common rocks and minerals, and we need gold to be naturally concentrated at least a thousandfold before we can mine it, then we need a concentrating mechanism. How can this happen if gold is inert, immobile, noble? The answer lies, clearly, in the fact that the Earth's surface, with abundant dilute water and oxygen, is only one geological/geochemical environment in which gold may be found. In the Earth's interior are hypersaline systems within which high temperature conspires with variations in depth and pressure to produce acidic brines most unlike ground or surface waters. The rotten-egg odors encountered in a walk over a geyser basin in Yellowstone Park or by a thermal spring in the Imperial Valley will convince you that there are abundant terrestrial environments rich in sulfur gasses. Everyone has seen the photos of the chimneys of silica, gypsum, pyrite, and other metal sulfides that precipitate on the ocean floor from fluids exiting the mid-ocean ridges. Tiny bubbles of fluid trapped in minerals demonstrate to us that the upper portions of the Earth's crust contain hot aqueous fluids rich in carbon dioxide, sodium chloride, and methane—and no oxygen.

It is within these environments that we need to test the stability/mobility of gold. In fact, when we sample fluids from some of these environments, we find that the nobility—the inertness—of gold is not quite what it seems at the Earth's surface. Geochemical modeling of brines from 21°N on the East Pacific Rise and from the Axial Seamount have shown that they are capable of precipitating sulfides with 200 to 4,900 ppb gold (Hannington et al., 1991); those sulfides would be ore were they not located on the bottom of the ocean. Experimental results, as well as such direct analyses, reveal that at elevated temperatures of from 200° to 450°C—temperatures readily accessible in deep basinal brines, in geothermal fields, and in aqueous circulation systems surrounding volcanoes and deeper igneous intrusions—and in the presence of low pH, abundant chloride, sulfur, and/or methane, gold is soluble (Seward, 1991). Gold is soluble as gold complexes with chloride and sulfur, the sulfur primarily as bisulfide (e.g., $Au(HS)_2^-$). And if gold is soluble, it can move with hot circulating, anoxic aqueous fluids—what we often call hydrothermal solutions—that flow through the cracks and pores of rocks in response to a host of dynamic geologic processes.

In general, then, we expect to see that highly dispersed gold and other metals that come in contact with hydrothermal and other solutions of the right composition are dissolved from the rocks in which they are found in very low concentrations and transported in solution along thermal, pressure, and chemical gradients. At some point along the way there occurs a change in fluid composition as a result of reactions with surrounding rocks or mixing of other fluids or a decline in temperature, or the occurrence of an irreversible physical process such as boiling, which alters the conditions that favor the solubility of gold. For example, boiling may strip off the sulfur gasses that keep pH low and enhance gold solubility (the rotten-egg smell), or oxygenated ground waters may mix with the hot gold-bearing fluid to cool it, producing oxygen-rich surface-like conditions where we know gold to be insoluble. If this occurs in a small enough volume of the Earth's crust, we have the capacity for concentrating relatively large amounts of previously dispersed gold into a valuable deposit.

If such an occurrence seems unlikely, it is. That is why we do not find gold deposits everywhere. But given 4.5 billion years of Earth history and the expanse of the Earth, unlikely things happen. If gold is so scarce, it also seems unlikely, perhaps, that we can in fact produce large volumes of it. But low concentrations in a large enough volume of rock can be a source if the extraction process is efficient. It is not difficult to show that the entire annual production of gold in the U.S. could be leached from a 20 km cube of rock that initially contained only 1 ppb of gold.

Where Do We Find Gold Deposits?

Geographically, gold is mined on every continent save Antarctica—and very likely would be mined there were it open to mineral entry. The Americas and Africa have traditionally dominated gold production, but gold is mined in Europe, Asia, and Australia and increasingly in the islands of Oceania as well. The widespread geographic extent of gold mining would imply to a geologist that economic concentrations of gold must be possible in a variety of geological settings, and that interpretation would be correct. With a few notable exceptions, such as rocks derived from passive continental margin sedimentation and intraplate volcanism, there are few geologic settings in which gold exploration has not been successfully pursued.

As has been noted, "[t]here are hundreds of thousands of reported gold occurrences around the world (Keays and Skinner, 1989)." Many are small; many are of little note. Every gold-hunter wants to be able to distill from these occurrences one or more generalizations that will allow him or her success in finding the next one. It's not that easy. While there are useful rules of thumb (for example, hydrothermal gold is almost always found with pyrite, and very little gold is found in rocks formed between the Archean and the Mesozoic), none of them are very helpful guides in specific instances. Pyrite is ubiquitous. Pre-Mesozoic but still Phanerozic gold deposits are just as valuable as others. If the latter were not true, we would not be publishing this volume in honor of a gold discovery in Cambrian host-rocks.

So the goal of this paper is not to provide the prospector's guide to finding more and richer gold deposits. Rather, in the remainder of this paper, I will briefly highlight six of the world's more important styles of gold occurrences in hopes that the reader might gain a sense of the diversity of geologic settings that have enriched gold-hunters over the centuries. Those seeking more information might start with the following references in the bibliography: Cox and Singer (1986), Bonham (1989), and Foster (1991). Out of consideration for the local audience, I have included examples of southern Appalachian gold deposits when in fact there are regional equivalents of the types I describe.

1. Quartz-lode Deposits

Veins of milky quartz with gold and metal sulfides are common in many metamorphic terranes worldwide. Often the precursors of the host rocks were volcanic deposits or volcanic sediments deposited in greenstone belts. An association with late granitic magmatism can result in veins that cut meta-igneous rocks as well. The veins themselves can be centimeters to meters thick, often found along fault sets or major structural breaks. Some veins can pinch and swell or form complex stockworks; others are regionally persistent. The veins often contain free-milling gold from dust to nugget size in association with base metal sulfides and sometimes tellurides. Richest lodes are often at vein intersections or where veins cut lithologic contacts. In general, fluid salinities are low and contain mixtures of CO_2-H_2O. These gold-bearing quartz lodes appear to have formed in the latest stages of post-orogenic metamorphic events. A good example of this type of deposit is the Mother Lode district of California or the gold deposits of the Abitibi greenstone belt in Ontario and Quebec. Small, gold-bearing quartz lodes in the South Mountain district of North Carolina and the Piedmont were mined in the southeastern U.S. as well.

2. Volcanic-hosted and Hot-spring Deposits

In the upper portions of many continental and continental margin stratovolcanoes, as well as in the geothermal fields along their flanks, the heating of meteoric and near-surface waters by the proximity of magma bodies sets up large-scale thermal circulation systems that drive mineralization processes. In these complex environments, the availability of heat, ongoing eruptive and post-eruption processes, a superstructure of abundant permeable volcaniclastic and epiclastic rocks, and the prevalence of descending and ascending waters result in a geologic setting rich in ore-forming potential. Complex stockworks, mineralized breccia pipes and veins, and disseminations of gold-rich siliceous and pyritic rock of complex variety have yielded gold in rocks as old as Archean, as young as Pliocene. Many of these relatively low-grade (2-10 ppm) deposits have become economic as a consequence of the refinement in heap-leaching methods in the past twenty years.

Goldfield and Round Mountain, Nevada, and Creede, Colorado, are examples of several of the various subtypes of volcanic-hosted gold deposits. Although still controversial because of post-mineralization metamorphic effects, the huge Homestake gold deposit of South Dakota, for many years the preeminent gold mine in the U.S., is likely a subtype in which submarine chemical sediments derived from volcanic-driven brines account for the gold deposition. The Haile, Brewer, and Ridgeway deposits of South Carolina certainly are of this class.

3. Granite-hosted Base Metal Deposits

A great deal of the gold produced in the United States, particularly prior to 1980, is, in fact, a by-product of the production of copper and other base metals in post-magmatic hydrothermal deposits collectively referred to as porphyry-style deposits. Associated in particular with syntectonic granitoid intrusive bodies in the Great Basin, these deposits are huge, disseminated ore bodies containing hundreds of millions of tons of ore. The gold is associated with copper sulfides in and surrounding the upper portions of these intrusive bodies, in which large-scale hydrothermal circulation driven by magma cooling and crystallization drives mineralization and regional scale wall-rock alteration. Minable only by open-pit bulk-mining methods, sufficient quantities of rock are treated to make possible the exploitation of very low gold values, on the order of parts per million, in addition to copper and, often, molybdenum. Any of the classic porphyry-style deposits such as Bingham Canyon, Utah, would serve as examples of this kind of deposit. None have been found in the Southeast.

4. Polymetallic, Massive Sulfide/Oxide Deposits

Since gold is associated with the modern mid-ocean-ridge vent systems, it is not surprising that gold can also be extracted from the polymetallic massive (more than 60 percent by weight) sulfide deposits associated with submarine extensional volcanism of all ages. These systems also produce zoned and massive ores of iron and manganese oxides (and, rarely, carbonates) and sulfates, as well as mineralized, sulfide-rich stockwork feeder systems and associated breccias and slump deposits. More commonly mined for copper, lead, zinc, and silver, these deposits often contain from 1 to 3 ppm gold. The Kuroko district of Japan is the best example of this group of deposits, but so too are Kidd Creek, Ontario; Mount Lyell, Australia; and perhaps Pueblo Viejo in the Dominican Republic. Smaller examples closer to home would be Silver Hill and Gold Hill in North Carolina. Some of the gold deposits in the Kings Mountain belt may well have been carbonate equivalents of these sulfide-rich types.

5. Carlin-type Deposits

Thinly bedded silty or argillaceous carbonaceous limestones and dolomites can host "no-see-um" (micron or smaller grains) gold mineralization in association with pyrite and arsenic minerals. Named after the classic deposit at Carlin, Nevada, these deposits are associated with jasperoidal silica replacements of the carbonate rocks. The deposits are found in proximity to regional structures. The temperatures of mineralization are low—125° to 300°C—and the fluids are low in salinity and relatively rich in CO_2 and methane. Gold is finely disseminated as selected replacements of the host carbonate rocks. Gold values range from 1 to 10 ppm. As noted, Carlin, Nevada, is the best example of these deposits. There are no examples in the Southeast.

6. Placer Deposits

Exceptions to the rule that gold mineralization must occur as a result of the transport of gold by anoxic fluids are residual or placer deposits. Here, the noble character of gold—its insolubility and nonreactivity in contact with the oxygen-rich atmosphere and surface hydrosphere—allows gold to persist in surface environments while all other components of the original host rock of the gold are removed by weathering, erosion, and surface transport. We can add to gold's nobility two additional characteristics that allow it to accumulate in active fluvial environments: (1) gold is malleable and thus resistant to fracture or other physical destruction while being transported in clast-laden streams and debris flows; and (2) gold is very dense, so that it collects in riffles and crevices of highly active and turbulent streams, where all but the coarsest of rocks and minerals are entrained. Thus deposited in high-energy systems with coarse sands and gravels in bars and along creek beds or in ancient terraces, free grains and rarely nuggets of gold are readily washed from modern sediments or, as in the case of ancient deposits or paleoplacers, must be liberated from the lithified equivalents of fluvial clastic sediments. The richest gold deposits are of this type, but the nugget effect means that average gold content is highly variable. In point of fact, every placer deposit, ancient or modern, is the result of the erosive destruction of a previous gold deposit—likely one of the above described types. Nearly every gold find starts as a placer deposit at some scale or other. Along the entire western slope of the Sierra Nevada Mountains in California, the finds that started the gold rush of 1848 are placer discoveries. The classic paleoplacer deposit (i.e., the lithified equivalent of a modern placer) is the Witwatersrand district of South Africa, which remains the richest gold district in the world.

Concluding Thoughts

It was one of the latter-described gold occurrences, a placer gold discovery, that set off the events we commemorate in this volume. For about the first forty years of gold mining in the Southeast, placer workings dominated gold production—and some of those workings produced legendary finds such as Conrad Reed's seventeen-pound nugget that started it all. How did those nuggets get so large? We really haven't a clue. Some think that the age of the Piedmont surface has allowed nuggets of large proportion to grow in the subsurface by slow, steady dissolution of gold in the presence, perhaps, of anoxic waters rich in organic acids and by slow migration in the permeable regolith, where those gold-rich fluids bathe stable nuggets in more oxygen-rich settings and nucleate into large nuggets (Feiss et al., 1991). It seems highly improbable that they eroded out of the original lodes at this size.

A reprise to the earlier discussion of the origin of gold: it is interesting to think of how large a volume of space was needed to make young Conrad's seventeen-pounder. Using the gold content of the universe above and the estimated density of the interstellar medium of one atom/cc, we can estimate that it would require a sphere of interstellar matter about 6 million kilometers in radius to produce the nugget of gold that started this all. That would be a sphere with a radius about fifteen times the distance from the Earth to the sun.

Acknowledgments

A paper of this breadth is not the work of any one person, though only the author should be held accountable for its errors and overgeneralizations. Nancy West is the source of much insight on the cosmogenesis of gold and of useful editorial assistance. What little I might know of the geochemistry and geology of gold is the by-product of thirty years of conversations, in and out of the field. Among the notable conversants are the late Bob Butler, John Guilbert, Dennis LaPoint, Sam Adams, Hunter Ware, and many, many students over the years. Most notable among these latter are John Scheetz, John Powers, Libba Hughes, and

Kent Ausburn. I should also note two people—Ulrich Petersen and John J. W. Rogers—who modeled for me the rewards of always thinking about the "big" picture.

References

Bonham, H. F., Jr. "Bulk Mineable Gold Deposits of the Western United States." In R. R. Keays, W. R. H. Ramsey, and D. I. Groves, eds., *The Geology of Gold Deposits*, Economic Geology Monograph 6. New Haven: Economic Geology Publishing Co., 1989. Pp. 193–207.

Cox, D. P., and D. A. Singer, eds. *Mineral Deposit Models*. United States Geological Survey Bulletin 1693. Washington: Government Printing Office, 1986.

Crocket, J. H. "Distribution of Gold in the Earth's Crust." In R. P. Foster, *Gold Metallogeny and Exploration*. New York: Chapman and Hall, 1991, 1–36.

Feiss, P. G., et al. "Mineral Resources of the Carolinas." In J. W. Wright Jr. and V. A. Zullo, eds., *The Geology of the Carolinas*. Knoxville: University of Tennessee Press, 1991, 319–345.

Foster, R. P. *Gold Metallogeny and Exploration*. Glasgow, Scotland: Blackie, 1991.

Harrington, M. D., P. M. Herzig, and S. D. Scott. "Auriferous Hydrothermal Precipitates on the Modern Seafloor." In Foster, *Gold Metallogeny and Exploration*, 249–282.

Keays, R. R., and B. J. Skinner. "Introduction." In Keays, Ramsey, and Groves, *The Geology of Gold Deposits*. Pp. 1–8.

Seward, T. M. "The Hydrothermal Geochemistry of Gold." In Foster, *Gold Metallogeny and Exploration*, 37–62.

World Gold Exploration: Discovering Earth's Gold Factories

Byron R. Berger

Byron R. Berger has been with the U.S. Geological Survey (based in Denver) since 1977 and has traveled around the world to study gold deposits in a number of countries. He has published a study of the world's largest lode deposit, Muruntau, in Uzbekistan, and has worked on other USGS projects in various parts of the world. He is currently directing a study of what localizes mineral systems and what further localizes ore deposits within mineral systems.

Introduction

Stars are remarkable cosmological bodies. The chemical elements of the periodic table are the "ashes of nuclear burning" within them (Kirshner, 1994). In the heart of the nuclear reactor we call the sun, hydrogen atoms are stripped of their electrons and, in a series of nuclear reactions, fused into helium. With time, helium nuclei fuse to form heavier elements, including carbon and oxygen, but rarer, more massive stars than the sun are required to allow the fusion reactions that result in elements such as silicon, sulfur, and iron. Constraints as to temperature and density require that elements heavier than iron (^{56}Fe), including gold (^{196}Au), are synthesized in even rarer celestial events (cf. Kragh, 1996). Thus, as matter coalesced 4.5 billion years ago to form the earth, the concentration of gold was intrinsically low. The structure of the gold atom lends to its preferential association with elements that are most abundant in the interior of the earth.

Demand for gold, favorable economics, dramatic political changes worldwide, and advances in knowledge and technology have fueled an enormous investment during the last two to three decades of the twentieth century in gold exploration and deposit development. Exploration is an expensive, high-risk business that, industry-wide, has a low success rate (*Mining Journal*, 1999a), but technical skill, persistence, and an entrepreneurial spirit have found many deposits hidden beneath the jungles, deserts, frozen tundra, and alpine environments on most of Earth's continents. Effectively, exploration is the search for evidence of the physical and chemical phenomena—the "gold factories" (Kerrich, 1999) that transfer gold found in low concentrations deep in the earth through the lower part of its outer crust and then deposit it in minable concentrations in the outer part of the crust as an *ore deposit*. It is the natural gold factories and the search for them that is the story told below.

The Geologic Environments of Gold Factories

The entrepreneurial spirit, serendipity, persistence, politics, and technology have always played critical roles in the success of the mining business. Exploration geology is but one of the investment strategies used by businesses to obtain mineral commodities. Explorationists depend on models of mineral deposits to guide their search in regions favorable for the occurrence of deposits. They construct models based on scientific knowledge. The single biggest conceptual advance in the earth sciences since World War II has been the widespread acceptance of the tenets of *sea-floor spreading* and *plate tectonics* as a model of how the earth works. This holistic view of an active earth integrates processes taking place deep within the earth with those taking place in shallow to surface regimes, and it is through those processes that gold is transferred to the shallow part of the earth's crust.

Sea-floor spreading and plate tectonics constitute a mixing mechanism wherein energy and mass are constantly being transferred from depth to the earth's surface and back into the earth's interior (figure 1). Hot, buoyant material from within the earth is added to the shallow regime at spreading centers, and, as it pushes previously added material laterally, cold, older crustal material at the outer margins of plates is pushed down at trenches into the earth beneath adjacent plates. The sinking process of one plate beneath another is known as *subduction*. In the subducting plate, the increase in pressure and temperature as the plate descends leads to the

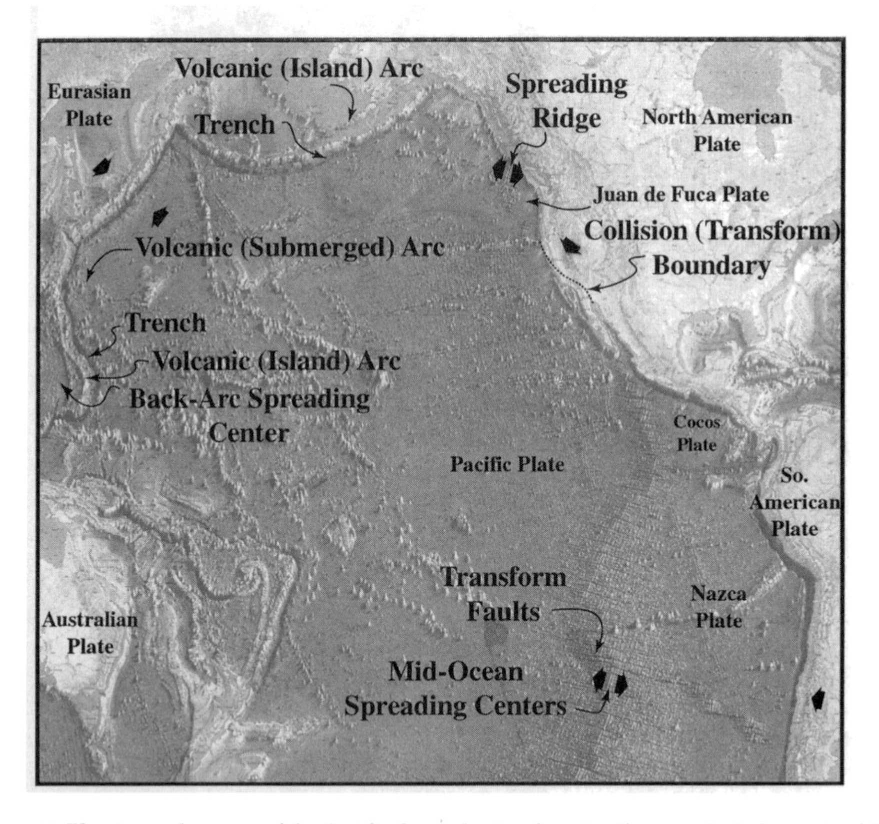

Figure 1. Physiographic map of the Pacific Ocean basin, showing the principal elements of the zones in which there is a significant transfer of heat and material from deeper in the earth into the shallow outer crust. These zones, where most gold deposits are formed, include the mid-ocean spreading centers, subduction complexes, and collision boundaries. The heavy, dark arrows show the relative motion of selected plates. Mid-ocean spreading centers are segmented by fracture systems known as *transform* faults. Subduction complexes, where one tectonic plate is moving beneath another, consist of oceanic *trenches*, *volcanic arcs* where volcanoes form on the overriding plate above the subducting plate and, in some circumstances, *back-arc spreading centers*, where new crust forms analogously to the mid-ocean spreading centers. Volcanic arcs may form island chains, be submerged, or form on land masses.

dehydration of minerals that are unstable at high pressures and temperatures, and the upward migration of those fluids contributes to the melting of rocks above the subducting plate to form *magma*. Some magmas rise to the surface above the subducting plate to form volcanoes in a string parallel to the overriding plate margin ("volcanic arcs"). In places, subduction may cease and adjacent plates collide along major fault zones ("collision boundaries"). Through time, older spreading centers, subduction-related environments, and collisional boundaries are preserved only as remnants attached together in collages that make up the earth's continental masses. Recognizing these ancient environments, unraveling the puzzle of geologic history, and knowing how gold factories work within the different environments are parts of the knowledge base necessary to conduct minerals exploration.

The gold factories for which the greatest exploration investment is made are those in which tectonic-plate interactions are most effective at concentrating gold into economically viable deposits. The most important of these environments are sea-floor spreading centers, the volcanic arcs that form above subducting plates (submarine or subaerial), and the fault zones that bound colliding plates (figure 2).

Sea-floor Spreading Center Environments

New material from deep within the earth's interior is added to the crust along fracture zones (figure 1). The most voluminous outpourings are along the mid-ocean fracture zones that ring the earth ("mid-ocean ridge spreading centers") and, to a lesser extent, in basins that are located behind ("back-arc spreading centers") volcanic arcs. The deposits that form in those environments typically contain large volumes of sulfide minerals (≥50 percent) and, therefore, are referred to as "volcanogenic massive sulfide deposits." Although there are modern hydrothermal vent areas on mid-ocean ridges producing massive sulfide deposits, chemical data indicate that ancient analog deposits that have been mined on the continents are remnants of back-arc spreading centers (cf. Franklin et al., 1998). While remnants of sea-floor spreading environments have been less of an exploration target in recent years than other geologic environments, a brief discussion of sea-floor gold factories illustrates how deposits in submarine environments differ from those in subaerial environments.

Mid-ocean Spreading Centers

The greatest volume of magma is added to the earth's crust from its interior mantle along long, sinuous ridges within the major ocean basins (figures 1 and 2a). These magmas transfer heat into the oceanic crust, heat that drives the convection of hydrothermal fluids along the ridge axes. Where these fluids vent at the ocean floor, sulfide minerals may accumulate in mounds around the hydrothermal vents. The sulfide minerals are primarily those containing iron, copper, and zinc, with gold as an important co-product at many localities (figure 3).

Although mid-ocean ridge mineral deposits are apparently not often preserved in the continental rock record in comparison to back-arc spreading center deposits, presently forming deposits are analogs of sea-floor spreading gold factories. Scientific study of them provides glimpses inside the doors of these gold factories. Repeated dives to the sea floor on two ridge segments with active sulfide precipitation have provided exciting new discoveries as to how mid-oceanic hydrothermal systems work, thereby revealing information that can be incorporated in mineral-deposit models used in exploration. Those sites are the TAG hydrothermal field

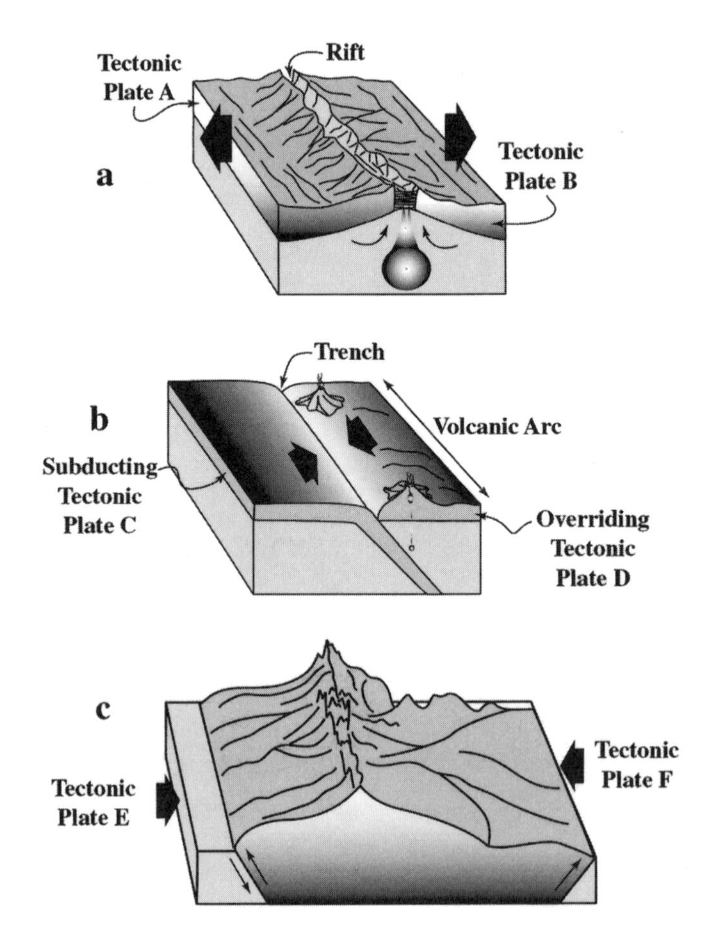

Figure 2. Schematic drawing of the three principal plate-tectonic environments in which earth's gold factories occur. **a**: Spreading centers, mid-ocean or back-arc, are localities in which new crust is formed through the extrusion of magma onto the surface along an axial rift, with the two plates (A and B) diverging away from this center. The related flow of high-temperature fluids, *hydrothermal fluids*, in the rift zones may result in deposits of sulfide-bearing mineral deposits with associated gold and silver. **b**: A subduction zone in which Plate C is moving beneath Plate D. Magmas form in the vicinity of the subducting plate and rise to the shallow crust, where volcanoes may form during periods of eruption. Hydrothermal fluids related to the volcanism may result in the formation of gold deposits in and around the volcanic rocks. **c**: A collision boundary at which Plate E and Plate F are converging at a sufficiently high angle to result in the uplift of a mountain range along the margin of Plate F, with oblique motion occurring along a plate-margin parallel set of faults. Gold deposits form within the plate where uplift is taking place.

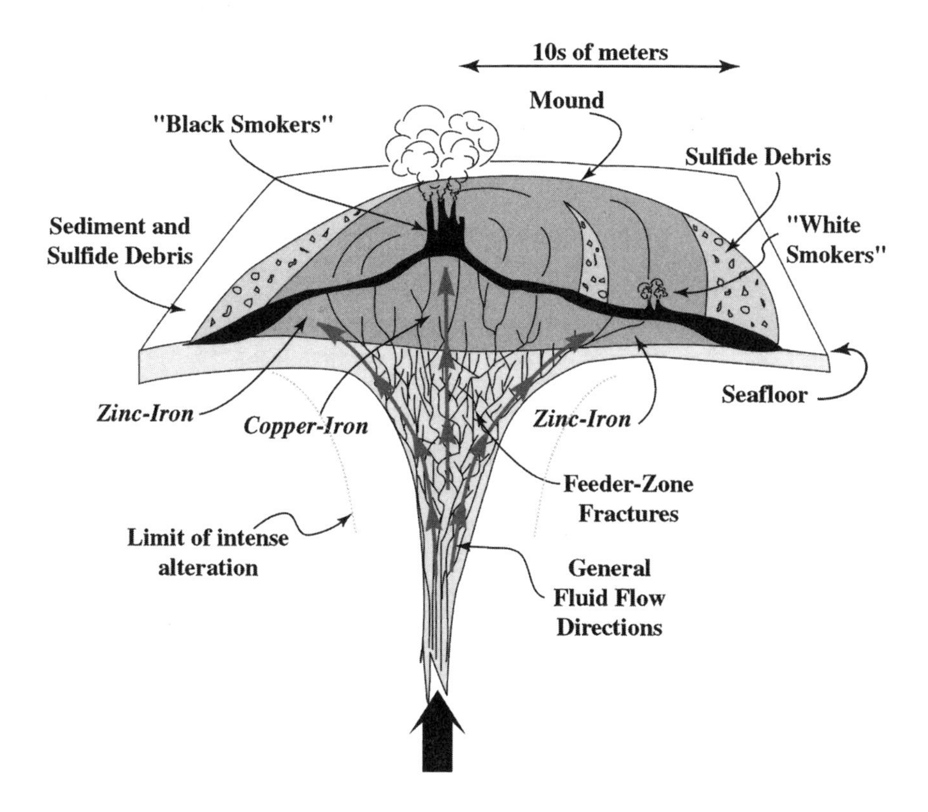

Figure 3. Schematic drawing of a typical deposit of sulfides on the sea floor within the rift zone of a spreading center (after Hannington et al., 1995) or around a deeply submerged volcanic arc. If the sulfides form approximately 50 percent or more of the mineralized mass, they are termed *massive sulfide deposits*. The deposits form by the emanation of hydrothermal fluids at and immediately beneath the seawater-sediment interface with dissolved metals precipitating as a consequence of the sudden cooling of the high-temperature fluid at the interface. The deposits commonly form a mound around the higher-temperature vents (*black smokers*), the mound being underlain by a zone of fracture-controlled (*stockwork*) veins. Lower-temperature, peripheral hydrothermal vents are called *white smokers*. Copper minerals are generally most abundant in the higher-temperature parts of the hydrothermal vent area, whereas zinc minerals are most abundant in the cooler parts of the discharge regime.

(26°N) along the Mid-Atlantic Ridge and the Middle Valley region (48°N) along the northern Juan de Fuca Ridge off the Pacific Northwestern coast of North America. The most important discovery has been of sulfide accumulations similar in size to ancient deposits found on the continents—for example, 7 to 15 million tons of copper- and zinc-bearing massive sulfide in Middle Valley (W. C. Shanks, U.S. Geological Survey, personal communication, 1999). Thus, scientists are truly studying active gold factories.

From the research on oceanic hydrothermal systems has come a realization and appreciation of the *essential* linkage, an interdependence among natural phenomena, including the transfer of heat to the surface (volcanism), active earthquake faults (deformation), hydrothermal fluid flow, and chemical exchanges between the oceans and ocean-floor rocks. Among the observations that may assist in the exploration for ancient sea-floor deposits are the apparent relationships among the (1) cross-sectional area of the ridge, magma budget, and the abundance of hydrothermal venting; and (2) rate of spreading, nature of fracture-related segmentation, and localization of hydrothermal activity (MacDonald, 1998). Such relationships may assist in evaluating which of two ostensibly similar geologic environments poses the lowest exploration risk. At the more detailed scale, studies of the mechanisms that focus fluid flow demonstrate the linkage of active faulting, magma intrusion, and rock permeabilities to hydrothermal flow, and understanding mound growth (Shipboard Scientific Party, 1998) has direct implications for prospect evaluation on land.

Back-arc Spreading Centers

Whether or not there is sea-floor spreading behind volcanic arcs depends on the angle at which the subducting plate dips beneath the overriding plate (figure 2b). When the dip of the subducting plate is relatively steep, extension occurs on the overriding plate side of the arc, forming back-arc spreading centers. The Lau, North Fiji, and Manus basins in the South Pacific and the Okinawa Trough in the West Pacific are sites where gold factories have been studied in back-arc environments (cf. Rona and Scott, 1993).

The preservation of ancient back-arc volcanogenic massive sulfide deposits may result from the collision of tectonic plates, which more often preserve volcanic-arc and back-arc remnants than mid-ocean-ridge spreading centers. Mid-ocean spreading centers are lost along with sea floor through subduction into the interior of the earth. Nevertheless, back-arc gold factories work in the same manner as the hydrothermal systems currently active along the mid-ocean ridges. An example of a descriptive model of a sea-floor spreading center mineral deposit is as follows:

A Gold Factory in Sea-floor Spreading Environments

The marine environment has covered a substantial proportion of the earth for most of earth history. At present the proportion is ≈70% (Scott, 1997). Much of the heat transfer from the earth's interior to the surface occurs in two environments: (1) along the interconnected, extensional ridges in the deep ocean basins, and (2) smaller zones of extension in the seas behind island arcs. The heat transfer is largely in the form of magma intruding the earth's crust and extruding in these environments, sea-floor spreading centers, and through related hydrothermal convection. The hydrothermal convection associated with sea-floor volcanism leaches metals from sea-floor rocks and may deposit them at the sites of spreading. At and immediately below the sea floor there may be high concentrations of metals, including zinc, copper, silver, and gold, because of the mixing of the hydrothermal fluids with cold seawater. That process has occurred intermittently through geologic time and is occurring at present.

Based on chemical data about the volcanic rocks associated with massive sulfide deposits, all of the known deposits formed in geologic environments related to subduction. Deposits formed in back-arc environments are known as "Cyprus type" because of the analogous deposits located on the island of Cyprus. The associated rock types are marine magnesium- and iron-rich volcanic flows and intrusions and minor deepwater sedimentary rocks. Cyprus-type ores are predominantly economic for copper but also contain some zinc and silver and minor gold and lead (Cox and Singer, 1986).

The heat sources that drive the spreading-center hydrothermal systems are magma reservoirs beneath the spreading centers at depths varying between ≈1.5 and 3 km below the ocean floor (Scott, 1997). The hydrothermal fluids are dominated by seawater that was heated at least to 350°C and commonly to over 400°C (Hannington et al., 1995).

To become established as steady-state phenomena, sea-floor hydrothermal systems must initially displace large volumes of cold seawater occupying the crack fissures in the ocean-floor rocks, replacing it with hot fluids, and also heat the rock surrounding the fractures to the ambient temperature of the fluid. That process alters the minerals in the rock beneath the sea floor to a new assemblage of minerals, *hydrothermal alteration*. The deepest and highest temperature alteration zones include the minerals epidote ($Ca_2(Al,Fe^{+3})_3(SiO_4)_3(OH)$) and quartz ($SiO_2$). Toward the surface, the central part of the upwelling plume of hot fluids is accompanied by quartz and chlorite ($(Mg,Fe)_5Al(Si_3Al)O_{10}(OH)_8$) alteration. The chlorite is iron-rich (Fe) in the most fractured rock and magnesium-rich (Mg) in the surrounding rock. This outermost zone paralleling the upwelling regime is an assemblage of mica-clay minerals, including illite $(K,H_3O)(Al,Mg,Fe)_2(Si,Al)_4O_{10}[(OH)_2,H_2O]$ and smectite (e.g., montmorillonite $(Na,Ca)_{0.3}(Al,Mg)_2Si_4O_{10}(OH)_2.nH_2O$)) (cf. Hannington et al., 1995; Scott, 1997).

The precipitation of ore minerals occurs as a result of boiling of the fluid beneath the sea floor and/or the mixing of the high-temperature fluid with cold seawater at the rock-seawater interface. The textures of the ore minerals are highly varied and, in the active oceanic systems, include bladed, fibrous, acicular, myrmekitic, and colloform forms (Scott, 1997). The abundance of individual minerals is highly variable, but chalcopyrite ($CuFeS_2$), sphalerite (ZnS), pyrite (FeS_2), marcasite (FeS_2), and pyrrhotite ($Fe_{1-x}S$) are most frequent. Less abundant are bornite (Cu_5FeS_4), covellite (CuS), digenite (Cu_9S_5), and galena (PbS). Gold is generally encased within sulfide minerals but may also occur as discrete, visible grains. Silver-bearing minerals may include tetrahedrite ($Cu_{12}Sb_4S_{13}$) and tennantite ($Cu_{12}As_4S_{13}$). At the highest temperatures, chalcopyrite and pyrrhotite are precipitated, followed by pyrite and sphalerite at lower temperatures (Hannington et al., 1995). Non-ore minerals associated with the sulfide minerals may include anhydrite ($CaSO_4$), barite ($BaSO_4$), calcite ($CaCO_3$), dolomite ($CaMg(CO_3)$, and siderite ($FeCO_3$).

Volcanic-arc Settings above Subduction Zones

Subduction of crustal materials into the earth has two important effects. One is that the stresses along the boundary of the colder subducting plate and the hotter overriding plate lead to heat transfer or advection in the wedge-shaped leading edge of the overriding plate (figure 2b). That process brings higher temperatures to shallower depths in the wedge. Second, the subduction of crustal materials into the earth's interior results, as pressure and temperature increase, in the recrystallization of minerals not chemically stable under those changed conditions. Water released from the crystal structures and fluids trapped in pore spaces are released and migrate upward. Those fluids may lower the melting temperature of some rock components in the overriding plate to form magmas. Because magmas are hot and buoyant, they rise in the earth, heating the surrounding rocks as they intrude the shallower, cooler crust. Eruptions of magma onto the earth's surface can form volcanoes and other volcanic landforms that make up the volcanic arcs. Modern examples such as the active volcanoes in the Andes of South America, the Aleutian Island chain of Alaska, and the Cascade Mountains of western Washington and Oregon circumscribe the Pacific Ocean basin. A model of gold factories in volcanic-arc environments is shown in figure 4.

Not all volcanoes in volcanic arcs are subaerial. Some form in the submarine environment—for example, along the Izu-Ogasawara Arc offshore of Japan and the Kermadec Ridge north of New Zealand. Hydrothermal convection related to the cooling of magma at depth in the submarine-arc volcanoes discharges fluids in the volcanic edifice on the sea floor. The Sunrise sulfide deposit on Suiyo seamount in the Izu-Ogasawara Arc (Iizasa, 1999) is a presently active hydrothermal system, and vents in the Brothers caldera on the southern Kermadec Ridge are also depositing sulfides (Stoffers et al., 1999). Those gold factories operate in the same manner as those in sea-floor spreading regimes, but the contained metals are more similar to deposits formed above sea level in that they often contain lead along with zinc and copper. The Kuroko deposits in Japan may be analogs to these sea-floor volcanic arcs.

Magmas commonly pond into "reservoirs" of magma beneath the earth's surface. The heat emanating from those reservoirs results in convecting hydrothermal systems above them that produce hot springs and geysers on the earth's surface (e.g., Yellowstone National Park). These hydrothermal systems, when enriched in gases such as hydrogen sulfide (H_2S) and carbon dioxide (CO_2), are a style of gold factory found in volcanic-arc environments. Some active geothermal systems in New Zealand and Japan are examples of such gold factories.

Another factory is the magma reservoirs themselves, which feed the volcanoes. Under certain conditions, the crystallized parts of reservoirs will be intensely broken into closely spaced, cross-cutting fractures along which gold may be precipitated as a consequence of the precipitation of sulfide minerals. Because of the short length and cross-cutting nature of the fractures, such deposits are referred to as "stockwork" fracture deposits or, more commonly, "porphyry-style" deposits. Remnants of ancient volcanic arcs, such as are found along the eastern and western oceanic margins of the United States, have been historically among the world's most significant gold resources. Because those geologic terranes were among the earliest to receive scientific scrutiny, they have served as templates for understanding similar volcanic arc settings elsewhere in the world. The Spanish called the gold ores with high unit value "bonanzas," and bonanzas continue to be highly sought prizes in

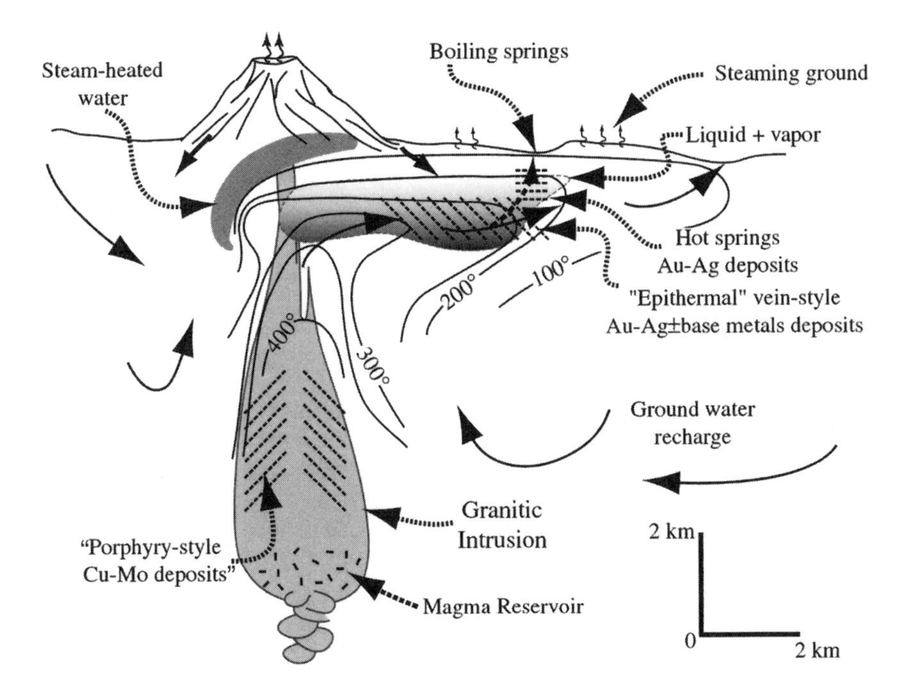

Figure 4. Schematic drawing of a hydrothermal flow system associated with a volcano within a volcanic arc. A ponding of magma in a *reservoir* beneath the volcanic edifice is the main source for lavas and ash that erupt on the surface. Solidified parts of this reservoir may be the site of intense fracturing and hydrothermal fluid flow, which may result in the formation of a *porphyry-style* copper deposit with significant quantities of gold. As the magma reservoir cools, it creates a convecting plume of water above it that flows upward and then laterally toward the margin of the volcanic landform because of gravitational forces. This plume of laterally flowing fluid may result in the formation of *epithermal-style* gold and silver deposits as veins in the deeper parts of the discharge regime or in broken masses of rock, *breccias*, and veinlets near the discharge vents at the ground surface (*hot-spring-style* deposits).

mineral exploration. Similarly, porphyry-style ores are an important exploration target because they are generally very large and amenable to inexpensive mining and milling methods. The most active areas for current exploration are the Andes of South America and the islands of the South Pacific. An example of a descriptive model of a volcanic-arc mineral deposit is as follows:

A Gold Factory in a Subduction Zone Environment

Subduction of material into the crust balances the addition of mass at the earth's surface along sea-floor spreading centers. Subducted materials may vary from cold sea-floor volcanic rocks with a veneer of sedimentary materials to younger, and therefore warmer, sea-floor volcanic rocks with a veneer of sediment. There are stresses along the upper surface of the subducting plate that lead to convection in the overriding plate. In time, this process brings higher-temperature materials closer to the earth's surface. Simultaneously, the dehydration of minerals in the subducting plate leads to the partial melting of material in the overlying region and the production of siliceous magmas, generally of diorite to granodiorite composition. Intrusion of magma into the shallow parts of the earth's crust leads to bodies of ponded magma, and plutons, as well as volcanic activity. Gold-bearing deposits may occur within the plutons after crystallizing, in veins lateral to them, or in the discharge regime of hot-spring systems.

Deposits formed in plutons are commonly called "porphyry-style" deposits because the deposits occur in porphyritic granitoids within a myriad of interconnected small veins ("stockwork veins"), predominantly within the plutonic rocks. Porphyry-style ores are primarily economic because of their copper content, with gold and silver being by-products. The gold is generally associated with the copper-bearing minerals bornite (Cu_5FeS_4) and chalcopyrite ($CuFeS_2$).

Deposits formed in veins are known as "epithermal"- or "polymetallic"-style deposits, depending on their ore-mineral assemblages, temperatures of formation, and ore-fluid composition. Epithermal veins are generally formed within volcanic rocks that are the eruptive equivalents of underlying plutonic rocks. Epithermal veins characteristically are laminated quartz (SiO_2) with ore minerals occurring in some laminations as disseminations or massive accumulations. Economic metal assemblages can vary from merely Au-Ag to Au-Ag-Zn-Pb-Cu. Fluid inclusion data indicate that ores were deposited from extremely dilute to moderately saline fluids, predominantly within the temperature range of 180°C to 280°C. Polymetallic veins may be considered a subclass of epithermal veins wherein the metal assemblage contains economic concentrations of the base metals zinc, lead, and copper, as well as silver. For those polymetallic veins formed in close proximity to related granitic intrusions, ore-deposition temperatures may range up to 350°C. The ore-depositing fluids are more saline than are those that form Au-Ag-only deposits.

Shallow, cooling siliceous magmas, dioritic to granodioritic in composition, are the heat sources that drive convection to make porphyry- and vein-style deposits and also provide some components of the convecting fluids. The magmas occur at depths varying between ≈1.5 and 3 km below the earth's surface. With the exception of porphyry copper-gold deposits, the hydrothermal fluids are predominantly of meteoric origin.

During the early stages of reaching a steady state, hydrothermal fluid flow is not focused and is out of thermal equilibrium with its surroundings, and the fluid/rock ratio is low. At low water/rock ratios, the wet, hot rock alters—whether basalt, andesite, or dacite—to a mineral assemblage including albite ($NaAlSi_3O_8$), chlorite ($(Mg,Fe)_5Al(Si_3Al)O_{10}(OH)_8$), and epidote ($Ca_2(Al,Fe^{+3})_3(SiO_4)_3(OH)$) (Reed, 1997). At high water/rock ratios, the fluid dominates the wall-rock alteration. Different fluid composition-temperature combinations produce different alteration types. The interior parts of porphyry Cu-Au deposits consist of quartz (SiO_2), potassium feldspar ($KAlSi_3O_8$) and/or biotite ($KFe_3AlSi_3O_{10}(OH)_2$), muscovite ($KAl_3Si_3O_{10}(OH)_2$), and chlorite. In most porphyry deposits there is an overprint of an assemblage that also occurs as the inner selvage of epithermal and polymetallic veins consisting of quartz, muscovite, and chlorite. Within quartz veins other non-ore minerals may include potassium feldspar, calcite ($CaCO_3$), barite ($BaSO_4$), and fluorite (CaF_2). Hydrothermal fluids that are moderately to highly acidic produce what are referred to as "intermediate argillic" and "advanced argillic" mineral assemblages. Intermediate-argillic assemblages are characterized by the minerals kaolinite and smectite (e.g., montmorillonite, $(Na,Ca)_{0.3}(Al,Mg)_2Si_4O_{10}(OH)_2.nH_2O$). Advanced-argillic assemblages include the minerals quartz, alunite ($KAl_3(SO_4)_2(OH)_6$), and kaolinite ($Al_2Si_2O_5(OH)_4$) ± pyrophyllite ($Al_2Si_4O_{10}(OH)_2$), zunyite, or topaz. Acidity in hydrothermal solutions is most probably related to the aqueous chemistry of CO_2 and HCl (Giggenbach, 1997). The presence of alunite, however, implies that oxidation-reduction reactions in the fluid were controlled by the H_2S/SO_2 gas buffer. The disproportionation of SO_2 in magmatic vapor in the presence of water forms highly acid, oxidized solutions altering surrounding rocks to an advanced-argillic alteration assemblage. Advanced-argillic alteration assemblages are known to occur with all styles of hydrothermal mineralization associated with subduction zone magmatism.

The precipitation of ore minerals in most cases is a result of phase separation or boiling and to a lesser extent through sulfidation reactions and fluid mixing. Ore mineral assemblages are relatively simple in porphyry Cu deposits and many types of epithermal deposits. The mineral assemblages can be quite complex, however, in all deposit styles with polymetallic mineral assemblages. Ore minerals in porphyry Cu-Au deposits are chalcopyrite ($CuFeS_2$), bornite (Cu_5FeS_4), molybdenite (MoS_2), native gold, and electrum. In Au-Ag vein deposits, native gold, electrum, argentite or acanthite (Ag_2S), tetrahedrite ($Cu_{12}Sb_4S_{13}$), and tennantite ($Cu_{12}As_4S_{13}$) are the principal ore minerals, with only accessory sphalerite (ZnS), galena (PbS), and chalcopyrite. Pyrite (FeS_2) is ubiquitous in all deposit styles and frequently contains inclusions of gold. Epithermal Cu-Au associated with advanced-argillic alteration contains native gold, electrum, enargite (CuAsS), and famatinite (CuSbS). Other less common phases include covellite (CuS), chalcocite (Cu_2S), and various tellurides, possibly including goldfieldite and calaverite. Marcasite (FeS_2), calcite ($CaCO_3$), barite ($BaSO_4$), and fluorite (CaF_2) are common accessory minerals. The abundance of minerals is highly variable.

Andes Mountains, South America

Keen exploration interest in the Andes followed the discovery of the El Indio, Chile, deposit (3.5 million ounces of gold) during the 1970s (figure 5a), and the Andean region receives about 29 percent of the world's exploration investment (*Mining Journal*, 1999a). The climate of the high Andes readily oxidizes sulfide minerals exposed in outcrop, yielding vividly colored outcrops that guided much of the exploration during the 1980s. That exploration led to the discovery of additional deposits in Chile, including Refugio, Lobo, Marte, La Coipa, and El Hueso (Suttill, 1991). Economic geologists often group deposits into linear trends and refer to them as "belts"; terrains with more than a single belt of deposits are referred to as "provinces." The Chilean discoveries following El Indio constitute the Maricunga belt.

The Maricunga belt deposits resulted from hydrothermal activity related to centers of volcanic eruptions that occurred above a subducting plate within the past sixty million years. At the northern end of the province, the El Hueso mine (19 Mt @ 1.68 grams Au/t) was the first mine in Chile to use heap-leach technology when it went into production in 1987. The gold is widely dispersed along numerous small fractures in the rock and, because of its association with arsenic and mercury, is

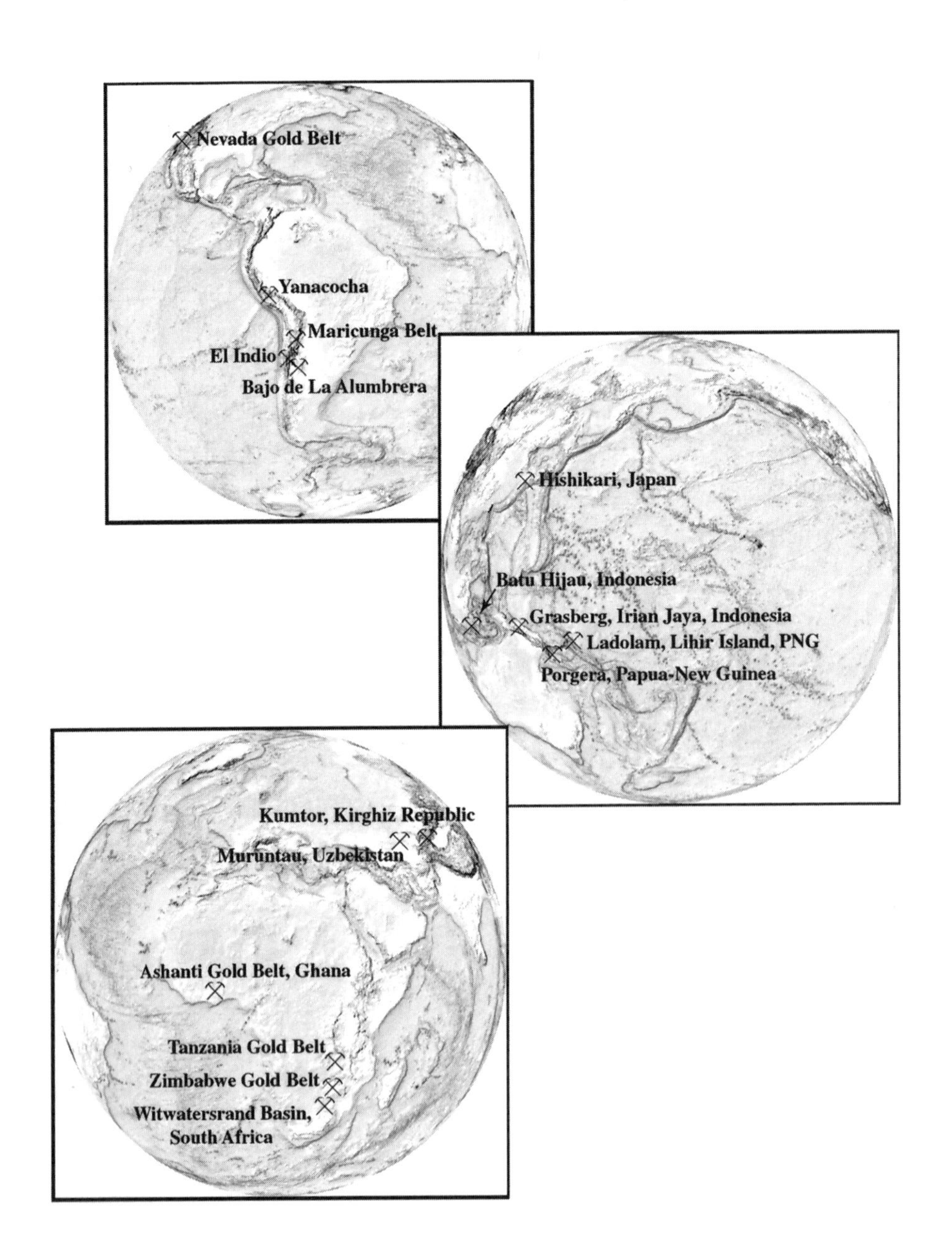

Figure 5. Selected regions wherein there is a considerable investment in gold exploration at the present time. **a**: North and South America; **b**: the southwest Pacific Ocean region; **c**: Africa and central Asia.

considered to have formed at very shallow levels in a volcanic center. Southeast of El Hueso is La Coipa (61.2 Mt @ 1.37 grams Au/t), the premier gold deposit in the province and also the world's largest silver mine (Suttill, 1991). La Coipa illustrates how old prospects (ca. 1911) that fall into obscurity can at some future date become important deposits.

Exploration of the high Andes has extended north into Peru and southeast into Argentina. Although political difficulties have made exploration in Peru a high-risk endeavor for much of the past two decades, discoveries such as Yanacocha, possibly the lowest-cost gold mine in the world (Phelps, 1996), have encouraged additional exploration investment in Peru. A potentially onerous royalty code in Argentina has stifled exploration for much of the twentieth century (*Mining Journal*, 1998a). The premier mining discovery in Argentina is the Bajo de la Alumbrera porphyry copper-gold deposit, where the crystallized magma reservoir that fed overlying volcanoes contains 767 million tons of ore averaging 0.51 percent copper and 0.64 grams Au/t.

South Pacific Ocean Region

If the rate of plate convergence is rapid, then mountain ranges in the overriding plate may gain altitude at a rapid rate. When there is a conjunction of rapid rates of uplift and high rates of erosion, recently formed mineral deposits can be exposed soon after their formation. Such is the case in the tropical southwestern Pacific Ocean basin (figure 5b). With the steep terrain and dense rain forest masking deposit locations, success in finding the most productive gold mine in the world, Grasberg, Irian Jaya, has spurred on the flow of exploration investment into this region during the past two decades. The proportion now constitutes more than 9 percent of the worldwide total (*Mining Journal*, 1999a).

Indonesia

There was a considerable effort by the Dutch in decades past to find mineral deposits on the many islands that make up Indonesia. With samples collected from stream sediments and geophysical data collected from aerial surveys, many new deposits have been found during the most recent spate of exploration activity. Grasberg on Irian Jaya, Batu Hijau on Sumbawa Island, and Mesel on Sulawesi are among the many new discoveries.

Irian Jaya is part of a large island divided between Indonesia and Papua-New Guinea. Although there are some active volcanoes on the island, most are extinct. The now extinct volcanoes have received considerable exploration attention and, consequently, are sites at which some of the economically most important gold deposits in the world have been discovered. The largest of these, Grasberg, Irian Jaya, was formed in the hearth of a volcano that became extinct ≈2 to 3 million years ago. Geologists often call this style of gold factory "porphyry copper-gold deposits" (figure 4). Fluids that escaped along the margins of the porphyry deposit rock and into the surrounding rocks reacted with carbonate-bearing formations to form deposits that geologists refer to as "skarns."

Papua-New Guinea

Just as the 1960s and 1970s were the decades of successful copper exploration in Papua-New Guinea, the last two decades have been years of a successful search for gold deposits in the same region. Porgera is the flagship gold mine of the country, having produced ≈7 1/2 million ounces of gold since 1990 (*Mining Journal*, 1998b). The Porgera veins display bands or laminations of quartz, some of which are encrusted with ore minerals, as a result of the dynamic processes that take place in near-surface, high-temperature hydrothermal systems.

The offshore islands of Papua-New Guinea have proved to be lucrative areas in which to search for gold deposits in island-arc environments. There is now production from Misima, Fergusson (Wapolu mine), and Lihir islands. The Lihir mine, Lihir Island, is one of the largest gold deposits found to date in the discharge regime of a hot-springs system. The ore reserves are approximately 104 million tons averaging 4.37 grams Au/ton (≈14.6 million ounces) (*Mining Journal*, 1998b). The main thermal source for this still active hydrothermal system is a volcanic center that is the backbone of the island. Because of clay alteration in the volcanic edifice, there was a prehistoric catastrophic collapse of one side of the volcano, which resulted in a refocusing of fluid flow to the zone lateral to the volcano that now constitutes the Lihir mine.

Collision-related Suture Zones

Tectonic plates collide (figure 6) when subduction can no longer accommodate the relative motion between them, when there is a change in the direction(s) of plate convergence, or when the mid-ocean spreading ridge itself is subducted. Those situations result in the compressional contact of plates along great fault systems called transform faults (such as the San Andreas fault in California) or the actual suturing of pieces of different tectonic plates together and onto one of the plates. The subduction of younger, and therefore warmer, ocean-floor volcanic rocks along with heat accumulated in the earth's crust during subduction are dissipated through the generation of post-collision magmas and, very importantly, the alteration or *metamorphism* of large regions of the earth's crust along collisional boundaries.

The most famous of the great suture zones in the United States is the Melones fault zone in California, where the "Mother Lode" deposits and related, derivative ancient riverbed and alluvial "placer" ores (which were the "elephants" sought by the forty-niners during the nineteenth-century California gold rush) were found. The suturing in California began on the order of 150 million years ago. The California goldfields have probably yielded most of their available wealth, but even more ancient collisional zones are the targets of current exploration in West Africa, Central Asia, and West Australia. An example of a descriptive model of a collision-zone mineral deposit is as follows:

A Gold Factory in a Collisional Environment

Continental growth occurs through the collision of plates upon the closing of ocean basins. Most of the large continental masses are amalgamations of different-sized blocks joined along major fault zones (e.g., Africa), and plate margins may reflect the amalgamation of numerous small pieces of different terranes (e.g., western Cordillera of North America). The variety of collisional relationships that have resulted in gold-bearing veins is illustrated by the following: (1) veins in the European Alps are a consequence of continent-to-continent collision; (2) veins in the Mother Lode in California are a consequence of subduction-zone-to-continent collision; (3) veins in West Africa appear to be a consequence of the collision of oceanic volcanic arcs.

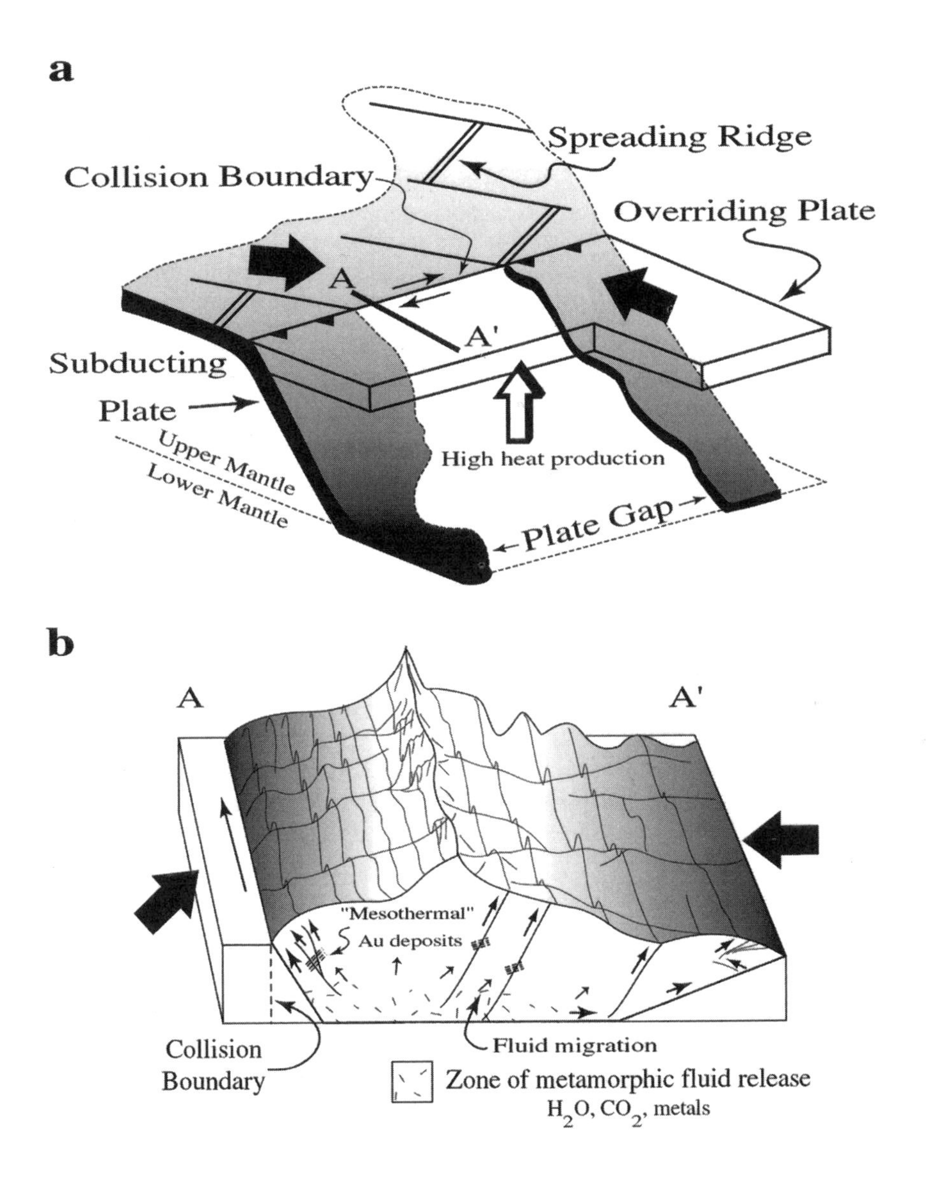

Figure 6. Schematic drawing of a collisional boundary zone. **a**: Illustration of the *plate gap* that occurs when subduction ceases along a portion of a convergent plate margin. The cessation of subduction above this gap results in collision, as illustrated in **b**. The gap allows the rise toward the surface of high-temperature material from deep within the earth. This high heat flow alters the mineralogy of the rocks in the shallower zone, *metamorphism*, releasing fluids from no longer stable hydrous minerals to form a hydrothermal fluid. **b**: The released fluid rises along fracture zones wherein quartz veins are formed that commonly contain important concentrations of gold. The veins are commonly called "mesothermal" gold deposits.

Crustal heating accompanies plate collisions. That process results from the subduction of young sea-floor, mid-ocean ridges, and radiogenic sedimentary materials. The heating leads to the formation of magmas and drives metamorphic reactions. Metamorphism and magmatism mobilize large quantities of hot, buoyant fluids, water and compounds such as CO_2 originally contained within minerals in the crust. These fluids scavenge small quantities of gold and flow along those faults, actively accommodating strain along the collisional boundaries. At depths of 8 to 10 km in the crust, quartz veins are formed during decompression of the fluids at the time of seismic events. The decompression sets in motion processes and chemical interactions in the fluid and adjacent wall rocks that lead to the precipitation of dissolved metals, including gold in the quartz veins. Those vein systems are variously referred to as "mesothermal"- (Lindgren, 1933) and "orogenic"- (Groves et al., 1998) type gold deposits.

Mesothermal veins are texturally distinct from those formed in the shallower crustal environments of sea-floor spreading and above subduction zones. At the depths of formation, fractures do not remain open for any duration. Thus, the nature of flow differs significantly from flow at shallower depths. During periods of seismic activity, fluid flow occurs in the matrix of the rock along grain boundaries and into and along shear zones. Fracture opening is intermittent, resulting in textures referred to as "crack seal." Thin slivers of wall rock may spall off during fracturing, resulting in bands or "ribbons" of quartz separated by seams of wall rock. Thus, mesothermal veins are referred to as "ribbon vein" deposits. Ore depositional temperatures range from 500°C to about 250°C.

Few deposits are economic for metals other than gold. Native gold and electrum are most common, but tellurides occur in minor abundance at most localities. Accompanying minerals include pyrite (FeS_2) and arsenopyrite (FeAsS), and minor amounts of sphalerite (ZnS), galena (PbS), and chalcopyrite ($CuFeS_2$). Occasional minerals include scheelite ($CaWO_3$), bornite (Cu_5FeS_4), tennantite ($Cu_{12}As_4S_{13}$), and molybdenite (MoS_2).

The ore-depositing fluids are generally low-salinity, H_2O-CO_2±CH_4 type. Because of the nature of flow at depth in the crust, the water/rock ratios are always low. Thus, wall rock alteration varies from one host-rock type to another. The relatively lower temperature assemblage generally consists of chlorite ($(Mg,Fe)_5Al(Si_3Al)O_{10}(OH)_8$), sericite, and ankerite. At somewhat higher temperatures biotite is part of the assemblage. The highest temperature assemblage includes amphibole, pyroxene, garnet, and calcite. When magnesium- and iron-rich rocks are parts of the host-rock suite, such as are found in subduction zone complexes, a chromian muscovite mineral called fuchsite or mariposite occurs. Potassium feldspar is commonly associated with ore; albite ($NaAlSi_3O_8$) is characteristically earlier than the K-feldspar and generally indicates higher temperatures.

Coseismic decompression is probably of considerable importance to gold deposition. In addition, wall rock exchange reactions (e.g., sulfidation) are probably of considerable importance for selected host-rock compositions such as iron-rich volcanic rocks and carbonaceous, iron-bearing metasedimentary rocks.

West Africa

The African continent contains some of the oldest rocks on earth, amalgamated several hundred million years ago into a geologically stable mass referred to as the

African *shield*. The nature of continental growth differed somewhat in the first half of earth history from that at present (Windley, 1992). Although there were continent-to-continent collisions, as occurred in more recent earth history, accretion involved the amalgamation of oceanic volcanic arcs and some fragments of ocean basins. Such an accretionary process took place in West Africa about two billion years ago where oceanic volcanic and sedimentary rocks make up a series of structural belts bounded by regionally extensive faults. That zone is found in eastern Guinea, southern Mali, Ivory Coast, western Ghana, Burkina Faso, and western Niger. Gold deposits (figure 5c) occur along a number of the fault systems, with the largest deposits occurring in Ghana, the second largest gold-producing nation in Africa. Although artisanal gold mining has occurred in West Africa since prehistoric times, inadequate infrastructure, onerous mining codes, and destabilizing social conditions have depressed international exploration investment during the latter half of the twentieth century. However, changing conditions during the past decade have brought the proportion of worldwide investment dollars flowing into African exploration to about 17 percent (*Mining Journal*, 1999a).

The Ashanti structural belt is the primary producing region in Ghana and includes the Obuasi, Prestea, Bogosu, Konongo, and Obenemase deposits (cf. Eisenlohr, 1992). Recent exploration has focused on the discovery of high-grade veins, lower-grade concentrations of gold amenable to open-pit mining and heap leaching, and the reprocessing of materials such as mill tailings from previous mining operations. In the most productive mines, the gold occurs for the most part in quartz veins.

Revised mining codes in Burkina Faso and Mali have stimulated gold exploration in these countries for deposits similar to those known in Ghana. The gold-mining industry in Burkina Faso is quite small, but exploration has brought several gold deposits to production decisions that should increase the relative importance of the industry during the next few years (*Mining Journal*, 1999b, p. 51). In Mali the Syama mine has a resource of ≈31 million tons of ore averaging 3.3 grams/ton (*Mining Magazine*, 1999), and the Sadiola Hill deposit is purported to be a major producer (*Mining Journal*, 1999b, p. 77).

Central Asia

The opening of the former Soviet Union to foreign investment in 1991 resulted in a flow of mining-related investment into the former empire, a sum that now constitutes about 7 to 8 percent of the world total (*Mining Journal*, 1999a). Large concessions to explore and develop have become available in Russia and central Asia. The principal investment targets for gold are the republics of Uzbekistan, Kazakhstan, and Khirgizia.

Uzbekistan in Central Asia is home to the world's single largest lode-gold deposit, Muruntau. In all likelihood, turquoise, copper, and gold were being mined in the vicinity of Muruntau at the time that caravans were traversing the Silk Route not too far to the south. During the late 1950s, renewed exploration in the region by Soviet Union geologists resulted in the rediscovery of gold. The great Muruntau mine with somewhere on the order of 130 million or more ounces of gold was the result. The deposit was formed following the collision of two tectonic plate masses about 280 to 250 million years ago (Drew et al., 1996).

The collision of the large tectonic plates that preceded formation of the Muruntau deposit produced a large number of major fault zones along the length of the suture zone, a zone referred to as the Tian Shan fault and fold belt. Many of those fault zones contain significant gold deposits similar in general character to Muruntau. One of them, Kumtor, occurs high in the mountains of the Kyrgyz Republic. The deposit, opened in 1997, has proved to be a technological challenge because parts of the ore body lie beneath mountain glaciers.

Western Australia

Gold was discovered in the Kalgoorlie region of West Australia during the nineteenth century. Success in finding ore bodies intermittently since has established the region as one of the world's premier gold producers. As a whole, Australia currently receives about 17.5 percent of the total investment in mining exploration (*Mining Journal*, 1999a). Much of that capital goes into Western Australia, which currently accounts for about 75 percent of the nation's total gold production (*Mining Journal*, 1999c). Important new operations in the western part of Australia include the Bronzewing and Jundee mines.

Erosion as a Mechanism of Gold Deposit Formation

The exposure of primary, lode-gold deposits to weathering results in the accumulation of gold particles in stream beds, alluvial plains, and other sorts of water-lain materials. The exploration for such deposits is not the focus of large sums of venture capital; however, because those "placer" accumulations are among the historically most important gold deposits, they warrant some mention. The most important placer deposits in the world occur in the Witwatersrand Basin in South Africa.

Summary

The challenge of mineral exploration for gold is to find the earth's gold factories—places in which heat, fluid, and precious metals are transferred from deeper to shallower regimes in the earth. The plate-tectonic environments of oceanic spreading centers, subduction zones, and collision zones (ancient and modern) are the primary locations in which gold deposits form. During the past three decades the increase in the price of gold has spawned a substantial investment in the search for new gold resources. Over the course of time, political changes around the world have opened up new opportunities to explore virtually any area in which the right geologic environments occur. Some of the most interesting developments have occurred in the western United States, western South America, the southwestern Pacific region, western Australia, West Africa, and central Asia.

Acknowledgments

I wish to thank Dennis LaPoint for affording me the opportunity to prepare this paper and to the American Gold Discovery Bicentennial Committee for supporting my participation in the projected symposium "Gold in Carolina and America: A Bicentennial Perspective."

References Cited

Appiah, Henry, David I. Norman, and Isaac Boadi. "The Geology of the Prestea and Ashanti Goldfields: A Comparative Study." In Eduardo A. Ladeira, ed., *Brazil Gold '91*. Rotterdam: A. A. Balkema, Rotterdam, 1991. Pp. 247–255.

Cox, Dennis P., and Donald A. Singer. *Mineral Deposit Models*. U.S. Geological Survey Bulletin 1693 (1986).

Drew, Lawrence J., Byron R. Berger, and Namik K. Kurbanov. "Geology and Structural Evolution of the Muruntau Gold Deposit, Kyzylkum Desert, Uzbekistan." *Ore Geology Reviews* 11 (1996):175–196.

Eisenlohr, Burkhard N. "Conflicting Evidence on the Timing of Mesothermal and Paleoplacer Gold Mineralisation in Early Proterozoic Rocks from Southwest Ghana, West Africa." *Mineralium Deposita* 27 (1992): 23–29.

Franklin, James M., et al. "Arc-Related Volcanogenic Massive Sulphide Deposits." British Columbia Geological Survey, short course on Metallogeny of Volcanic Arcs (1998), Open-File 1998–5, Section B, p. N-1–N-32.

Giggenbach, Werner F. "The Origin and Evolution of Fluids in Magmatic-Hydrothermal Systems." In Hubert L. Barnes, ed., *Geochemistry of Hydrothermal Ore Deposits*. New York: John Wiley and Sons, 1997. Pp. 737–796.

Groves, D. I., et al. "Orogenic Gold Deposits: A Proposed Classification in the Context of Their Crustal Distribution and Relationship to Other Gold Deposit Types." *Ore Geology Reviews* 13 (1998): 7–27.

Hannington, M. D., et al. "Physical and Chemical Processes of Sea-floor Mineralization at Mid-ocean Ridges." In Susan E. Humphris et al., eds., *Seafloor Hydrothermal Systems*. Geophysical Monograph 91. Washington, D.C.: American Geophysical Union, 1995. Pp. 115–157.

Lizasa, Kokichi, et al. "A Kuroko-type Polymetallic Sulfide Deposit in a Submarine Silicic Caldera." *Science* 283 (1999): 975–977.

Kerrich, Robert. "Nature's Gold Factory." *Science* 284 (1999): 2101–2102.

Kirshner, Robert P. "The Earth's Elements." *Scientific American* 271 (1994): 58–65.

Kragh, Helge. *Cosmology and Controversy*. Princeton: Princeton University Press, 1996.

Lindgren, Waldemar. *Mineral Deposits*. New York: McGraw-Hill, 1933.

MacDonald, Ken C. "Linkages between Faulting, Volcanism, Hydrothermal Activity and Segmentation on Fast Spreading Centers." In W. Roger Buck et al., eds., *Faulting and Magmatism at Mid-Ocean Ridges*. Geophysical Monograph 106. Washington, D.C.: American Geophysical Union, 1998. Pp. 27–58.

Mining Journal, 1998a. Central and South America: Annual Review Supplement. *Mining Journal* 331, no. 8493.

Mining Journal, 1998b. Asia and Australasia: Annual Review Supplement. *Mining Journal* 331, no. 8489.

Mining Journal, 1999a. Exploration Supplement: Slump in Activity. *Mining Journal* 332, no. 8519.

Mining Journal, 1999b. Open Spaces: Investing in Africa. *Mining Journal* 332, no. 8515.

Mining Journal, 1999c. "Uncertainty Remains: Asia & Australasia." *Mining Journal* 333, no. 8553: 141–142.

Mining Magazine, 1999. "Randgold Resources' Mali." *Mining Magazine* 180: 12–15.

Oberthür, Thomas, et al. "Gold Mineralization at the Ashanti Mine, Obuasi, Ghana: Preliminary Mineralogical and Geochemical Data." In Eduardo A. Ladeira, ed., *Brazil Gold '91*. Rotterdam: A. A. Balkema, 1991. Pp. 533–537.

Phelps, Richard W. "Newmont Mining Corp.: An Emphasis on the New." *Engineering and Mining Journal* 197 (1996): 34–47.

Reed, Mark H. "Hydrothermal Alteration and its Relationship to Ore Fluid Composition." In Barnes, *Geochemistry of Hydrothermal Ore Deposits*, 303–365.

Rona, Peter A., and Steven D. Scott. Preface to "A Special Issue on Sea-Floor Hydrothermal Mineralization: New Perspectives." *Economic Geology and the Bulletin of the Society of Economic Geologists* 88 (December 1993): 1935–1976.

Scott, Steven D. "Submarine Hydrothermal Systems and Deposits." In Barnes, *Geochemistry of Hydrothermal Ore Deposits*, 797–875.

Shipboard Scientific Party. "Middle Valley: Bent Hill Area (Site 1035)." In Y. Fouquet et al., *Proceedings of the Ocean Drilling Program, Initial Reports* 169 (1998) College Station, Texas. Pp. 35–152.

Stoffers, Peter, et al. "Little-studied Arc-Backarc System in the Spotlight." *EOS, Transactions, American Geophysical Union* 80 (August 10, 1999). Pp. 353.

Suttill, Keith R. "Maricunga—The World's Next Great Gold 'Province?'" *Engineering and Mining Journal* 192 (1991): 33–37.

Windley, Brian. "Proterozoic Collisional and Accretionary Orogens." In Kent C. Condie, ed., *Proterozoic Crustal Evolution*. New York: Elsevier, 1992. Pp. 419–446.

Pre–Columbian Gold

I. S. Parrish

Irwin S. Parrish is president of DMBW, Inc., a mining consulting firm in Golden, Colorado. With degrees from Brooklyn College and Indiana University, he has been active in mining since 1953 and specializes in appraisal and evaluation of mineral properties, compilation of ore reserves, grade control, and mining geology.

Introduction

From the time when our earliest ancestors first picked up and kept a shining golden pebble, mankind has collected on the order of 151,000 metric tons of gold. Because of its virtual indestructibility, most gold produced is still extant. Some has been buried, lost at sea, or is otherwise not accounted for, but it has been estimated that "lost" gold would account for considerably less than 10 percent of past production. Of that estimated 151,000 tons of gold produced over the past, say eight thousand years, close to 19 percent, or 28,500 tons, were mined prior to A.D. 1500. The following table summarizes that production by period and notes the principal sources during each period:

PRODUCTION OF GOLD PRIOR TO A.D. 1500

Period	Tons Au	Cumulative	% of Total	Principal Sources
Pre-4000 B.C.	100	100	0.4	Romania, Egypt
4000–3000 B.C.	900	1,000	3.2	Egypt
3000–2000 B.C.	2,000	3,000	7.0	Nubia (Sudan), Egypt, Altai (central Asia), Balkans
2000–1000 B.C.	12,000	15,000	42.1	Nubia, Arabia, Ireland
1000–500 B.C.	3,500	18,500	12.3	Lydia (Turkey), Bactria, Altai
500–1 B.C.	4,500	23,000	15.8	Spain
1 B.C.–A.D. 500	3,000	26,000	10.5	Spain, Balkans, Bactria (Afghanistan)
A.D. 500–A.D. 1000	1,000	27,000	3.5	West Africa
A.D. 1000–A.D. 1500	1,500	28,500	5.3	West Africa

The Pre-Columbian History of Gold

Tracing the early history of gold production parallels, but is not an exact replica of, the growth of civilization. Civilization developed in the lands bordering the Mediterranean Sea (map 1). Ancient gold production came from modern-day Egypt, Arabia, Turkey, Romania, Greece, and Spain. Production elsewhere moved over early trade routes to the Mediterranean powers from distant Ireland, the Sino-Russian Altai Mountains, Afghanistan, and far-off India. With the conquering of the African deserts, the gold of Ghana, Mali, and Nigeria was carried by camel caravan to the Mediterranean empires. Finally, by 1500 the great Atlantic Ocean was traversed and Spanish ships brought gold from the New World to the kingdoms of the Mediterranean and Europe.

Map 1 ***I. S. Parrish - DMBW, Inc.***

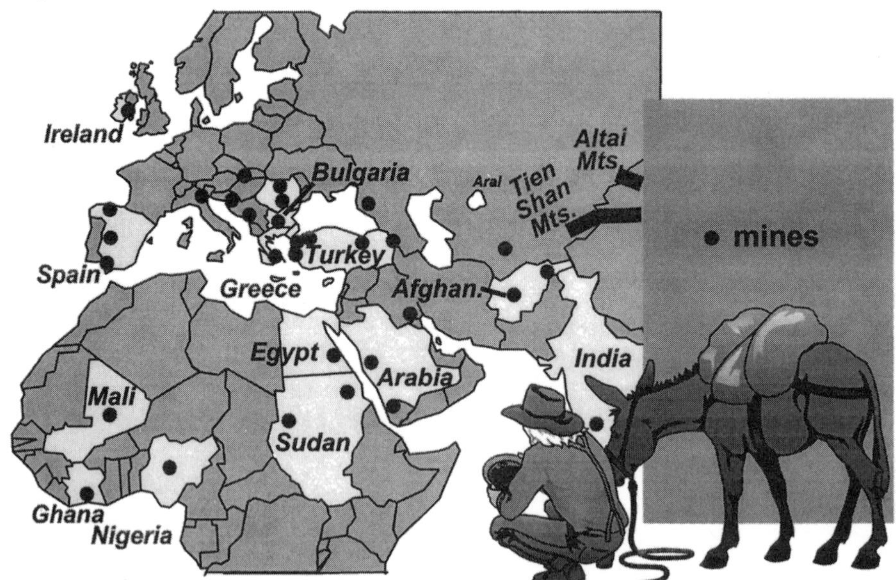

Pre-Columbian Principal Gold Sources

Before 4000 B.C.

There are no records of the very earliest discoveries of gold. The first collecting of gold must have been ancient, inasmuch as gold-adorned stone knives have been noted in numerous neolithic burials. The first evidence of gold having been actively sought, collected, shaped, and treasured comes from a 1972 archaeological discovery made in the Balkans on the western shore of the Black Sea. The site, near Varna, Bulgaria (map 2), yielded 5.5 kg of gold, which had been hammered and shaped into ornaments to be sewed onto clothing or worn as bracelets and so on. Archaeologists date the burial site as being pre-4000 B.C. The gold found at Varna is postulated to have come from sources in the Transylvanian Alps of Romania (map 2). It is probable that gold was also being "mined" from Wadi Allaqi in southern Egypt at about the same time.

Gold "mining" during this period probably entailed little more than gathering of exposed nuggets or flakes. Metallurgy was restricted to hammering and basic smelting by fusion using charcoal fires. Total gold produced in these two thousand or more years would have been no more than a hundred tons.

Map 2

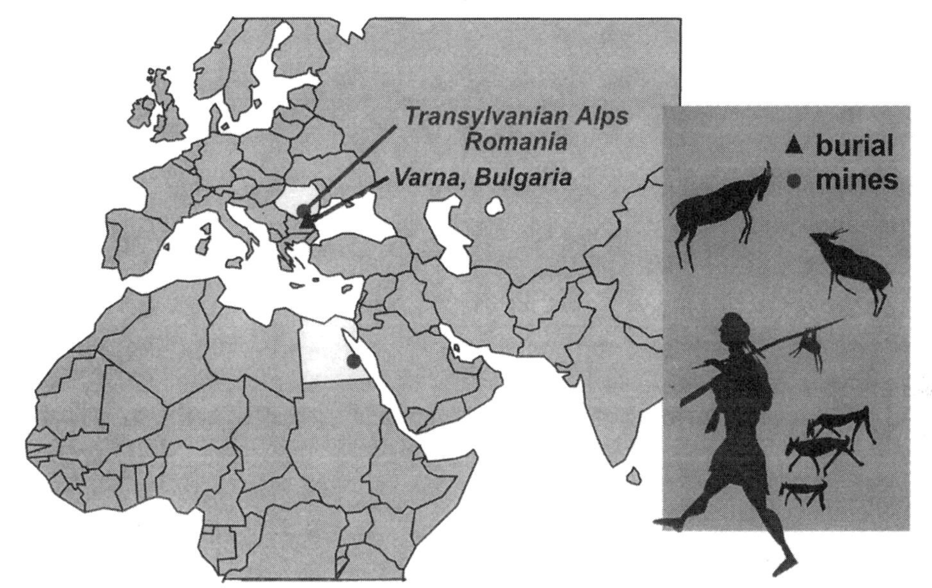

4000 B.C. to 3000 B.C.

The major source of gold in the fourth millennium B.C. was in upper (southern) Egypt. At least thirty-six individual bedrock gold mines have been identified as being in production between the Red Sea and the Nile River in southeastern Egypt (map 3). Seven of the ancient mines were to be reopened in the twentieth century. In addition, the sands and gravels of the Nile River yielded much placer gold to the early Egyptians.

Gold ornaments have been found in archaeological digs near the ancient Sumerian "city" of Uruk on the Euphrates River near the head of the Persian Gulf. That gold may have come from Egypt, or possible from Dilmun on the northeastern coast of Arabia near Bahrain. Other potential sources are in western Turkey near Lake Van or in eastern Iran near Hamadan. Gold was also being mined in the Kolar-Hutti area of India, where underground workings date back to the fourth millennium B.C.

Total production from 4000 to 3000 B.C. would have been on the order of nine hundred tons. Mining included placering using gourds, batea, and other primitive pans, as well as pitting or trenching and unsupported shallow underground "shafts" and "drifts" excavated by using stone hammers, antler picks, and bone and wooden shovels. Metallurgy was similarly primitive, entailing crushing and hand sorting

Map 3

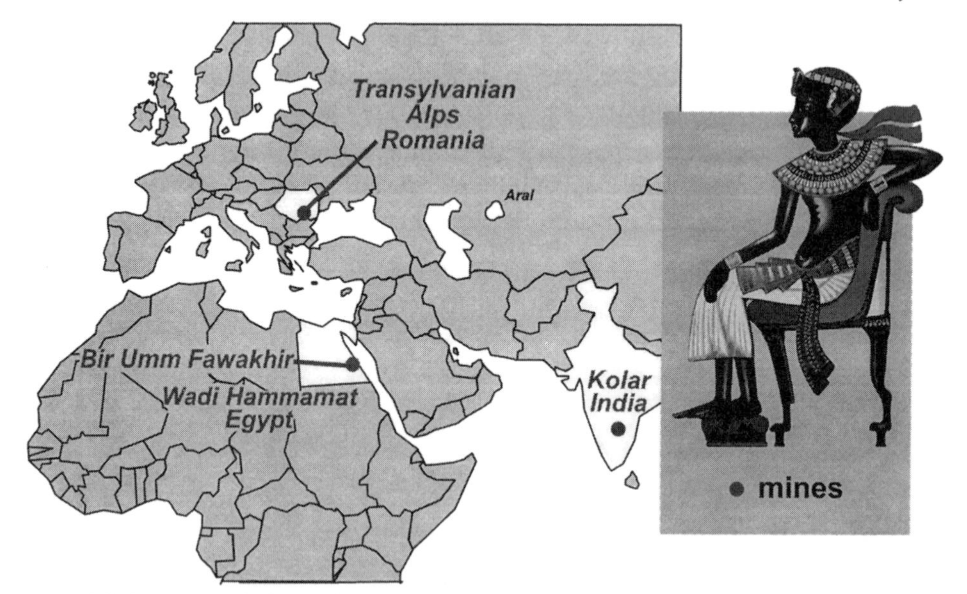

and possibly some washing of debris. Refined gold was unknown, so gold was 800-850 fine. Ancient Egyptian gold is often speckled with white because of inclusions of silver, platinum, and copper. Gold ornaments were cut and hammered.

The growing significance of gold resulted in the "Code of Menes," issued in 3500 B.C. by Menes, the first pharaoh. The code, which declared that "one part gold was equal to two and one-half parts of silver in value," suggests not only that man accepted and considered gold a basis for measuring worth but also that silver was more highly valued in ancient times than at present, when one part of gold is equal to fifty-three parts of silver.

3000 B.C. to 2000 B.C.

Most of the gold produced during the third millennium B.C. originated or found its way to Egypt. Much of it was produced in Egypt east of the Nile at places like Wadi Hammamat and Wadi Fawakhir. Production in that area began about 3200 B.C. and wound down about 1793 B.C.—a span of fifteen hundred years. Centuries later mining of those deposits would begin anew. Although Egypt was an important producer during the third millennium, the greatest production was from Nubia south of Egypt (map 4). Nubia, named from *Nub*, the Egyptian word for gold, had seen sporadic production in the fourth millennium, but in 2980 B.C. Egyptian forces invaded and captured Nubia. Mining was systematized. Slaves, prisoners of war, and others out of political favor were sentenced to work the Nubian mines, then controlled by the military.

An archaeological find at Ur in Sumeria at the head of the Persian Gulf yielded many exquisite gold ornaments from this period. The source of the gold may have been Arabia, where about 2450 B.C. gold was mined near the Persian Gulf at Rumma, Yabrin and Dawasin (map 4), or to the north in Anatolia (ancient Turkey). Gold was also being produced in some quantity from placers in the Avoca valley (Crogan Kinsella) in County Wicklow in eastern Ireland (map 4). This gold was exported east and showed up in Scotland about 2250 B.C. The Irish deposits were worked intermittently until the eighteenth century.

About 2500 B.C. additional players entered the scene as the first of the Scythians moved into eastern Europe. The Scythians were a nomadic, warlike people being driven westward from the steppes of central Asia. The Scythians brought with them gold from the Urals, the Altai region, and other sources east and north of the Aral Sea (map 4). Another European gold supplier in the third millennium B.C. was France. Neolithic monoliths, possibly of Celtic origin, contain beads of gold considered by archaeologists to be locally derived.

In the Americas, mankind was increasing, but there was no record of gold workings. The North American Indian never developed any particular attraction to gold—the Amerindians and the Aborigines in Australia being unique among

Map 4 *I. S. Parrish - DMBW, Inc.*

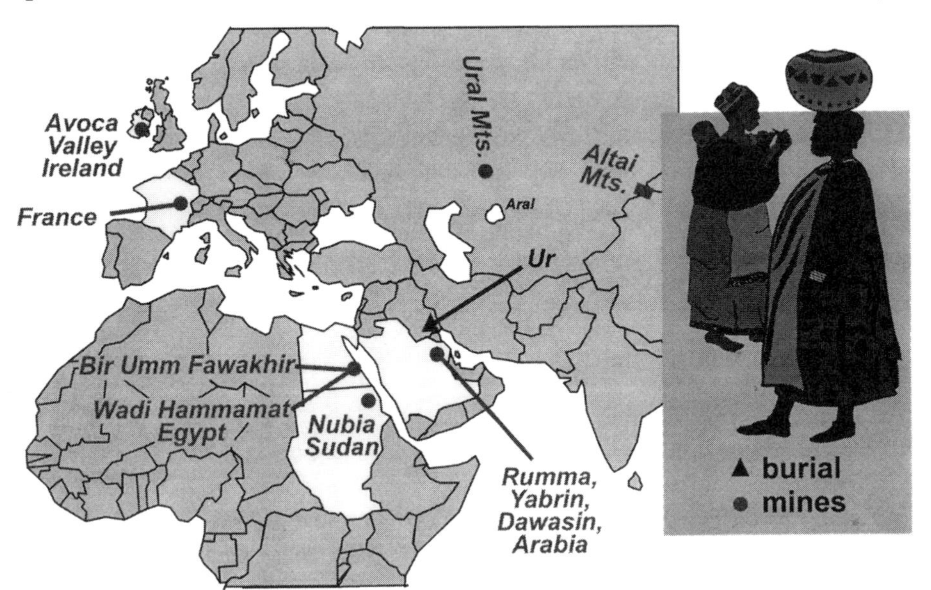

3000 - 2000 BC Principal Gold Sources

major human groups in that regard. Worldwide production of gold from 3000 to 2000 B.C. was on the order of two thousand tons, with well over half coming from Egypt and Nubia.

2000 B.C. to 1000 B.C.

From 2000 to 1000 B.C. the total gold produced increased several fold over that accumulated to the start of the period. Mining during this "Golden Age" would account for 42 percent of all gold mined prior to A.D. 1500. The principal source remained in Nubia, but production had begun from deposits in Europe, Asia, and the Americas. In Europe the Phoenicians were mining at Huelva and Los Millares in southern Spain by 1240 B.C. (map 5). Ladles and crucibles for working gold have been found along with gold ornaments in both Limerick and Tipperary Counties in Ireland. Presently, the gold collection of the National Museum in Dublin is unequaled in western Europe.

Gold, probably of Irish origin, was fashioned into elaborate ornaments as uncovered in Trundholm, Denmark (dated 1650 B.C.), and Leubingen, Germany (1500 B.C.). Gold from the Balkans and the Carpathian Mountains near the Black Sea was dealt in trade to Crete, where the Minoan culture was accumulating gold up to the sack of Knossus in 1375 B.C. Crete was second only to Egypt in its holdings of gold.

The gold found by archaeologist Heinrich Schliemann at Troy (map 5) may well have arrived there via Crete. A small gold mine is known to have existed twenty miles from Troy at Abydos in Turkey. Another possible source of the Trojan treasure was in the Caucasus Mountains at Colchis (map 5). It was that Caucasian area which was visited about 1200 B.C. by Jason and his Argonauts in search of the Golden Fleece. (The fleece was literally sheepskin, which even up to the present time is used as a primitive trap to ensnare gold being washed from a stream.)

In Asia gold was being produced from placers in China and by 1122 B.C. from placers in Korea. Production continued in India and is described in Hindu scriptures. In the Americas the early Peruvians were placering for gold by 1200 B.C. In southern Africa the oxidized deposits of Zimbabwe and Mozambique were yielding gold from a myriad of small pits, trenches, and shallow shafts. Some authors believe that that area was the fabled land of Punt, whose gold found its way to Egypt, Sumeria, and Babylon. The exact location of Punt was an ancient secret. Its location is still in question. All that is known is that a round trip from Egypt to Punt took two to three years by sea.

Most of the gold produced in the second millennium B.C. came from Nubia (modern-day Sudan and Ethiopia). Nubian gold was used for the golden coffin of King Tut about 1450 B.C. It was mined from underground drifts and stopes. Mining was by a crude room-and-pillar method. Usually pillars were of rock, but at the Atallah mine relics of both stone and wooden supports have been found. Loose rocks, or walls, were held up by crude masonry. Lighting was afforded by fire sticks and later by oil lamps. In some cases the lamps were affixed to the heads of slaves.

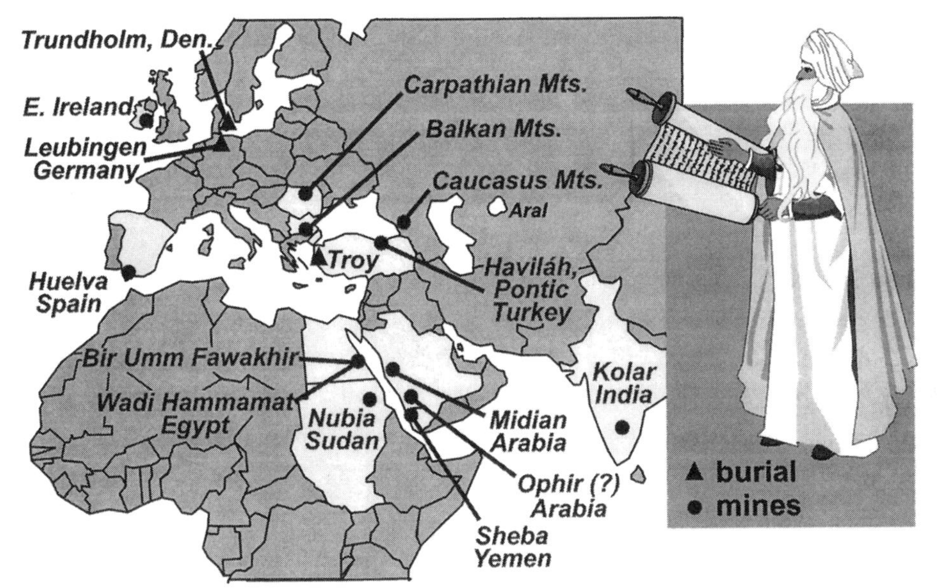

Ore was carried from the mine in baskets in much the same way it is today in parts of South America and Asia. Similar conditions prevailed in eastern Egypt at Bir Umm Fawakhir (map 5). That area housed more than one thousand people from about 1307 to 1070 B.C. It was worked again between A.D. 100 and 200 and from A.D. 500 to 600. At Am Rus on the Red Sea one mine was twelve hundred feet long and three hundred feet deep.

The oldest map known to man is the Turin Papyrus, which depicts an Egyptian gold camp of about 1150 B.C. The map shows the gold "mountain," the workers' huts, temples, and roads to other areas. The unidentified camp was probably Wadi Hammamat (map 5). As many as ten thousand people were employed there.

Gold concentrate was sent from the mines to outposts such as Kublen for smelting and refining. Whereas gold of earlier times was only 800 to 850 fine, gold now was 950 to 960 fine and of a rich golden yellow hue as man learned the art of refining. By 1580 B.C. goldsmithing was considered a profession of great stature and importance in Egypt.

One remaining gold province during the second millennium B.C. and probably the best known was the Biblical (Arabian) gold mines at Ophir, Sheba, Havilah, and Midian (map 5). "Havilah" probably refers to placers along the Pison River near Trabzon, Turkey—the Pontic gold field. Midian, a small mine in Saudi Arabia at Hejaz near the Red Sea, was likely the Eldorado of the Hebrews. There gold was found in oxidized quartz veins and was mined to a considerable depth. Ophir was likely in Yemen and was probably the source of the gold the queen of Sheba brought to King Solomon. Other sites have been proposed for Ophir, but Yemen is currently the preferred one. One mine was located 120 km north of Mareh.

The queen of Sheba's gift to King Solomon was on the order of 32.5 tons of gold. Solomon's tribute from his kingdom was eighteen tons per year. These are exceptional quantities. The best estimate for the amount of gold produced worldwide from 2000 to 1000 B.C. is on the order of twelve thousand tons. The increase, in part, reflects major metallurgical advances. "Amalgamation," or the collection of gold by the use of mercury, allowed the recovery of very fine gold flakes or particles. "Liquation" introduced lead into the smelting process to improve gold recoveries. "Salt parting" permitted silver to be separated from gold during the refining process. The net result was that the second millennium B.C. produced more than six times the gold produced in the preceding millennium.

1000 B.C. to 500 B.C.

The first gold-bearing coins were made toward the end of the second millennium B.C. By 687 B.C. crude coins were widely used. Electrum coins were available by 640 B.C. in Asia Minor, and in 550 B.C. King Croesus of Lydia minted the first pure gold coins. He gave large amounts of gold to the temple at Delphi. His wealth led to the expression "rich as Croesus." The source of the Lydian gold was the Pactolus River in modern-day Turkey (map 6). The Pactolus, today called the Gediz or Sarahat, derived its gold from Mount Timolus and drained west through Smyrna to the sea. In mythology King Midas of Phrygia (Lydia), to rid himself of the ill-considered Golden Touch, was told to immerse himself in the Pactolus River. From that time on the river sands and gravels contained gold. The Pactolus was a major source of gold for three centuries. It ceased to bear gold after 550 B.C. as the headwaters eroded through the Mount Timolus source area.

Other sources of gold important to the Greek city-states were the steppes of central Asia (the Altai region), through trade with the Scythians, and the islands of the Aegean Sea. Herodotus, writing about 450 B.C., said that the Altai gold came from an area near present-day Semiplatinsk in Kazakhstan (map 6). Herodotus wrote that the Altai gold was mined by the Chudi, a people of Finnish origin. The Scythians, trying to protect their sources, said that the gold was from a land with huge snowflakes and winter that persisted for more than three-fourths of the year. They said that the gold was mined by one-eyed men-monsters and protected by griffins and other large beasts. The winters thus described fit the north Kazakhstan region of Semiplatinsk. The "one-eyed monsters" were derived from underground miners using oil head lamps or candles. The "other large beasts" may relate to frozen mammoths or saber-toothed tigers. The aggressive Scythians, known for their horsemanship, were also talented goldsmiths.

Aegean islands with significant gold deposits included Siphnos; Thasos, offshore from Thrace; and Samos (map 6). The tiny island of Thassopolis, near the north

Map 6 *I. S. Parrish - DMBW, Inc.*

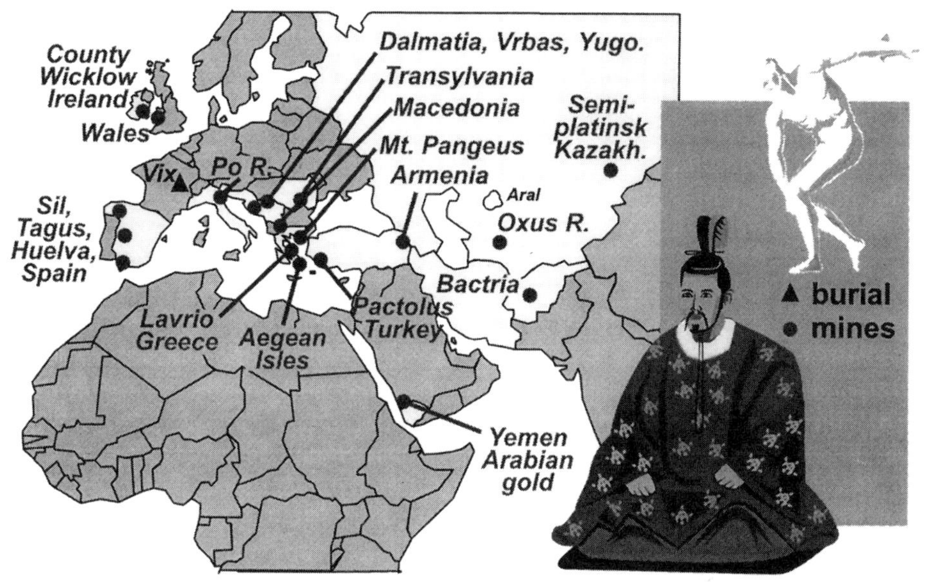

1000 - 1 BC Principal Gold Sources

shore of the Aegean Sea, had a small gold mine. The island was owned by the wife of the Greek poet/historian Thucydides. The mine yielded the equivalent of one hundred thousand dollars per year, allowing Thucydides to follow his life of letters. Siphnos contained the richest and most extensive gold deposits of all Greece. These were mined from shortly before 700 B.C. until about 500 B.C., when the underground workings were flooded. For a short time Thasos produced two tons of gold per year, according to Herodotus.

To the Greek city-states, gold was of less importance than silver. Significant, if not major, silver deposits were mined on the mainland at Lavrio (map 6) and in the north at Mount Pangeus. Gold would have been a by-product of both areas. Placer gold was recovered near Pangeus in Macedonia at Styrynon and near Serres at Amphipolis.

Greek mining, although it relied upon the efforts of slaves and prisoners, was quite sophisticated. Stoping was performed both overhead and underfoot. Timber supports were used sparingly, but rock pillars were common. Short crosscuts were put in not only to connect ore drifts but also to provide ventilation. Fires were set at shafts to assist air circulation. Workings were so placed and divided as to improve air movement. Fires were also used to assist in breaking rock. Heated rock faces, when quenched with water, would split, making them easier to mine. It has been estimated that daily advance by fire-setting would have been only on the order of 12 cm (5 to 6 inches) in a one-meter-wide drift. The Greek mining was, nonetheless, effective. At Lavrio quartz veins in schist within diorite were mined for lengths up to seven hundred feet. Virtually no ore was left. Waste heaps now contain only traces of gold.

Unfortunately, there are few written descriptions of mining during the first millennium B.C. since mining was performed by slaves and was not considered a "profession" or an endeavor worthy of the cultured intelligentsia, who viewed it as being beneath them. Thus, mining was among the most poorly described human pursuits. The nearby Etruscan people received their gold from central France and northern Italy's Po River (map 6). Celtic (Irish) gold continued to be produced. A form of gold coinage may have been in use in Ireland prior to 500 B.C. Spanish gold was being mined in Andalusia by the Phoenicians and was a principal source of gold to the Mediterranean. The noun "Spain" is derived from *Spania,* the old Phoenician word for burrowing or mining. Production in Asia Minor would have come from Yemen and Arabia, but the Biblical mines of Ophir, Midian, and so on would have been exhausted before 500 B.C.

Most of the Persian gold was accumulated through conquest of Egypt, Babylon, Judea, and other lands. From about 750 B.C., additional gold would have come from trade with the Scythians and their Altai gold. At about that time Scythians settled in present-day Armenia, and gold production increased from that area. Production from the Oxus River and the nearby Bactria region (today's Iran, Tajikistan, Afghanistan, et al.) was probably the chief source of newly mined gold to the civilizations of Asia Minor (map 6). Bactria provided ten tons of gold per year to Persia by 500 B.C.

Elsewhere, China and Korea in Asia probably continued to recover placer gold, but the record is incomplete. Japan's first recorded production occurred about 660 B.C. Indian traders, much as the Scythians, tried to hide the sources of their gold. A Sanskrit myth calls the Indian gold "Piplika"—meaning "collected by ants." The story goes that the gold was collected in hills by giant ants with exceptional powers of smell. The ants would track by smell any man who carried away gold. In the Americas, Peruvian goldsmiths were hammering and shaping gold ornaments near Nazca, but gold mining had not advanced as rapidly as it had in Africa, Asia, and Europe.

500 B.C. to 1 B.C.

Gold was being produced in record quantities. The principal source was now Spain. Macedonia, Egypt, Bactria, and the Balkans were of secondary importance. A detailed description of the Egyptian gold mines was given by Agartharchides, who visited the mines in 170 B.C. His description was quoted in 50 B.C. by Diodorus Siculus. The actual area visited was on the same latitude as Aswan in Egypt in the Atahi Mountains between the Nile River and the Red Sea. By 50 B.C. production from Asia Minor, including Arabia, dwindled as the deposits were no longer economical to work. Gold mining continued in India. A fourth-century B.C. Hindu text describes gold as occurring as nodules and needles and being either yellow, red, or bluish. The yellow ore was probably pyritic, the red limonitic; the bluish likely referred to copper ores bearing gold.

Of particular significance and historical importance were the gold and silver mines of northern Greece or, more accurately, Thrace and Macedonia (map 6). The mines were principally near Mount Pangeus, at Paeoria in Thrace, along the Struma River east of Pella, and north of modern Salonika. The Strimonas River was worked as early as 436 B.C. Those mines originally financed Alexander the Great. Alexander later confiscated one thousand tons of gold from the Persians. He took over the mines of Turkey, the Oxus, Bactria, and the Indus valley. Wherever he went, Alexander brought along in his entourage experts in the art of gold prospecting and extraction. By 335 B.C. Alexander had accumulated most of the world's gold. By the time of his death in 323 B.C., the Thracian mines had produced nearly seventeen hundred tons of gold.

With the depletion of resources in Egypt, Asia Minor, and Greece (including Thrace and Macedonia), the chief source of gold from 500 B.C. to 1 B.C. was Spain. The gold of Spain was known and worked as early as the time of King Solomon. Later it was worked by the Phoenicians and their vassals. By 206 B.C. the Romans had taken over the mines. For sixty years Rome battled to quell revolts and repel invaders covetous of the Spanish deposits. By 146 B.C. Rome had completed and solidified its domain. To work the mines, Rome enslaved, among others, sixty thousand Carthaginians. Roman engineers began to improve and organize mining on a scale previously unknown. The amount of gold produced was such that Roman gold holdings in 150 B.C. were four times those of 218 B.C. despite the cost of almost constant warfare. In 58 B.C. Julius Caesar brought enough gold from Spain to pay all of Rome's debts and to grant two hundred gold coins to every Roman soldier. By the start of the Christian era, Spain was producing ten tons of gold per year.

The first deposits mined were near the Mediterranean coast, by Huelva (map 6). The mineralized trend was followed and extended into central and then northern Spain, a distance of more than two hundred miles. In the last century before the Christian era, the Sil River area of northwest Spain was the principal source of gold.

Rome methodically prospected for gold. Records indicate that the Romans searched for gossans and quartz pebbles as guides to deposits. In mining they introduced hydraulicking, using monitors to wear down auriferous gravel cliffs. Where the pay gravels were above water levels, the Romans used aqueducts and waterwheels powered by slaves and up to 14 1/2 feet in diameter to raise waters to huge dammed ponds, or cisterns, as a means of supplying their monitors. "Booming," or allowing waters to flood over boulder fields to concentrate gold in boulder traps or upon mats of Ulex plants, was introduced. Riffles were likewise used to concentrate gold.

Underground workings were extended below the water table by using force pumps and Archimedes' screws to dewater the workings. Two- to three-foot-wide circular shafts with footholds in the wall accessed some deposits. Level workings were wider at the top than at the bottom in order to accommodate workers carrying baskets of rock on their backs. Slaves used stone hatchets and pointed moils of sheep bone to break the ore. Centuries later, in A.D. 1845 at Potosi, Spain, remains of seventeen Roman slaves and their utensils were uncovered; these men were victims of some unrecorded mine disaster.

Roman mining was widespread, large in scale, and effective. As noted earlier, Roman workings could be found in almost all of Spain (and Portugal). In northwestern Spain, one mine in Asturias treated thirty to forty million tons of ore, and in Galicia five million tons of quartz were crushed. Mines were placers, open pits, open trenches, and underground, entered by adits or shafts. The Tagus River was a particularly productive placer area. Pliny wrote of Tagus River gold nuggets weighing half a pound. Relics and remnants of the Roman mines can still be seen in Spain. From 500 B.C. to 1 B.C. Spain was the principal supplier of gold to Rome, but it was not the only one (map 6). The Roman Empire had taken over the mines of Egypt and Macedonia, and those sources were worked to virtual depletion. In 54 B.C. Caesar brought Celtic gold back from Dolgelly and Golofau, Wales, in England. Gold was mined in Dalmatia (west of the Balkans) from 150 B.C. to A.D. 80. The old Etruscan gold sources in northern Italy, the western Alps, and along the Po River were reworked. In Transylvania and in Yugoslavia (Vrbas) the Romans introduced "booming" to recover alluvial gold. Gold was mined at Salassi/Victmuli and Vencllae in northern Italy (map 6), forty miles east of Milan, and at Aguilesa, twenty-five miles north of Trieste. Although the energy and engineering skills of the Romans dramatically increased the yearly supply of gold, their efforts affected only the final years of the first millennium B.C.

From 1000 B.C. to 1 B.C. it is estimated that eight thousand tons of gold were produced. That total represents only 67 percent of that estimated for the 2000 B.C.-1000 B.C. period. The decrease in production reflects the depletion of the mines in Arabia, Turkey, and, most significantly, Nubia.

1 B.C. to A.D. 500

At the start of the millennium, Rome actively sought gold from any source. Rome demanded that all taxes be paid in gold and silver, resulting in rebellions in northern Germany, where gold was scarce. By A.D. 120 Rome controlled every known gold mine in the Western world.

The Macedonian mines, reopened in the last decades B.C., were actively mined until about A.D. 150. The emperor Nero mined gold from the Dalmatia area of Yugoslavia (map 7). One mine yielded 22 kg of gold per day. The emperor Trajan captured and mined gold from Transylvania, Romania, in an area near Mount Apuseni on the Oltul River. In A.D. 101 there was a "gold rush" to Dacia (Romania) equal to later-day rushes to California or the Klondike. From A.D. 97 to 106 Rome removed 250 tons of gold from Dacia (Romania). Gold was also recovered from mines between the Danube River and the Carpathian Mountains (map 7). Most of these European mines were, however, worked out by A.D. 250.

Map 7 ***I. S. Parrish - DMBW, Inc.***

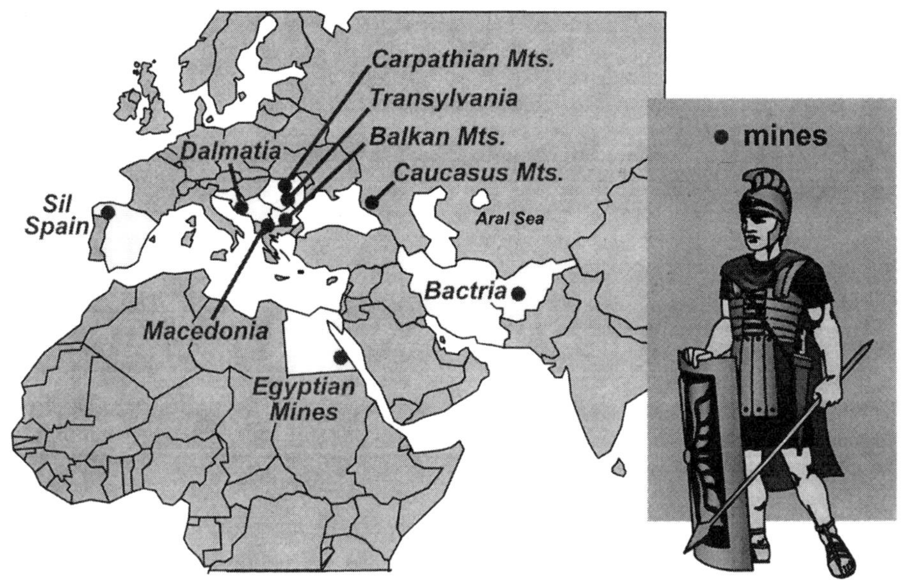

1 BC - 500 AD Principal Gold Sources

For the most part, the mines were operated using slaves. Changes were occurring, however, and records from the second century A.D. describe a "freeman" contracted to work six months as a miner in exchange for board and the equivalent of two hundred eighty dollars.

The Romans reopened the mines of upper Egypt and kept them in operation until A.D. 600. The Egyptian mines, although small in comparison to the Spanish, were still of good size. One underground operation followed a quartz vein in granite for more than three hundred feet in a six-foot-high drift. Air shafts and side "galleries," or stopes, were also driven. More than a thousand people were housed at one Roman mine in Egypt.

Nonetheless, Rome's prime source of gold was Spain, particularly the Sil area of northwest Spain. Production remained at nearly ten tons per year for the first century of the new era but dropped afterward. All production ceased in the fourth century, following invasions by the Visigoths, Vandals, and other barbarian tribes.

From about A.D. 200 on, Rome's search for gold became less and less successful. Known mines were being depleted. New mines were not being discovered. The situation resulted in part from a stagnation in Rome's growth. New lands were not being conquered; indeed, the opposite was happening. Lands were being lost to the marauding barbarians. Rather than collecting gold as taxes, Rome found itself paying gold as tribute, or ransom, to keep the barbarians at bay. When the Goths sacked

Rome in A.D. 408, a ransom of 2.25 tons of gold (and 13.5 tons of silver and 1.3 tons of pepper) had to be paid to the victors. In A.D. 461 the Ostrogoths were bought off with three hundred pounds of gold. From A.D. 447 on, the Huns received a tribute of one ton of gold per year. Gold also flowed east to the Orient in exchange for spices and silks coveted by the Roman gentry.

As Roman power and gold dwindled, Byzantium's grew. Much of the Byzantium gold came from the Balkans. Turkey, the Caucasus, and Bactria (map 7), where there was active exploration, were other sources. More came from gold seized from pagan temples from A.D. 363 on, after Constantinople officially declared Christianity the state religion. Some gold, no doubt, came to the Byzantines from Africa, including Egypt. But as the Vandal invasions extended to the Nile by A.D. 500, even those sources must have been intermittent at best.

India, for centuries a steady producer, saw its gold production become insignificant after A.D. 300 as slavery waned in the subcontinent's culture. Gold production from India did not revive until A.D. 1500. On the other side of the world, Peruvian and Mexican cultures were developing, and gold was used for ornamental purposes, but not as "money" or as a measure of wealth but rather more as a symbol of status. Total gold production during the first five centuries of the Christian era was on the order of three thousand tons, and most of it came in the first few hundred years.

A.D. 500 to A.D. 1000

In Europe almost no gold mining (less than one ton per year) was taking place from A.D. 500 to 1000. Gold was so scarce that no gold coins were minted in Europe from A.D. 850 to 1243. It was not until A.D. 900 that gold mining was renewed with discoveries in then Hungary, now modern-day Czechoslovakia. In A.D. 938 there was a major gold rush to the Harz Mountains in Germany. Spain was producing silver and a little gold, but by the seventh century it was controlled by the Arabs, and the metals flowed east. There was no shortage of gold in Byzantium in the fifth and sixth centuries. Most of it came from Timbuktu, the "City of Gold," and likely was from West Africa (map 8). Byzantines operated gold mines at Wadi Bir Umm Fawakhir in eastern Egypt and in Syria and Persia. By the ninth century all of those sources fell to the Islamic Arabs, and Byzantium's gold came only from mines in the Balkans, around the Black Sea, the steppes of central Asia, and the Pontic field of Turkey; none of those equaled the African production.

With the rise of Islamism in the Arab world after A.D. 630, the balance of power shifted to Mecca and into Arabia. Gold mines from Spain to India were controlled by the young Moslem empire. The main mines were in Syria, Arabia, Afghanistan (Bactria), and Armenia. The Nubian and Egyptian gold fields were once again reopened. The principal source of Muslim gold, however, was Ghana and the surrounding West African area. The Arabs were careful not to describe the location of the source. It was said to be in a fabled land called *Waranga*. Several centuries passed before the Portuguese finally located and identified *Waranga* as the Gold Coast of West Africa. The Arabs brought the Waranga gold over desert trade routes using

Map 8 *I. S. Parrish - DMBW, Inc.*

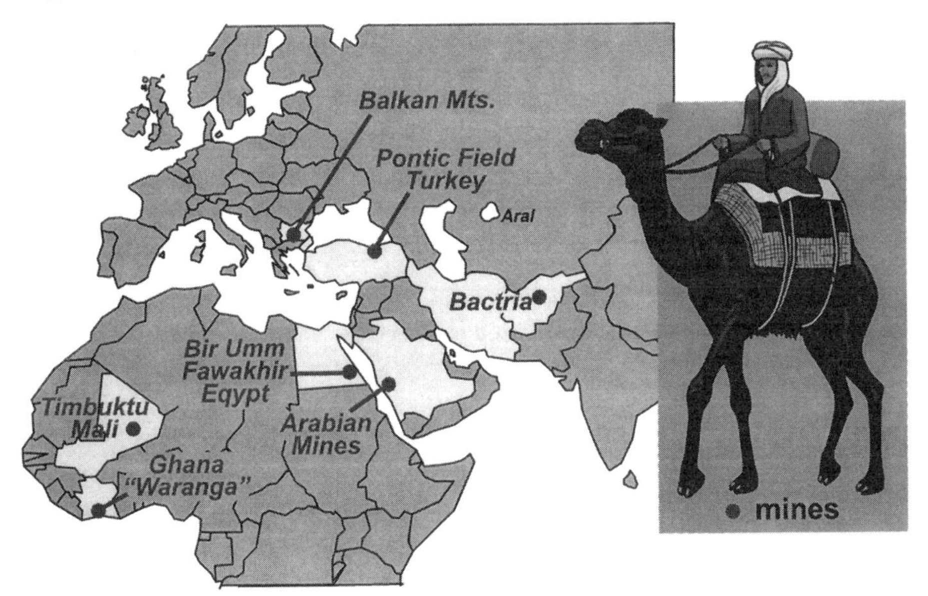

500 - 1000 AD Principal Gold Sources

trains of camels (as large as twenty-five thousand in a train) to transport gold north and east and manufactured goods and salt south and west. Elsewhere in Africa the Bantu people of Zimbabwe were panning gold along the Limpopo River. In the Americas, gold metallurgy advanced in Ecuador and Peru, with the alloying of gold and platinum as evidence.

Gold production from A.D. 500 to 1000 did not compare to immediately earlier periods. Estimated production was only on the order of one thousand tons and mostly from West Africa. The first millennium of the Christian era saw gold production of somewhat over four thousand tons, just over half that of the previous millennium and only one-third of that produced in the Golden Age of the second millennium B.C. The principal sources were Spain for the first several centuries, the Balkans in the early years of the Byzantine Empire, and West Africa in the final century of the first millennium A.D.

A.D. 1000 to A.D. 1500

Gold mining virtually ceased in Europe with the fall of the Roman Empire. It was estimated that whereas nearly ten tons of gold were produced each year during Roman time, less than two tons were produced each year during the Middle Ages. The European mining revival began in about A.D. 900 with the discovery of deposits north of Budapest and east of Vienna in an area of the Hungarian Empire called

Kremnica, or Kremnitz (now Czechoslovakia) (map 9). Mines were also located at Klagenfurt, or Magdalenburg. Virtually all of the gold produced in Europe from 900 to 1500 originated in the Bohemia area of the Hungarian Empire. Bohemia's resources made it the wealthiest of the medieval nations. The gold of the mountains south of Prague attracted a major gold rush in A.D. 1260. By 1400 some gold was being mined underground in France at Vosges and Alvergne in the Pyrenees and in the Alps at Badgastien. Placers were being reworked in Germany along the Rhine, Rhone, and Caronne Rivers. In Italy the Po River was again being worked.

During the early part of the Middle Ages, slaves, serfs, and prisoners were used to work the mines. Then, from about A.D. 1300, "professional" miners began to appear. Special privileges were accorded to those career miners. In Bohemia miners were excluded from taxation and military service. Mining communities were allowed to be self-governing and to brew their own beer. The art, or science, of mining became socially acceptable. Books were published dealing with the previously spurned endeavor. Georg Agricola's *De Re Metallica*, the leading classic text on mining, was published in 1556.

Until about 1500 the mines themselves still belonged to princes, but with the rise of capitalism, privately owned mines became more common. With the privatization of mines the miner became an employee and lost the stature, privileges, and perquisites accorded to the Bohemian miners of the fourteenth and fifteenth centuries. The rise of the Bohemian and Saxon "professional" miners also resulted in

Map 9

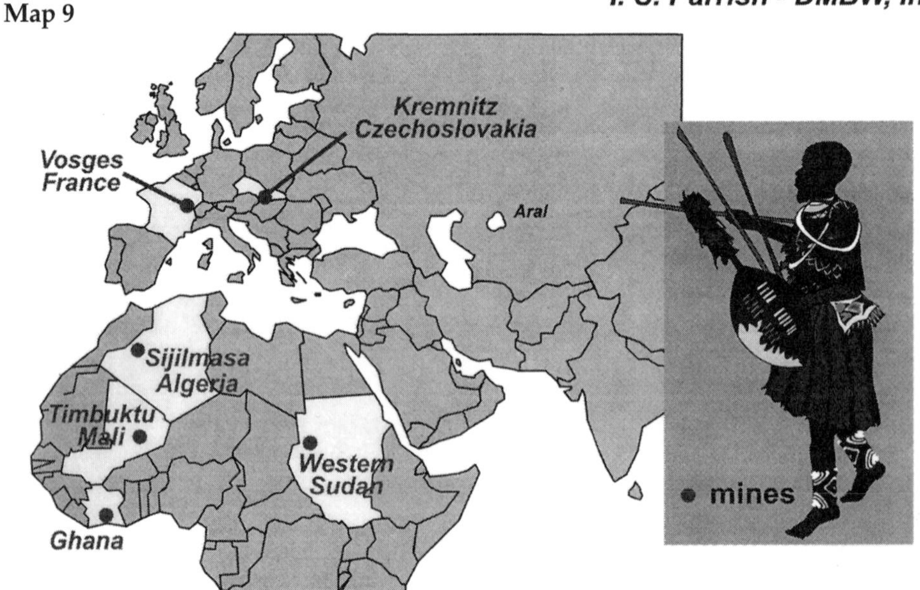

improved mining methods and technology. For example, the piston pump was introduced in the 1400s to dewater Bohemian mines.

The rebirth of gold mining in Europe was significant in that it mirrored the growth of technology, of capitalism, and of the ability of freedmen to work as they chose. The rebirth of gold mining in Europe was more significant as a social phenomenon than for its economic import, however. For in terms of gold produced, Africa from A.D. 1000 to 1500 was by far the largest producer of gold in the world. Sadly, that fact reflects as much the continuing use of slave labor as it does the tenor of the deposits. For some decades, more than eight tons of gold each year passed from Ghana, Mali, and Nigeria through Timbuktu, over the Sahara to Sijilmasa (map 9), and thence to the Mediterranean. Up to the thirteenth century most of that gold went to Arabia. Italy began to receive it in later years. In the 1400s the Portuguese established trading posts on the Atlantic shore of the Gold Coast. Other parts of Africa were likewise producing gold. Western Sudan was turning out about one ton per year; Zimbabwe was another producer.

Production from the Orient was principally from the Chinese provinces of Shandong and Zhejiang. The quantity produced, although of great importance to China, was, nonetheless, small when measured worldwide. Most of the gold that graced the Chinese dynasties was gained through trade with the West. For at least a thousand years China and the Far East traded their spices and silks for the gold of the West.

In Peru gold was being recovered from placers near Urchana, Sandia, and Cajamarca. Mining of gold-bearing veins was evidenced at Carabaya and Cajamarca. Gold was utilized to ornament temples, to adorn nobility and priests, and as offerings. It was also used for temple ornamentation and offerings by the Toltec, Aztec, and Olmec civilizations in Mexico. In 1502 Columbus noted that natives of Honduras were wearing gold ornaments. Early Portuguese explorers of Colombia and Brazil found that the natives were using fish hooks made of gold—one of the very rare instances of gold being used in antiquity for some purpose other than ornamentation, offerings, or currency. The gold from local rivers was present as a natural alloy of 19.5 percent copper, 1.4 percent platinum, and 79 percent gold. The alloy was malleable enough to be shaped into fish hooks but strong enough to maintain its shape for its end use.

The amount of gold amassed over the centuries in the Americas, although substantial, was small in comparison to the gold mined and collected in the Old World. The roomful of gold paid in ransom by the Incas to Francisco Pizarro in Peru contained about six tons of gold. It has been estimated that the total amount of gold sent from Peru to Spain was 7.4 tons, which would represent the greatest part of all the gold mined prior to A.D. 1500 along the Andes and west coast of South America. Similarly, in Mexico much of the gold amassed by all of the native civilizations prior to 1500 was shipped to Spain in the early 1500s. This treasure was estimated to have been on the order of thirty tons, including the seven tons taken by Cortez during the sack of Tenochtitlán (Mexico City). Much of the gold was mined from a site fifty miles west of Mexico City.

The total amount of gold transferred to Europe from the plundering of the New World was on the order of forty to fifty tons. That quantity, delivered to Europe over the period of a few decades, represented a tremendous influx, but in light of the total production of close to two thousand years, it is not exceptional. The Romans mined as much gold in Spain in less than a decade and plundered five times as much from Dacia (Romania) between A.D. 97 and 106. The New World's gold was significant in that it reached Europe at a time when the once prolific gold mining areas (Spain, Romania, Macedonia, and other areas) had been depleted. From A.D. 1000 to 1200 Africa was the only significant producer of gold. Production was probably about one ton per year, or approximately two hundred tons altogether. From 1200 to 1500, with the increased activity in Europe and the opening of trade routes into western Sudan and the Gold Coast, production increased steadily to nearly ten tons per year by 1500, averaging about four tons per year. For the period A.D. 1000-1500, total worldwide production is estimated at fifteen hundred tons.

Epilogue

Tracing and locating the sources of the gold that graced the ancient pre-Columbian world is not only of archaeological or historical interest. There is also an economic benefit. Modern-day prospectors and miners are finding that those pre-Columbian gold fields are bona fide exploration targets. In 1999 a license was granted to enable prospectors in eastern Egypt to explore for gold deposits in the area shown on the Turin Papyrus. Since 1997 more than 153,000 ounces of gold have been recovered from waste piles around the ancient Egyptian mines east of the Nile River. On the Aegean isle of Minos, one company is currently drilling out a gold resource in anticipation of a future mining operation.

The single largest gold-producing mine in Europe at present is the El Valle mine in northern Spain. The El Valle pit, opened in 1998, is producing just over three tons of gold per year from a district dotted with old Roman workings.

In Romania, at Rosia Montana near Deva, a mining museum includes an old Roman adit and drive. Nearby mining produces about one-third of a ton of gold per year. The West African Ghanaian gold deposits are currently being worked. The single largest gold mine in South America at present is north of Cajamarca, Peru. Mining at Yanacocha has located old "Incan-age" adits and an altar with gold offerings. There is perhaps no better testimonial to our respect for the ancestral prospectors than the attention being given by today's geologists to dusty, erudite archaeological writings describing ancient trade routes and gold mines.

Selected References

Australian Minerals Economics Pty. Ltd. Letter to author (undated).

Binneman, Johan. Letter to author, 1998.

Boyle, Robert W. *The Geochemistry of Gold and Its Deposits*. Geological Society of Canada Bulletin 280. Ottawa: Geological Society of Canada, 1979.

_______. *Gold: History and Genesis of Deposits*. New York: Van Nostrand Reinhold, 1987.

Chadwick, John. "Rio Tinto—Still Improving." *Mining Magazine* 179 (October 1998): 195.

Cremin, Aedeen. *The Celts in Europe*. Sydney, Australia: Center for Celtic Studies, University of Sydney, 1992.

Cunliffe, Barry, ed.*The Oxford Illustrated Prehistory of Europe*. Oxford and New York: Oxford University Press, 1994.

Davies, Glyn. *A History of Money: From Ancient Times to the Present Day*. Cardiff: University of Wales Press, 1996.

Dekalb, Courtenay. "Pyrite in the Huelva District, Spain." *Mining and Scientific Press* 122 (January 22, 1921): 125–130.

Dominian, Leon. "History of the Geology of the Ancient Gold Fields of Turkey," *AIME Transactions* 42 (1911): 569–589.

Habashi, Fathi. "Historical Metallurgy: Gold through the Ages," in *CIM* [Canadian Institute of Mining and Metallurgy] *Bulletin* 88 (May 1995): 60–69.

Healy, John F. "The Processing of Gold Ores in the Ancient World." *CIM Bulletin* 78 (May 1985): 84–88.

Josephy, Alvin M., Jr., ed. in charge. *The Horizon History of Africa*. New York: American Heritage, 1971.

Kavanagh, Paul. "Have 6000 Years of Gold Mining Exhausted the World's Gold Reserves?" *CIM Bulletin* 61 (April 1968): 553–558.

Kutz, Kenneth. *Gold Fever*. Darien, Conn.: Gold Fever Publications, 1988.

Martino, Orlando, et al. *Mineral Industries of Latin America*. U.S. Bureau of Mines Mineral Perspectives. Washington: Bureau of Mines, U.S. Department of the Interior, 1981.

Meyer, Carol. "Bir Umm Fawakhir: Insights into Ancient Egyptian Mining." *JOM* [*Journal of Metals*] 49 (March 1997): 64–67.

Mohide, Thomas P. *Gold*. Mineral Policy Background Paper 12. Toronto: Ontario Ministry of Natural Resources, 1981.

_______. *Silver*. Mineral Policy Background Paper 20. Toronto: Ontario Ministry of Natural Resources, 1985.

Mullen, Thomas V., Jr., and I. S. Parrish. "A Short History of Man and Gold." *Mining Engineering* 50 (January 1998): 50–56. Also in *California Geology* 51 (July/August 1998): 3–13.

Murdock, George P. *Africa—Its Peoples and Their Culture History*. New York: McGraw-Hill Book Company, 1959.

Oliphant, Margaret. *The Atlas of the Ancient World: Charting the Great Civilizations of the Past*. New York and London: Simon and Schuster, 1992.

Poss, John R. *Stones of Destiny. Keystones of Civilization: A Story of Man's Quest for Riches*. Houghton, Mich.: Michigan Technological University, 1975.

Purington, C. W. "Ancient Gold Mining." *Engineering and Mining Journal* 75 (1903): 437.

Rickard, Thomas A. "With the Geologists in Spain." *Engineering and Mining Journal* 124 (July 16, 1927): 91–95.

_______. *Man and Metals: A History of Mining in Relation to the Development of Civilization*, 2 vols. New York: McGraw-Hill Book Company, 1932.

Samamé Boggio, Mario. *Peruvian Mining: Biography and Strategy of a Decisive Activity*. 2d ed. Lima, Peru: Editorial Gráfica Labor, 1974.

Silverman, David P., gen. ed. *Ancient Egypt*. New York: Oxford University Press, 1997.

Strauss, Simon D. "Gold and Silver as Stores of Value." *Mining Congress Journal* 67 (August 1981): 23–30.

Sutherland, Carol H. V. *Gold: Its Beauty, Power, and Allure*. London: Thames and Hudson, 1969.

Tien, J. C. *The Gold Industries of China*. Chicago: Intertec-Group IV, 1994.

Vanished Kingdoms of the Nile (centennial celebration exhibit, University of Chicago, February 4–December 13, 1995).

Wells, William V. *Explorations and Adventures in Honduras*. New York: Harper and Brothers, 1857.

Wernick, Robert, and Pierre Boulat. "Who Was the Lady of Vix?" *Smithsonian* 16 (March 1986): 140.

William, Jonathan, with Joe Cribb and Elizabeth Errington, eds. *Money: A History*. New York: St. Martins Press, 1997.

World Gold Council. *Gold History* (gold information sheets No. 1 and 2), 1998, at *www.gold.org*.

Gold and the Discovery of America

Peter Bakewell

Peter Bakewell is professor of history at Southern Methodist University in Dallas. Among his publications are Silver Mining and Society in Colonial Mexico *(1971); Miners of the Red Mountain: Indian Labor in Potosí, 1545–1650 *(1984);* A History of Latin America: Empires and Sequels, 1450–1930 *(1997); and* Mines of Silver and Gold in the Americas *(1997).*

On the morning of Saturday, October 13, 1492, native people of the island of San Salvador, as Columbus had named it, came to the beach to continue the bartering with the Spanish that had begun the day before. The Spaniards had offered glass beads and hawk's bells; and now the natives brought more of the balls of cotton thread, parrots, and spears that they had produced the previous day, and, Columbus wrote in his log, "exchanged them for anything that was given to them." But Columbus had higher hopes. "I watched carefully," he wrote,"to discover whether they had gold and saw that some of them carried a small piece hanging from a hole pierced in the nose. I was able to understand from their signs that to the south . . . there was a king who had large vessels made of it and possessed a great deal. . . . I decided to remain till the afternoon of the next day and then to sail south-west, for according to the signs which many of them made there was land to the south, south-west and north-west. They indicated that men from the north-west often came to attack them. So I resolved to go south-west to seek the gold and precious stones."[1]

Up to that point there had been no sign of precious stones, but Columbus may have thought that where there was gold, there must be gems too. Both, after all, were forms of wealth long associated with the rich Asian lands on whose eastern margins he now believed himself to be. As he passed southward through the Bahamas, he found more gold decorations among the native people but no firmer information about the source of the metal. His uncertainty ended only when, on October 28, he reached the northeastern coast of "Colba" (Cuba, which he took to be Cipangu—that is, Japan), where native people told him that great quantities of gold, pearls, and spices existed in another land to the east, called Bohio, or Babeque.[2] Columbus immediately abandoned his reconnaissance of the Cuban coast, turned eastward, and on December 6 made landfall on the northwest shores of that new land, which he decided to name *Española* (Hispaniola, the island now shared by Haiti and the Dominican Republic). Now, information about gold from the local people was more exact and encouraging: the source of the precious metal lay further east, along the north coast. Columbus therefore proceeded eastward, enquiring after gold, until late on Christmas Eve, when his largest ship, the *Santa María*, was wrecked on a reef. Columbus's dismay at this disaster was soon replaced by a certainty that, given the date, the event was in fact a divine sign that he should stop at that point and found a settlement. This he proceeded to do, building, partly from the timbers of the grounded ship, a small fort, which he named "La Navidad" (Christmas). He appealed for help to a local ruler, or cacique, with whom he had made contact shortly before and whom he took to be king of the whole land. The cacique provided buildings in which to store items removed from the *Santa María*; but, better, when he "saw that the Admiral liked gold he promised to have a great quantity brought to him from Cibao, the place that had the most."[3]

Thus, finally—though in fact quite quickly, within only ten weeks of his American landfall—Columbus learned of a large source of gold. It was indeed to prove the largest source of gold not only on Hispaniola but anywhere in the Caribbean islands. Cibao was the native name for the middle section of the central cordillera of Hispaniola, which consists of an igneous intrusion into Cretaceous sedimentary rock. The sedimentary covering has been eroded away except along the flanks of the cordillera. Parts of it have metamorphosed into slate, in which exist many quartz veins. The veins closest to the underlying intrusion often bear gold. And streams flowing down the north slopes of the range and cutting into the quartz veins typically had particles and nuggets of gold in their beds and in the sand and gravel bars that they formed. There, some thirty miles inland from the north coast of Hispaniola, about halfway along the island, the first American gold rush took place.[4]

That scramble for the gold of Cibao did not materialize, however, until the spring of 1494, after Columbus's return to Hispaniola on his second voyage. That expedition was far less an exploratory venture than an attempt at settlement. Seventeen ships carrying fifteen hundred men had crossed the Atlantic. Having found Navidad destroyed and those he had left there killed by the native people (probably in reaction to ill-treatment by the Spanish), Columbus founded a new base and port further east along the north coast, naming it Isabela, after the queen. The site was directly north of the Cibao gold fields, which was one of Columbus's reasons for choosing it. From it in March 1494 Columbus himself went inland to see Cibao and there founded a fort to solidify the Spanish presence in the area. That building he named Santo Tomás (Saint Thomas) as a response to those who had doubted the presence of gold.[5]

Whether in the dark watches of the night Columbus himself was ever among those doubters, we cannot say. Certainly his logbook for the weeks after the American landfall is full of sightings of gold—in the form of bodily ornaments among the Bahamians and the natives of Cuba. And Columbus was quite capable of believing that divine guidance would lead him to the metal, as indeed finally seemed to have happened with the wrecking of the *Santa María*.

Columbus was a man not above a materialistic hankering for wealth, or so the immensely generous terms of his initial agreement with the Spanish crown in 1492 would suggest. The most striking of them granted to him a tenth of any gold, silver, pearls, gems, spices and other goods obtained by barter, or produced as a result of his efforts, in the lands he might discover. Inasmuch as Columbus truly believed in early 1492 that he was about to undertake a direct voyage westward to the Far East, then that concession might well have given him wealth beyond imagination. His tenacious pursuit of gold in the Caribbean, though, had other motives as well. He was clearly anxious to justify, both to the crown and to his own men, his exuberant predictions and promises about the outcome of his exploratory project; and particularly so once he had proclaimed, in glowing terms, the Caribbean islands' beauty and potential after his return from the first voyage.[6] The fifteen hundred men in the outbound ships of 1493 were there as a result of that message and could not be disappointed (although in fact most finally were).

But beyond even these questions of Columbus's own credibility and honor with the crown and his followers, there were deeper and grander reasons for his pursuit of gold. His log entry for December 26, 1492, after expressing his realization that the loss of the *Santa María* had been divinely ordained, continues with his hope that the gold and spices to be found by the men he is leaving at Navidad will enable the Spanish monarchs—Ferdinand and Isabella—to undertake the reconquest of the Holy Sepulchre within three years, since "I declared to Your Highnesses that all the gain of this my enterprise should be spent in the conquest of Jerusalem. . . ."[7] The recovery of Jerusalem from Islam, a project of western Christianity for centuries past, was something particularly likely to appeal to the Spanish king and queen, since 1492 was for them notable less for Columbus's ventures than for their final success, after ten years of fighting, in overcoming the last Islamic stronghold in Spain itself: the Emirate of Granada.

But Columbus was not merely addressing royal enthusiasms of the day. The notion of Jerusalem, and of the connection of his own efforts with it, was constantly in his mind. Peter Martyr, an Italian writer who in Spain chronicled developments in America almost as they happened, relates that in the late 1490s the Spanish discovered in Hispaniola some gold workings in the form of deep trenches. "The Admiral [Columbus] asserts," Martyr wrote, "that Solomon, King of Jerusalem, obtained there, via the Persian Gulf, the enormous treasures of which the Old Testament speaks." Martyr added: "It is not for me to decide whether this is true or not, though in my view it is very far from being so."[8] It was an idea in Columbus of a piece with his famous notion, as he crossed the mouth of the Orinoco on the outward leg of his third voyage in 1498, that a river large enough to send so much fresh water so far out to sea must certainly be one of the four streams that had their source in the Garden of Eden, the earthly paradise. Columbus was a lateral thinker, gifted (or afflicted) with an associative mind to an unusual degree even for his time. It was a mental quality that often added persuasive force, though not the force of deductive logic, to his arguments.

The role of gold in the initial European discovery of America, then, was central. Columbus himself was inspired by the prospect of finding it in the fabled lands of the Orient, to which he hoped to give Europeans quicker and easier access than they had ever had. He desired gold both for usual and unusual reasons. His promises, and then reports, of gold induced the Spanish monarchs to support him and men to join and follow him.

Having drawn people across the Atlantic, gold was for several decades equally powerful in luring them on to explore the vast regions to which Columbus had opened the door. The deposits in Hispaniola were soon—within twenty-five years—

worked down to the limit of profitability. And the other large islands to which the Spaniards soon spread from Hispaniola (Puerto Rico, Jamaica, and Cuba) proved to have far less gold than Hispaniola.[9] Attention then naturally turned to the mainland, precisely, in fact, to what was already known, as a result of reconnoitering voyages, as Tierra Firme ("Solid Land"). The English later called that area the Spanish Main. It included much of the coast of Venezuela and also the Caribbean shores of modern Colombia and Panama. In that venture of expanding settlement the influence of Columbus was still profound. On his fourth and final voyage, of 1502-1504, he had sailed down the Caribbean coast of Central America to a point a little east of where the Panama Canal now enters the sea. He had named the Panamanian coast Veragua, and in his usual optimistic manner had inflated quite minor traces of gold there into vast fortunes awaiting both crown and explorers.

Thus, in 1508 the home government issued two contracts for the closer exploration, with possible settlement, of the Main. The central aim of both contracts was royal tax income from mines (of gold, silver, and *quañín*, an alloy of gold and copper that the Spaniards had encountered in native ornaments gathered along the north coast of South America). The contracts went to two experienced Caribbean hands, both residents of Hispaniola: Diego de Nicuesa, for Veragua; and Alonso de Hojeda, for a long stretch of South American coast from east of Caracas to the Gulf of Urabá (at the base of the Isthmus of Panama).

Nicuesa's expedition, the larger of the two, was a disaster. He, without organizing ability, allowed it to divide into three, in his headlong rush to find the supposed gold of Veragua. Three-quarters, or more, of his initial force of 785 men perished, including himself. Hojeda did little better in securing wealth, but his effort did have a strategic result of enormous importance for further Spanish expansion on the mainland. Following various combats with bellicose natives who used poisoned arrows (considered wildly unfair by the Spanish) and Hojeda's own retreat after being wounded and taken to Hispaniola (where he died soon after), the remaining sixty of his original three hundred (along with reinforcements from Hispaniola) fetched up in November 1510 at a more welcoming native settlement on the west side of the Gulf of Urabá. That place they soon named Santa María la Antigua de Darién. It was the first permanent Spanish town on the American mainland and served as bridgehead for the exploration and wider settlement of the Isthmus. The importance of Darién is suggested by the presence there from its founding of two particular men. One was Francisco Pizarro, the conqueror of Incaic Peru twenty years later; the other was Vasco Núñez de Balboa, who in 1513 led a party of men across the Isthmus, thereby "discovering" the "South Sea" (the Pacific). Pizarro was the second most senior man in Balboa's group. Over the following years he prospered in that first mainland colony, emerging as a leading figure in the city of Panama, founded on the Pacific shore of the Isthmus in 1519.[10]

Gold, then, provided a crucial incentive for Spanish expansion to the mainland. And though Panama never yielded gold on the scale suggested by Columbus's reports of 1504, enough of it was found there to anchor Spanish settlement while further explorations took place from the new base. Enough was found, in fact, to finance those explorations in part. Pizarro had gold workings in Panama. Their product supported the seaborne expeditions that he organized down the northwest coast of South America in the 1520s—expeditions that finally took him to the northern edges of the Inca empire. But before we move on to that story, another of equal historical weight demands attention.

While the Isthmus was being explored and settled, the conquest and occupation of Cuba were proceeding, from Hispaniola. The island, however, presented few obvious opportunities for gain—it had less gold than Hispaniola, for instance—and as a result the westward urge soon reasserted itself among the Spanish. A result, in 1517, was a small, one-ship expedition that brought the Spanish for the first time into significant contact with Maya people on the west coast of Yucatan. A little gold was found in native temples on that coast, which was promising. But just as alluring to the Europeans was the obviously higher cultural level of the Maya in comparison with the natives of the islands. Something more advanced in the way of human society seemed to be appearing on the American horizon as this small force pressed southward down the coast of Yucatan.

And so, even though the Spaniards had a number of tough scuffles with the Maya on that 1517 foray—their leader in fact dying after the return to Cuba of wounds he had received—the governor of the island, Diego Velázquez, immediately decided to send off another expedition. Four ships left Santiago de Cuba in April 1518. They first retraced the track of the previous year's party, then moved onward to the base of the Gulf of Mexico and from there followed the coastline as it turned north. This brought the Spanish to the shores of the Aztec empire and to confirmation of the previous year's suspicions about the presence of a more advanced native culture to the west. Here there were large towns on or near the coast, temples with a professional priesthood (though performing human sacrifice, which gave the Spaniards pause), and a greater quantity and quality of gold items than had been seen before. The Spaniards' total haul of gold, through barter, from this second voyage was twenty thousand pesos. It is hard to know, at that early date, what weight of gold that sum corresponded to; but it was an impressive amount—enough to persuade Velázquez to start organizing a third expedition for exploration and trade in Mexico.

To lead that third venture, Velázquez chose a respectable, but hardly outstanding, settler of Cuba—a man in his mid-thirties by the name of Hernán Cortés. Cortés had gold workings in Cuba, but they had certainly not made him rich.[11] He obviously sensed a possibility of greater wealth, and perhaps more than mere wealth, in the leadership of the new expedition. The story of his making off with Velázquez's ships and men, sailing away as the governor shouted after him from the shore, is one of the classics of Latin American history.

Cortés set up a base on the Gulf Coast of Mexico, near Veracruz, on Holy Thursday, 1519. He made contact with the people of nearby native towns and from them learned of the power of the Aztecs and their ruler, Moctezuma II. The Aztec empire extended to that coast, so that Moctezuma soon had news of the presence of these peculiar outsiders (as he had had, in fact, of the expedition of 1518). Gold was in evidence on the coast from the start as a barter item. But the Spaniards were truly impressed when large gifts of gold objects arrived from Moctezuma: a golden disk, the size of a cart wheel, intricately engraved; twenty golden ducks; many other gold ornaments shaped like animals; staffs, necklaces, and pendants; and, demonstrating that gold placers were currently yielding, a helmet full of small grains of the metal.[12]

In all likelihood, these and other rich gifts were intended to impress the Spanish with the Aztecs' wealth and power[13] and hence to scare the outsiders away. But, of course, they had precisely the contrary effect. And Cortés sent Moctezuma a notorious, and apparently cynical, message that the Spaniards suffered from a disease of the heart that gold alone could cure.

Thus, gold drew the Spanish on to the Aztec capital, Tenochtitlan, where they arrived in November 1519. It was not the only attraction, however, since the closer they drew to Tenochtitlan, the more obvious it became that the Aztecs were something qualitatively quite superior to any other native people encountered to date in the Americas; and for the first time these Europeans had the sense that a chance awaited them there to earn fame comparable to that of the heroes of classical antiquity. If they could overcome this never before suspected empire, their names would go down in glory; and that was an inspiring ambition, especially since most of them were men of humble origins.

Wealth, though, preferably in the form of gold, was certainly an equal incentive, and Cortés's Spaniards pursued it avidly. Even after they had entered Tenochtitlan, Moctezuma continued to be generous to them with gifts of gold. And they themselves found more. They had been housed in the palace built by Moctezuma's predecessor, Axayácatl. According to the conquistador-chronicler Bernal Díaz del Castillo, the Spanish were looking around for a suitable place for an altar in the building when they noticed a plastered-up doorway. "Now as we had heard that Moctezuma kept his father's treasure in this building, we immediately suspected that it must be in this room. . . . So the door was secretly opened, and Cortés went in first with certain captains. When the men saw the quantity of golden objects—jewels and plates and ingots—which lay in that chamber, they were quite transported. They did not know what to think of such riches. . . . Very secretly we all went in to see. The sight of all that wealth dumbfounded me. Being only a youth at the time and never having seen such riches before, I felt certain that there could not be a store like it in the whole world." The Spaniards left the hoard in place for the time being.[14] A few months later they tried to remove it from the city, at great cost in injuries and lives to themselves.

Early in 1520 Cortés, perhaps looking ahead to Spanish settlement, asked Moctezuma about the sources of the Aztecs' gold and then sent off small investigating parties (of Spaniards with native guides) to two of the sites he was told about: one was at Zacatula on the Pacific coast, southwest of Tenochtitlan, the other around Tuxtepec, a little inland from the Gulf coast and not far from Veracruz, where the Spaniards had landed. Both parties duly returned with reports that alluvial gold was washed from the sands of rivers in the two areas; both brought encouraging amounts of granular gold as confirmation.[15]

In the spring of 1520 the relationship between the Spanish and the Aztecs deteriorated as the invaders' demands on the native people grew heavier and the Aztecs began to recover from the initial surprise at the appearance of this totally unfamiliar variety of human beings. The outcome was that by midsummer Cortés and his followers found themselves trapped in the center of Tenochtitlan under siege in their lodging. The city occupied a large island, surrounded by lakes that were crossed only by three or four causeways. For defensive purposes, the causeways had gaps in them that were bridged by removable planks. Cortés decided to try to break out by night along the shortest causeway—one that led to dry land on the west. According to Bernal Díaz, the attempt was made on July 10, 1520. The Aztecs were on the alert, however, and what the Spaniards hoped would be a stealthy exit turned into a near disaster in which several hundred of them were killed and most of their horses and arms lost. The Aztecs, naturally, opened the gaps in the western causeway so that the escapers had to swim across them. Many, it seems, failed to make those crossings because they were weighed down with gold removed from the hoard discovered in Axayácatl's palace.[16]

That "Noche Triste" ("Gloomy Night"), as it is known, was, though, only a setback in the overall Spanish assault on the Aztecs. Thirteen months later, in August 1521, Tenochtitlan surrendered after a long and terrible siege in which most of it was destroyed. The Spanish quickly took charge of the Aztec empire—most of central and southern Mexico—and of all the sources of wealth it contained, including gold in the forms both of booty and alluvial deposits. A number of the conquistadores, Cortés among them, exploited those deposits, using native workers, in the early decades of the new colony, a land to which they gave the name New Spain.

With the Aztec conquest, Spanish expectations in America rose several notches. For much of the rest of that century, hopes of finding other high native civilizations propelled a series of forays, mainly into the interior of South America. Only one comparable culture was found, however—that of the Incas, centered in the Peruvian Andes.

Panama was the initial base for exploration of western South America. Rumors of interesting peoples to the south inspired a first expedition down the Pacific coast of Colombia in 1522, led by one Pascual de Andagoya. The venture did not go far enough south to encounter anything of promise. Francisco Pizarro followed with a second voyage, again fruitless, in 1524-1525. But a third exploration, again under Pizarro, in 1526-1527, had more luck. Having crossed the equator, the Spaniards met up with a seagoing sailing raft carrying Inca products: "many pieces of silver and gold as personal ornaments . . . including crowns and diadems, belts and bracelets, armor for the legs and breastplates; tweezers and rattles and strings and clusters of beads and rubies; mirrors decorated with silver, and cups and other drinking vessels. . . ." These and other articles the raft was carrying to be exchanged for "fish shells from which they make counters, coral-colored scarlet and white."[17] Encouraged by that encounter, Pizarro pressed further south late in 1527 and near the present Ecuadorian-Peruvian border came to the Incan town of Túmbez. Advancing still further down the coast, he picked up varied evidence of Peruvian high culture: samples of coastal pottery, metal vessels, llamas, fine fabrics, and a few boys to be trained as interpreters.[18] With such evidence, Pizarro returned to Spain in 1528 and obtained from the king a contract for the exploration and conquest of Peru.

That is another dazzling tale. But time and topic limit us here to gold. Following what was by now the standard Spanish pattern of attack, in November 1532 Pizarro seized the Inca leader Atahualpa in an ambush at the northern Peruvian town of Cajamarca. Gold was immediately forthcoming—eighty thousand pesos' worth, with seven thousand marks of silver besides, both metals in the form of "monstrous effigies, large and small dishes, pitchers, jugs, basins and large drinking vessels. . . ."[19] Atahualpa, noting the Spaniards' hunger for gold and being told by Pizarro that they were simply seeking gold for themselves and their emperor, then quickly made his famous offer to fill a large room (reportedly 22 by 17 feet) with gold objects halfway up its height of more than eight feet in exchange for his liberty. The Spanish, needing time for reinforcements to arrive and seeing nothing to be lost in such an arrangement, agreed. Gold arrived at Cajamarca, at Atahualpa's orders, for the next several months, much of it in the form of jars, pitchers, and plates. But the most spectacular contribution came from Cusco, the Incan capital in the Andes of southern Peru, where the walls of the Coricancha, the temple to the sun, were sheathed in gold plates (some seven hundred, each of 4 1/2 pounds). These were pried off by Spaniards sent to Cusco to investigate its wealth and the metal taken to Cajamarca.

Whether the room was filled with gold up to the height that Atahualpa had promised is not certain. But in June 1533 Pizarro decided that it was time to melt down and assay the gold that had arrived. The total that passed through the furnaces was counted at 13,240 pounds (along with 26,000 pounds of silver). Great numbers of objects of high artistic and cultural value were lost in this casting of the two metals into ingots for transport to Spain or distribution among the conquistadores.[20] (Atahualpa was not freed. On the suspicion that he had been plotting an attack on the Spaniards, he was garrotted on July 26, 1533.)

The Peruvian conquest certainly yielded the largest initial haul of gold booty that the Spaniards enjoyed in America. The presence of so much gold (and silver) also suggested to them that the Andes abounded in precious metals. And so they immediately launched new expeditions to look for similar loot, or the sources of the metals in the ground. One such was a remarkable excursion to Chile between 1535 and 1537 that returned northward to Peru up the coastal desert but missed the large gold deposits present there. Explorers had better luck, though, north of Peru, in northwestern South America.

Many expeditions pressed into that large and geographically resistant region from 1530 onward. One of the first was led by Diego de Ordaz, who had been with Cortés in the Mexican conquest. His inspiration at that time had been Aztec rather than Inca gold. Acting on the common belief of the time that gold was engendered by heat and so was most likely to be found near the equator, Ordaz took an expedition up the Orinoco in 1531. Some six hundred miles upstream he learned (or thought he learned) from local native people that there was gold in the province of Meta—the basin of the River Meta, a tributary of the Orinoco that rises in Colombia not far east of Bogotá.[21] The rivers were, however, in flood, and the party was exhausted after the long haul up the Orinoco. Ordaz decided to retreat and to mount another expedition; he died, however, before he could do so.

The prospect of gold in the Meta basin that he reported was, though, the origin of a very long-lived notion that great golden wealth lay awaiting discovery east of the Andes in northwestern South America. The next to pursue that gold were (uniquely in Spanish American exploration) Germans. The emperor Charles V, king of Spain, had in 1528 contracted out the governing, defense, and exploration of Venezuela to the Welser banking company. The Welsers hoped mainly for gold and sent six expeditions into the interior to look for it, with very little success. One of those parties, however, in 1539 pressed further westward than its predecessors and crossed the Andes (with great difficulty) into Colombia. And there it ran into two Spanish expeditions, also seeking gold, that had arrived from north and south. The northern one, under Gonzalo Jiménez de Quesada, had started out from the Caribbean coast of Colombia in 1536, impelled by large finds of gold in native tombs not far inland. And the southern one was a prolongation of the Spanish force that in the mid-thirties had overcome the northern tier of the Inca empire in Quito (present Ecuador).

Thus, three European intrusions converged from different directions in Colombia in 1539, all driven by the ambition for gold. They came together in the area of the Muisca, or Chibcha, culture, which centered on a place that is now Bogotá.[22] Though the Muisca were a far smaller and less organized group than the Incas or Aztecs, they did possess a stratified society, a political structure, and a considerable area under their control. They were skilled workers of emeralds, which they mined, and of gold, which they imported from surrounding regions. They can be ranked, then, among the high cultures of Middle and South America, though at the bottom of the scale. They were, in fact, the last of the high cultures that the Spanish overcame in America, and, once more, the search for gold had a large part in their discovery and conquest.

Colombia finally proved to be the part of Spanish America best endowed with gold deposits; its placers yielded particularly well in the eighteenth century. It was Colombian gold that led to the first use by the Spanish of the term "indio dorado" ("gilded Indian") when in 1534 or 1535 they captured a local native leader from the Popayán area of southern Colombia who had evidently gathered much of the metal from the rich deposits of the area.[23] But it was further south, in Quito, that the enduring legend of El Dorado seems to have had its origin, about 1541. That, at least, is the date that emerges from the first reference made to it, by the chronicler Gonzalo Fernández de Oviedo, then living in Hispaniola: "I asked Spaniards who have been in Quito and have come here to Santo Domingo . . . why they call that prince the 'Golden Chief or King.' They tell me that what they have learned from the Indians is that that great lord or prince goes about continually covered in gold dust as fine as ground salt. He feels that it would be less beautiful to wear any other ornament.

It would be crude and common to put on armor plate of hammered or stamped gold, for other rich lords wear these when they wish. But to powder oneself with gold is something exotic, unusual, novel and more costly—for he washes away at night what he puts on each morning, so that it is discarded and lost; and this he does every day of the year."[24] That vision—along with the parallel idea that the Muisca had periodically thrown offerings of gold into a lake in the province of Bogotá—was the inspiration for repeated expeditions into the interior of northwestern South America. But El Dorado was never found; nor much else of interest or profit to the Spanish in the mountains and plains of eastern Colombia and Ecuador. For the Spanish in America, those early decades of fecund rumors—such as had led to the finding of the Incas, for example—were drawing to a close. At much the same time as El Dorado began to be talked of in Quito, news of the fabled seven cities of Cíbola reached official Spanish ears in central Mexico. But that story yielded little more tangible result than did the notion of the gold-dusted native lord.

What was in fact happening was that the initial phase of looting—of scooping up the accumulated wealth of the native past—was ending. Much of that wealth had been in gold; and the search for that gold, from Columbus onward, had been an immense stimulus to Spanish exploration, discovery, and settlement. The lure of gold was, indeed, a prime cause of the amazing rapidity of Spanish expansion in America. But now, in the 1540s, there was to begin another phase of expansion, more mundane but finally far more profitable still. That was expansion driven by the search for silver, which in the middle of the decade yielded its first two great prizes: Potosí, in present Bolivia, in 1545; and Zacatecas, in northern Mexico, in 1546. It was silver—the metal that made Spain powerful in Europe for a century and the envy of its European rivals for almost three—that quickly became the most valuable export of Spanish America. But that is another, a longer, and a more complicated, story.

Notes

1. Columbus's log entry for October 13, 1492, in Christopher Columbus, *The Four Voyages*, ed. and trans. J. M. Cohen (Harmondsworth, United Kingdom: Penguin Books, 1969), 57.

2. Columbus, *The Four Voyages*, 75–76, 80.

3. Columbus, *The Four Voyages*, 93, chapter 34 of the *Life* of Columbus by his son Hernando.

4. Carl O. Sauer, *The Early Spanish Main* (Berkeley and Los Angeles: University of California Press, 1966), 80–81. The geological information given here is from William M. Gabb, "On the Topography and Geology of Santo Domingo," *Transactions of the American Philosophical Society* 15 (new series, 1881): 66 (cited by Sauer, 80–81).

5. Sauer, *The Early Spanish Main*, 80.

6. See, for example, the letter that Columbus wrote "to various persons" as he approached Spain on the return voyage in February 1493 (in Columbus, *The Four Voyages*, 115–123).

7. The entry is cited in Samuel E. Morison, *Admiral of the Ocean Sea: A Life of Christopher Columbus* (Boston: Little, Brown and Co., 1942, and later reprints), 304.

8. Pedro Mártir de Anglería, *Décadas del Nuevo Mundo*, trans. Agustín Millares Carlo (Mexico City: José Porrúa e Hijos, Sucesores, 1964), vol. 1, p. 150 (First Decade, Book 4).

9. Sauer, *The Early Spanish Main*, 197–198.

10. For Nicuesa, Hojeda, and in general the occupation of Tierra Firme, see Sauer, *The Early Spanish Main*, chapters 8 and 11 (on which the present summary is drawn).

11. Bernal Díaz del Castillo, *The Conquest of New Spain*, trans. J. M. Cohen (Harmondsworth, United Kingdom: Penguin Books, 1963), 47. The references to the two earlier voyages to Mexico are also taken from Bernal Díaz, as is most of the information about the Aztec conquest given here.

12. Díaz, *The Conquest of New Spain*, 93.

13. Inga Clendinnen, *Aztecs: An Interpretation* (Cambridge: Cambridge University Press, 1991), 268–269.

14. Díaz, *The Conquest of New Spain*, 241–242.

15. Díaz, *The Conquest of New Spain*, 265–269.

16. Díaz, *The Conquest of New Spain*, 306.

17. A report to Charles V, probably by Francisco de Xerez, quoted in John Hemming, *The Conquest of the Incas* (New York: Harcourt, Brace, Jovanovich, 1970), 24–25.

18. Hemming, *The Conquest of the Incas*, 26.

19. Francisco de Xerez, quoted by Hemming, *The Conquest of the Incas*, 47.

20. For Atahualpa's ransom, see Hemming, *The Conquest of the Incas*, 63–64, 72, 74.

21. John Hemming, *The Search for El Dorado* (New York: E. P. Dutton, 1979), 11–16.

22. For all of these expeditions, see Hemming, *The Search for El Dorado*, chapters 1–3.

23. Hemming, *The Search for El Dorado*, 53.

24. Quoted by Hemming, *The Search for El Dorado*, 98. For the original, see Oviedo, *Historia General y Natural de las Indias*, book 49, chapter 3 (Madrid: Ediciones Atlas, 1959) [*Biblioteca de Autores Españoles*, vol. 121, p. 236].

Gold Mining in North Carolina to 1840

Richard F. Knapp

Dr. Knapp is research curator for the Historic Sites Section of the North Carolina Division of Archives and History. His very first assignment in that position was to conduct historical research on gold mining in North Carolina and help develop Reed Gold Mine into a state historic site. Among his recent publications are a revised edition (1999) of Golden Promise in the Piedmont: The Story of John Reed's Mine *and (with Brent D. Glass)* Gold Mining in North Carolina: A Bicentennial History *(1999).*

For millennia gold has attracted men and empires. In the New World, gold mining in the United States arose after an accidental discovery in Cabarrus County, North Carolina, in 1799. Initially local farmers washed creek gold with pans and rockers. A genuine rush emerged after 1825 when farmers discovered gold in veins. Thousands sought riches in creeks and outcroppings of gold-bearing quartz. The bubble burst by the late 1830s, but some mining revived later. The story of North Carolina gold and Reed Gold Mine reflects the drama and irony that, with variations, occurred in many gold rushes. It is a saga of illegal immigrants, rich mines, slave labor, and incredible speculation.[1]

Hunters of gold in North Carolina usually found free gold on or near the earth's surface. Gold's physical properties made identification easy. It is nineteen times heavier than water and the most imperishable and malleable metal. For nearly two centuries, geologists have offered explanations of how gold deposits formed in North Carolina. I leave that topic to more qualified speakers here today. Visible coarse gold from eroded quartz veins was the first gold mined in the state. Weathering freed the gold, which concentrated in soil and creek bottoms. Some associated sulfides (or pyrites) of iron and copper also weathered and became easy to separate from the quartz by mechanical means. Bedrock with fine-grained disseminated gold also weathered into easily excavated and crushed saprolite (rotten, disintegrated rock) near the surface of the earth. Weathered rock usually extends down some fifty feet to the water table. That ore contrasted with deeper, unweathered ore, frequently combined with unoxidized pyrites. Deeper quartz veins and disseminated deposits were hard to excavate and crush. In addition, pyrites made separation of gold more difficult. Refractory pyritic ore, with sulfur compounds, was called sulfuret ore. Recovering gold from sulfuret ore required chemical procedures unavailable in the early 1800s.[2]

Gold occurs in six regions in North Carolina that correspond roughly with geologic belts identified by specific suites of rocks. The eastern gold region comprises at least 300 square miles northeast of Raleigh and features abundant weathered quartz veinlets. To the west is the Carolina slate belt, 10 to 60 miles wide from Virginia to South Carolina. Next is the Charlotte (Carolina igneous) belt, 15 to 30 miles wide, with gold in Guilford, Davidson, Rowan, Cabarrus, and Mecklenburg Counties. The slate and Charlotte belts contained most of the largest gold mines. Next, to the west, is the Kings Mountain belt in Cleveland, Gaston, Lincoln, and Catawba Counties. The South Mountain region is 10 to 11 miles wide in Burke, McDowell, and Rutherford Counties, with almost exclusively placers—gold in creek beds or near the surface in soils. Finally, the metal is found in most counties west of the Blue Ridge. Nearly half of the state is in a gold region.[3]

Various people encountered gold before 1799. Did pre-Columbian Indians pan gold for ornamental use? Spanish knowledge of gold and silver stimulated Europeans in North America to seek gold. Ponce de Leon in 1513 observed Indian gold in Florida. Later, Spaniards wrote of riches in the southern Appalachians. In 1663 English King Charles II reserved for himself one-fourth of any gold or silver found in Carolina. In 1774 Gov. Josiah Martin received gold from Guilford County. Thomas Jefferson reported a small nugget found in Virginia. Neither event created much excitement. In contrast, mining at John Reed's farm after 1799 led to the nation's first gold rush.[4]

Reed was born into poverty in Germany about 1758 and joined the Hessian militia as a reservist. He reached America as a soldier sent to aid the British in the American Revolution and helped occupy Savannah for several years. In 1782 he deserted and went to Cabarrus County. Germans there resided in log houses, spoke German, and seldom saw the outside world. Reed became an ordinary farmer, married Sarah Kiser, and raised nine children. Little is known of his personal qualities; an obituary called him a Christian and friend of the poor.[5]

One Sunday in 1799 his twelve-year-old son Conrad found an odd rock in Little Meadow Creek. The rock, the size of a small flatiron, weighed about seventeen pounds and became a doorstop. In 1802 a jeweler in Fayetteville identified the metal, fluxed it into a bar of gold, and bought it from Reed, paying $3.50 for $3,600 worth of gold. Reed bought gifts for his family, including coffee beans for his wife. (Sarah threw the odd beans out when they failed to make a good stew.)[6] The Reeds found more valuable rocks. In 1803 Reed took three partners, who (in late summer, after crops were planted and the stream had nearly dried up) supplied a few slaves to dig in the creek. Slave Peter unearthed a twenty-eight-pound nugget (the largest found at the mine). Offered a portion of the nugget, Peter refused to damage his dinner fork by trying to pry off the piece.[7]

Word of the discovery leaked out. In 1804 the partners made fourteen thousand dollars in six weeks, and the U.S. mint in Philadelphia received eleven thousand dollars' worth of Cabarrus gold. The news reached Washington and New York. Meanwhile Reed's miners sought finer particles of gold by washing sand from the creek in pans and rockers. A frying pan could swirl gravel and water and separate heavier gold. The rocker was a box, half of a barrel, or half of a log. Other creek mines—the Parker and Harris mines—opened about 1805. Businessmen purchased mining land to the south of Reed. Travelers found gold as far as seventy miles away.[8]

William Thornton, designer of the U.S. Capitol, in 1806 saw workers at Reed's mine using mercury—known since ancient times to bond chemically with gold—in pans to amalgamate with gold. Heating the amalgam vaporized the mercury, leaving

Little Meadow Creek flowed through the property of John Reed and for years was the principal source of placer gold discovered in North Carolina. It was in Little Meadow Creek in 1799 that Conrad Reed, John Reed's twelve-year-old son, discovered "a yellow substance shining in the water"—a seventeen-pound, wedge-shaped rock that subsequently proved to be the first discovery of gold in the United States. *Photograph from the files of the Division of Archives and History.*

gold behind. The partners obtained six or seven ounces of gold every few days, but Reed continued farming. Thornton formed the North Carolina Gold Mine Company (with a former governor of Maryland and the treasurer of the United States as directors) to mine land near Reed. Despite the prospect of achieving profits through slave labor, the company shortly became defunct.[9]

Gradually surface mining with pans and rockers grew as a seasonal, amateurish business. By 1810 the U.S. mint had assayed 1,300 ounces of Cabarrus gold; each sample exceeded the standard for coins. The creek that ran through John Reed's property, where the weight of recovered nuggets in excess of one pound each eventually passed 150 pounds, was the chief source of domestic gold. By 1820 gold was mined in four or five counties. One fellow found a five-pound nugget at the Parker mine; soon a hundred men were digging there. Yet, technology had scarcely changed since 1803. Local farmers became inefficient part-time miners; some (like Reed) forbade mining on cultivated lands. Geologist Denison Olmsted found Reed's mine a mass of small pits and piles of rubbish in a floodplain.[10]

About 1824 gold output began rising; five thousand dollars' worth of Carolina gold arrived at the mint, well above the previous average. The mint received one-third to one-half of gold mined; much of the rest was circulated or sold to artisans.

A decreasing amount of money in use (only five dollars per capita by 1819) enhanced the relative value of gold. In the gold region, Olmsted observed: "Almost every man carries with him a goose quill or two of it [gold], and a small pair of scales. . . . I saw a pint of whiskey paid for by weighing of 3 1/2 grains of gold." Merchants traded supplies for miners' gold.[11]

Yale-educated Olmsted in 1823 convinced the state government to fund geological fieldwork. He hired German mineralogist Charles E. Rothe to conduct a study of the slate belt, an endeavor completed by geologist Elisha Mitchell. Both Rothe and Mitchell produced articles in the new *American Journal of Science*, as well as official reports, the nation's first professional (and published) state-funded geological work. The geologists also recognized the economic value of agriculture.[12]

During periods of slack agricultural activity, notably those months in which crops did not require attention, there was a surplus of labor. As a consequence, numerous farmhands joined the search of gold. At the same time, a massive increase in the output of cotton (by 1821 North Carolina alone produced five times the national cotton crop of 1790, and Mecklenburg County grew a considerable portion of that crop) resulted in a chronic decline in the price of the staple. Those two factors significantly influenced mining for decades. As cotton prices declined after 1818, mining emerged as an attractive alternative. Landowners normally leased mining rights for one-fourth to one-half of any gold that might result. Owners ran a great risk: trespassers, miners, and slaves pocketed up to perhaps one-third of the precious metal found on their lands.[13]

By 1824 mines were still worked haphazardly. *Niles' Weekly Register* compared mining to a lottery. A traveler declared that the miners' scant machinery "looked as though it had been made by children, with a penknife." Panners cleared twenty-five cents to five dollars a day. As mining evolved into a business with employees, wages often were fifty cents per day. Mining was a low-technology, slow-growing enterprise.[14] In 1825, however, the mint received seventeen thousand dollars' worth of Carolina gold, three times the total in 1824, and in the next decade mining became a socioeconomic factor in the state. A few miners attained enduring stature, and a mining culture blossomed. By the early 1830s, portions of the state experienced a genuine gold rush, and Salisbury's *Western Carolinian* exulted that mining was "only second to the farming interest."[15]

The precipitating event was unplanned. In 1825 farmer Matthias Barringer, who resided twenty miles from Reed's mine, was searching for creek gold when he noticed a *vein* of quartz running into a hillside. He found nearly pure gold in the quartz. Soon dozens of men dug along the vein, and Barringer's profits reached eight thousand dollars. The miners followed the 4-inch-wide vein more than 30 feet to a depth up to 18 feet. Other prospectors sought quartz veins throughout the region. Mecklenburg became the leading gold county. Underground deposits proved more challenging than placers, but golden news spread afar. Gold hunters began to be regarded as "people that begin already to be accounted a distinct race," and backward North Carolina was called the "Golden State." In 1826 drivers hauled ore along Charlotte streets, and Yankee hucksters flocked to the region in wagons filled with goods.[16]

Landowners discovered gold in Guilford, Davidson, Randolph, and Rowan Counties. Citizens reported almost daily discoveries, and the estimated value of weekly production in Mecklenburg reached two thousand dollars. In 1828 Elisha Mitchell published a map with mines such as the Reed, Alexander, Capps, McComb, Parker, Wilson, Beaver Dam, Barringer, Anson, Cox, and Montgomery. Some new mines were placer operations: mine owner Nathaniel Bosworth reported eighty laborers at his site. The Chisholm mine daily yielded up to five pennyweights' worth of gold per man. By early 1827 the venture, which had employed dozens, used a rare steam engine and sought hands to work for wages.[17]

Mecklenburg County and Charlotte—as late as 1837 "a quiet little village . . . kept up . . . by mining"—claimed most of the handful of important vein mines. James Capps, a poor man, discovered gold on his land in 1827. By 1828 his rich mine had about eleven pits, some of them eighty feet deep. Capps plunged into extravagance, became an alcoholic, and died; his finely attired widow soon remarried. The mine, under new management, continued fruitful. The value of its annual output, with a labor force of twenty men, approached fifty thousand dollars.[18]

Near Charlotte, landowner Samuel McComb became an industry leader. In 1828 he sold part of his mine for six thousand dollars to J. Humphrey Bissell, a Charleston lawyer who had studied mining in Europe. Observers found Bissell "a gentleman" and "the most . . . scientific man in the [mining] business" and said he "lives the life of a lord and his manners can very well sustain him. . . ." Bissell brought to the mine new technology and men experienced in South American mining. He drove a new tunnel more than four hundred feet in a year, used carts on a mine railway, and raised production to more than one pound of gold weekly. Located on the same vein was the promising Rudisill mine.[19]

In 1828 placers sparked a gold rush in mountainous Burke County, which had few veins. Samuel Martin of Connecticut stopped at the log house of cobbler Bob Anderson and noticed flecks of gold in the mud on Anderson's cabin. The following day they became partners for six months. Martin made thousands of dollars and departed; Anderson frittered away his riches. By spring 1829 "gold fever" had become the chief ailment in the county. For months, one mine averaged an outturn valued at more two thousand dollars weekly. Some hands made five dollars a day. Several thousand men, including slaves, worked nearby creeks.[20]

Publicity fueled the excitement. Periodicals and local newspapers promoted mining. From 1830 to 1835 the *Miners' and Farmers' Journal* of Charlotte printed countless mining columns. Other papers copied such articles. Out-of-state papers (the *New York Observer* sent a correspondent) also helped. The *Western Carolinian* warned of a "thousand silly reports . . . daily set afloat." Faraway papers published outlandish stories, such as one about the Capps mine yielding sixteen hundred *pounds* of gold weekly. Papers advertised mines for sale as speculation mounted. *Niles' Register* of Baltimore claimed that the price of Mecklenburg land rose whenever a farmer saw quartz. Investors paid extravagant prices. A Guilford tract, offered at $300, sold for $3,000. Burke parcels once worth under $1 per acre brought $30. A paper offered advice for "salting" mines: "Melt up a silver dollar or a small gold piece. . . . Divide them into small particles . . . scatter them about your spring, or in a branch. . . . Let some of your neighbors discover them by accident."[21]

As production increased, observers suggested state subsidies, as with railroads. Instead, in 1827 the state chartered another North Carolina Gold Mining Company. Legislator Charles Fisher of Salisbury secured the charter and sought funds in Charleston. The firm planned to operate several mines and leased tracts in three counties. The state's few wealthy citizens, however, generally preferred agriculture. For venture capital, textile mills were cheaper and less risky than mines. Only a corporation might succeed in large-scale deep mining, with its costly preparatory work and machinery. For a time, corporate mining apparently surpassed the small textile industry in luring outside capital and technology, but farming dwarfed both industries.[22]

Incorporation was irresistible to some. In 1835 the state granted ten corporate charters. In 1836 one charter authorized capital up to $1.5 million. Soon legislators approved broader activities beyond mining. Much new capital originated in northern cities such as Baltimore, New York, and Philadelphia. In the 1830s significant funds also came from Europe. Editor Hezekiah Niles estimated a 10 percent return on foreign money. He noted a local benefit too: "Every 100 dollars worth of gold . . . represents . . . 75 dollars worth of American corn, beef, pork, etc consumed by the various laborers employed, and 10, or perhaps 15 percent more . . . into . . . American wealth. . . ."[23]

Most foreign funds and laborers were British, but miners came from numerous nations by the late 1820s. One mine supposedly employed nearly a thousand workmen, who spoke thirteen languages—undoubtedly an exaggeration. The most colorful foreign manager was Count Vincent de Rivafinoli, flamboyant head of British interests in Charlotte. The count was an Italian aristocrat, mining engineer, and promoter with backing and grandiose plans. A cultured man, he visited New York to direct an opera. In 1830 he imported expert miners from England, Germany, Wales, Scotland, Ireland, Switzerland, Italy, and France; of Charlotte's 717 residents at that time, 61 were foreigners. Some natives refused to work under Europeans and rioted, but generally nativism did not hamper mining.[24]

Among Europeans, Cornish miners stood out. By 1830 more than thirty thousand adults and children worked in Cornish deep mines in southwestern England—which produced much of the world's copper and tin—under conditions "barely tolerable even when times were good." Some miners emigrated; one claimed that miners in America earned a week's wages in a day. The Cornishman had a reputation as skilled, superstitious, clannish, and Methodist. The *Western Carolinian* said:

> He never was given to swearing or drinking,
> Yet got all his money by damming and sinking;
> He buried himself below all his life,
> And when dead he was buried up here by his wife.[25]

The number of Cornish in Carolina remains unknown, as does the overall mining population. Native whites, blacks, and perhaps women outnumbered Cornish miners. Miners and related workers were natives, foreigners, males, females, whites, blacks, adults, and children. One company employed hundreds; most mines operated with a handful. By the early 1830s up to thirty thousand people, most part-time placer miners, hunted for gold, but Mecklenburg had only one hundred miners, according to the 1840 census. Women worked in family mines; some were experts with pans and rockers. A few ladies in rude mining towns were surprisingly cultured—one taught a Sunday school class for miners.[26]

African Americans, one-fourth of the one hundred thousand people in the gold region in 1830, participated in mining and related pursuits. Most skilled miners were white, but hundreds, perhaps thousands, of slaves labored in and near mines and often cut timber and grew food for miners. At the height of Burke's rush, several thousand slaves reportedly toiled at placer mines. Some slaves, like free farmers, mined during slack farming seasons. Companies used slaves, although only 5 percent of slaves worked in *any* industry. About 1830 the Capps mine employed 33 African Americans (probably all slaves and including 10 women), and partners Bissell and Samuel Barker employed some 60 slaves. Slave tasks varied with owners' attitudes concerning the reliability and capabilities of blacks; most slaves provided unskilled labor. Owners felt that industrial mining, with opportunities for time off and extra wages, weakened their authority. By 1826 slaves hunted for gold on their own. Some acquired money with which to buy liberty, but few remained miners.[27]

By 1830 society in Mecklenburg mirrored mining life, often hard and given to excess. Observers noted "deplorably bad" morals around mines, blaming unregulated society and easy gold. One northerner could "hardly conceive of a more immoral community. . . . Drunkenness, gambling, fighting, lewdness, and every other vice. . . . men, by working three days in the week, make several dollars, and then devote the remaining four [days] to every species of vice."[28]

Miners abused alcohol; some worked merely to buy bread and whiskey. Charlotte drunkards abused peaceable citizens. One writer complained: "At every camp whiskey shanties were plentiful. . . ." Yet some miners were gentlemen and reformers. Some managers forbade liquor; others arranged for preaching of the Gospel to miners. One miner said that "any dissipation . . . would fail to come up to the reality" yet spent his idle hours reading and playing the violin. Liquor was an escape from a harsh life of disease and untimely death. A placer miner might live outdoors and carry only provisions and a few tools. Burke miners, even if wealthy, lived in small cabins. Mining "towns" were often bare frontier villages.[29]

Yet life had some merriment. Christmas and the Fourth of July might have a mining flavor. On Christmas 1825 miner Charles Rothe amused Charlotte by visiting and drinking eggnog in his "miner's uniform": a cape, cocked hat, and red "round coat with three rows of buttons having a hammer and pick stamped on them." One Fourth of July miners enjoyed toasts, such as "May the farmers around us be encouraged by the Miners of Burke." "May the gold miners meet with shallow pitting and rich grit for their reward." "Health of body, peace of mind, and four pennyweights to the hand." Mecklenburg mining captain John Penman and sixty miners once marched into Charlotte with fifes and drums for a banquet.[30]

Wages varied by skill and luck. Few grew rich. Miners increasingly became laborers toiling for others. A miner working for himself in the 1830s averaged a dollar a day or less. A dollar bought a pound of quicksilver or a shovel. A so-called "machine" cost fifteen dollars; a hand pump, two dollars. Horse-driven crushing mills cost as much as $2,500. Water or steam power added thousands more. Complex devices and some staples were imported. Burke miners purchased up to $20,000 worth of goods each week, including Kentucky pork and Philadelphia shovels. Many placer owners preferred pans and rockers to costlier machinery. Even important vein mines were backward: at Barringer's, men wastefully sank random pits.[31]

Nonetheless, technology crept into the gold fields after the mid-1820s. Charles Rothe promoted machinery, and other experts appeared. Papers printed technical articles on mines in Europe and Mecklenburg. By late 1830 a leading newspaper referred to rockers as "the old mode." One new machine was the stamp mill, which dated to sixteenth-century Germany and broke quartz into fine pieces so gold could be separated. In 1829 Humphrey Bissell installed at the Capps mine the first stamp mill in the United States. Stamp mills pounded ore (already broken, first by gunpowder and then by hand, to fragments measuring several inches across) into fine gravel. The wooden machine was covered in key places with iron. Ore flowed through a trough, or mortar, and received blows from stamps falling via a camshaft arrangement.[32]

Miners used other machines to further crush and amalgamate the golden particles with mercury. The common Chilean mill had two crushing wheels (made of stone) five or six feet in diameter that rotated within a circular stone base containing ore, water, and mercury. The arrastra substituted drag stones for the wheels and a rock pavement for the base. Both mills were slow compared to stamps. By the early 1830s mines near Charlotte had such machines and pumps, driven by horse, water, or steam power. Other mines, notably the Burke placers, continued with pans and rockers.[33]

Workers at Reed finally went underground. In 1831 the first pit yielded ore worth five dollars per bushel. Family members deepened pits into shafts up to ninety feet deep but continued creek mining. The men raised ore with horse-powered whims and erected an arrastra to crush it. The whim was a vertical drum around which a rope wound to raise or lower an ore bucket. Timothy Reed (grandson of John) supposedly recovered twenty thousand dollars' worth of gold from one shaft. One man claimed that he found fourteen pounds of gold one day before breakfast. Miners dug gold "like potatoes" from one trench. Slaves were said to pay for drinks of liquor with egg-sized hunks of gold and quartz. In 1834 conflict arose over a ten-pound nugget. John Reed sought family unity, but selfishness prevailed. One son obtained an injunction that closed the mine for a decade; it did not reopen until shortly after old John died in 1845.[34]

One exceptional mine in the early 1830s was the Capps, with up to a hundred laborers, black and white. A steam engine pumped water from a one-hundred-foot shaft, but horses raised ore. Miners used their muscles, hand tools, and gunpowder underground but only muscles and tools on the surface to break ore for milling. The Capps used stamp and Chilean mills. Soon it was being operated by the Mecklenburg Gold Mining Company, a corporation purportedly employing six hundred persons and managed by Count Rivafinoli. With entrepreneur Bissell, Rivafinoli leased the Capps, Rudisill, and McComb (renamed St. Catherine) mines. With dozens of European miners, he installed three steam-powered pumping engines. Spending large sums, Rivafinoli deepened the Rudisill's engine shaft and added timbered galleries and machine-driven bellows.[35]

At the opposite end of the same ore mass and one-half mile from the Rudisill was the St. Catherine mine. When Rivafinoli arrived, Bissell and Barker were converting a water-powered flour mill to Chilean mills. A 110-acre pond supplied water. Rivafinoli convinced them to use arrastras and added a steam engine, an engine-house, and buildings to house twelve arrastras. The eighteen water and steam-powered arrastras processed four hundred bushels of ore daily. The St. Catherine mill also had twelve water-powered, seventy-five-pound stamps requiring eight workers for twenty-four-hour operation. Such machinery represented only the handful of advanced mines.[36]

In the early 1830s North Carolina had numerous active mines. The Dunn and Alexander mine, near Charlotte, had surface and vein gold and a horse-powered Chilean mill. The Harris mine produced remarkable concentrations of gold. In Cabarrus, work continued at Pioneer Mills. Daily placer output of Burke County was valued at twenty-four hundred dollars. Prospectors discovered a large vein near Kings Mountain, a Revolutionary War battlefield. In Davidson County a corporation leased five thousand acres and installed water-powered stamps and Chilean mills. For Conrad Hill (likewise in Davidson County), at which workers dug to a depth of seventy feet and operated a mill four miles away on an undependable stream, Englishmen proposed steam machinery. Less fruitful gold fields were discovered forty-five miles northeast of Raleigh in the early 1830s. Never comparable in scale to Mecklenburg mining, intermittent operations there lasted a century. The most noted of about twenty eastern sites was the Portis mine in Franklin County.[37]

Western local promoters extolled the benefits of gold. Mecklenburg County reportedly produced $1,500 worth weekly. Mining counties enjoyed greater prosperity than other counties. A future governor exclaimed: "Those who have been esteemed prudent and cautious embark in speculation with the greatest enthusiasm—bankrupts have been restored to affluence, and paupers turned to nabobs." While few made fortunes, the industry aided regional economies. A Charlotte weekly cited growing business, new residents, more circulation of money, and rising property values. Northern and foreign capital created jobs and had a multiplier effect on local economies. Trade prospered with gold as currency. Gold reduced national dependence on foreign bullion and supposedly bolstered social values. Editors

Leading Gold Mines in North Carolina

1. Fisher Hill
2. Gardner Hill
3. North State
4. Fentress
5. Hoover Hill
6. Uwharrie
7. Conrad Hill
8. Silver Hill (Washington)
9. Ward
10. Gold Hill
11. Union Copper
12. Coggins (Appalachian)
13. Star
14. Tebe Saunders
15. Iola
16. Sam Christian
17. Barringer
18. Parker
19. Cotton Patch
20. Ingram (Crawford)
21. Phoenix
22. Reed
23. Pioneer Mills
24. Howie
25. Dunn
26. Capps
27. Alexander
28. McComb (St. Catherine)
29. Harris
30. Rudisill
31. Kings Mountain

Map by Michael T. Southern, Division of Archives and History.

declared that gold prompted "lazy, lounging fellows, who once hung as a weight upon society, into active and manly exertion." The industry fostered technology, enterprise, and orderly society, declared miner Charles Fisher.[38]

Mining also had negative effects. Observers cited alcoholism, degeneracy from easy money, neglected farms, a decaying work ethic, and declining thrift. Speculation made some rich and others poor: the *Charleston Courier* attacked the "corrupting treasure." Men were maimed or killed by blasting, cave-ins, and other mishaps. Editors often glossed over such news. A few observers noted a negative environmental impact. One predicted that Burke placers would ruin land, a situation that materialized in that county and elsewhere. Other effects, such as water pollution, often went unrecorded.[39]

Yet, production rose. North Carolina supplied all domestic gold sent to the U.S. mint ($110,000) prior to 1828. In 1827 the mint received $21,000 (11 percent of receipts) in such gold. In 1828 deposits from the state reached $46,000. In 1829 Virginia sent $2,500 worth and South Carolina $3,500, while North Carolina's sum ballooned to $134,000. Georgia and North Carolina each shipped more than $200,000 worth in 1830, providing gold for 80 percent of the mint's gold coinage. North Carolina regained the lead with $294,000 in 1831 and shipped $475,000 worth in 1833 (a record that endured for a decade). In 1834 the mint coined more gold than in the preceding nine years.[40]

The mint accounted for only one-third to one-half of North Carolina's output, and the exact total is unknown. The short life of most companies hindered record-keeping. Perhaps individuals secretly pocketed choice nuggets. Much gold went to domestic and European artisans, manufacturers, and other users. Some mines sent output directly to Europe, where American nuggets were popular. Jewelers in New York paid a premium for Carolina gold. Charlotte jewelers such as Victor Blandin and Thomas Trotter specialized in local gold. The metal passed from country merchants through market towns to Charleston, New York, and Europe. Perhaps one-fifth of output in 1832 went into the arts.[41]

Southerners began promoting branch mints in 1830. The long trip to Philadelphia was dangerous, and gold of questionable purity circulated in the gold region. Congressman Samuel P. Carson from Burke and the state legislature endorsed a regional mint, but the matter bogged down in Congress. Private initiative met the need. In 1831 German immigrant Christopher Bechtler began minting gold coins in Rutherford County. Templeton Reid had coined Georgia gold in 1830, but Bechtler was more successful. The middle-aged goldsmith, gunsmith, and clockmaker produced coins with homemade, hand-operated equipment. Besides $2.50 and $5.00 coins, he produced in 1832 the first gold dollar minted in the nation. Bechtler's integrity assured success. In four years he coined $109,000 worth of gold (a figure that grew twentyfold by 1840) and fluxed nearly 400,000 additional pennyweights' worth. His coins circulated for years. In 1834 North Carolina deposits at the U.S. mint reached $380,000, but Georgia sent $415,000. Congress voted in 1835 to establish a New Orleans mint to coin gold and silver and branch mints at Charlotte and Dahlonega, Georgia, to coin only gold. Philadelphia architect William Strickland designed the Charlotte facility.[42]

By the mid-1830s, key aspects of the mining culture had appeared. Yet—despite foreign capital and expertise; slaves, stamp mills, and steam engines; and continuing output—mining soon lost its vigor temporarily. Various factors, some long lasting, weakened the industry. North Carolina deposits at the U.S. mint fell sharply by 1837 to $116,900. Burke's placer rush was over. Agriculture (including cotton in Alabama) and railroad construction drew labor and capital from picked-over creeks.

Vein mines also faced challenges. After the national financial panic of 1837, most large firms were short of funds or bankrupt. Remaining "miners" were primarily farmers. The federal census enumerated only 589 miners in 1840, one hundred of them in Mecklenburg County.[43]

Mining had its own peculiar problems. Vein mining required major capital outlays in advance of production. Errors in management and technology plagued mines. Rivafinoli received criticism for hiring extra administrators and losing half of his gold during milling. Various mines closed as a result of ineffective direction. Once miners reached the water level, generally fifty to sixty feet underground, they needed pumps and dealt with undecomposed refractory pyritic ore. Without good pumps and a means to house such ore, most mining ceased where pumping began. Many efforts to drive off sulfur, such as by "roasting" ore, failed: dealing with sulfurets required a chemical or metallurgical process not available at that time. Deep mines frequently yielded low-grade ore. Some commentators derided as indolent those natives unable to develop mines skillfully.[44]

The late 1830s were difficult times. The Cabarrus Gold Mining Company, despite northern money, a promising mine, and a steam engine, battled bankruptcy. Rivafinoli's development outstripped returns, and the Mecklenburg sheriff auctioned his mines. John Penman bought the Rudisill, cleaned out old workings, and tried to dewater the engine shaft, but Rivafinoli's worn-out pumps failed. Penman abandoned that mine but also worked the St. Catherine for several years. Similar failures occurred at other mines.[45]

Yet the Charlotte mint opened in December 1837—in a town of fewer than one thousand souls and a county with fifteen mines and fifteen distilleries in operation. Twenty-five thousand dollars' worth of gold arrived that month. Depositors received gold bars or certificates for gold coins. Soon the mint received $130,000 in gold annually. Overall the mint coined about five million dollars in half eagles ($5.00—by far its most common product), quarter eagles ($2.50), and gold dollars.[46]

By the mid-1840s mining had largely recovered. Antebellum mines produced a significant amount of gold, livelihood for miners, a foretaste of coming industrialism, and markets for food and supplies in a rural, relatively backward state. Was mining profitable? Perhaps at best southern mines scarcely broke even overall.

Notes

1. The most recent published writings on gold mining in North Carolina are P. Albert Carpenter III, *Gold in North Carolina* (Raleigh: Department of Environmental, Health and Natural Resources, [North Carolina Geological Survey Information Circular 29], 1993; reprinted 1999); Jeffrey P. Forret, "Slave Labor in North Carolina's Antebellum Gold Mines," *North Carolina Historical Review* 76 (April 1999): 135–162; Richard F. Knapp, *Golden Promise in the Piedmont: The Story of John Reed's Mine* (Raleigh, Historic Sites Section, Division of Archives and History, Department of Cultural Resources, rev. ed., 1999); and Richard F. Knapp and Brent D. Glass, *Gold Mining in North Carolina: A Bicentennial History* (Raleigh:

Division of Archives and History, Department of Cultural Resources, 1999), which includes the most exhaustive bibliography on the subject. Older but still valuable are several articles in the *North Carolina Historical Review*, particularly Brent D. Glass, "The Miner's World: Life and Labor at Gold Hill," *NCHR* 62 (October 1985): 420–447; Brent D. Glass, "'Poor Men with Rude Machinery': The Formative Years of the Gold Hill Mining District, 1842–1853," *NCHR* 61 (January 1984): 1–35; and Fletcher M. Green, "Gold Mining: A Forgotten Industry in Ante-Bellum North Carolina," *NCHR* 14 (January, April 1937): 1–19, 135–155. Useful materials at repositories in North Carolina include nineteenth-century newspapers, especially the *Miners' and Farmers' Journal* (Charlotte), various public records and private collections at the North Carolina State Archives in Raleigh, and manuscript collections at such places as the Southern Historical Collection in the University of North Carolina Library at Chapel Hill and the Special Collections Department of the Duke University Library in Durham.

2. Geologists Henry S. Brown and Dennis LaPoint have provided these summary descriptions of geologic processes.

3. Herman J. Bryson, *Gold Deposits in North Carolina* (Raleigh: North Carolina Department of Conservation and Development [Bulletin 38], 1936), 43, 64–65, 103–104, 127, 133–136, 144. A more recent summary is in Carpenter, *Gold in North Carolina*.

4. Joseph Seawell Jones, *Memorials of North Carolina* (New York: Scatcherd and Adams, 1838), 84; Green, "Gold Mining," 6.

5. Mark A. Schwalm, *A Hessian Immigrant Finds Gold: The Story of John Reed* (Stanfield, N.C.: Reed Gold Mine, 1996), 1–12, 21–23.

6. John H. Wheeler, *Historical Sketches of North Carolina, from 1584 to 1851* (Philadelphia: Lippincott, Grambo, and Co., 1851), 63–64; August Partz, "Examinations and Explorations of the Gold-Bearing Belts of the Atlantic States: The Reid Mines, North Carolina," *Mining Magazine* 3 (August 1854): 162–164 (hereafter cited as Partz, "The Reid Mines").

7. Denison Olmsted, "Report on the Geology of North Carolina . . . ," *Southern Review* 1 (February 1828): 251–252n; William Thornton, *North Carolina Gold-Mine Company* [Washington, D.C.: n.p., 1806], 3; Partz, "The Reid Mines," 163; James F. Shinn, "Discovery of Gold in North Carolina," *Trinity Archive* 6 (1893), 336.

8. Thornton, *North Carolina Gold-Mine Company*, 3; U.S. Department of the Treasury, Bureau of the Mint, Elias Boudinot, "Report of the Director of the Mint for 1804," in *American State Papers: Finance* 2: 119; Stephen Ayres, "A Description of the Region in North Carolina Where Gold Has Been Found," *Medical Repository* 10 (1807): 149–151.

9. Thornton, *North Carolina Gold-Mine Company*, 8–17; W. Thornton, "Letter from W. Thornton, Esq. to the members of the North Carolina Gold Mine Company," *London, Edinburgh, and Dublin Philosophical Magazine* 27 (1807): 261–264.

10. Green, "Gold Mining," 9; "Progress of Finding Gold in North Carolina," *Medical Repository* 12 (1809): 193; *Western Carolinian* (Salisbury), September 21, 1824; Denison Olmsted, *Report on the Geology of North Carolina*, 3 vols. (Raleigh: J. Gales and Sons, 1824–1827), 1:34.

11. *Miners' and Farmers' Journal*, March 19, 1835; U.S. Congress, House, *Assay Offices—Gold Regions, South*, H. Rept. 391 to Accompany H.R. 407, 23d Cong., 1st sess., 1834, 1; *Western Carolinian*, February 7, 1826, October 13, 1829; Denison Olmsted, "On the Gold Mines of North Carolina," *American Journal of Science* 9 (1825): 12–13; Green, "Gold Mining," 137.

12. Jasper L. Stuckey, *North Carolina: Its Geology and Mineral Resources* (Raleigh: Department of Conservation and Development, 1965), 225–228; Charles E. Rothe, "Remarks on the Gold of North Carolina," *American Journal of Science and Arts* 13 (1828): 201–217.

13. Curtis P. Nettels, *The Emergence of a National Economy, 1775–1815* (New York: Holt, Rinehart, and Winston, 1962), 183; Cornelius O. Cathey, *Agriculture in North Carolina before the Civil War* (Raleigh: North Carolina Department of Archives and History, 1966), 34; Olmsted, "On the Gold Mines of North Carolina," 12; Rothe, "Remarks on the Gold of North Carolina," 36–37.

14. *Niles' Weekly Register* (Baltimore), April 18, 1829; *Western Carolinian*, June 23, 1829; *Miners' and Farmers' Journal*, April 28, November 16, 1831; *Mining Magazine* 10 (January 1858): 28–29; Phillip T. Tyson, *Report on the Gold Deposits and Works of the Manteo Mining Company in North Carolina* (Baltimore: John D. Toy, 1853), 23; Charles U. Shepard, *Description of Gold and Copper Mines at Gold Hill, Rowan County, North Carolina* (N.p., [1853]), 8.

15. *Miners' and Farmers' Journal*, March 2, 1833; "Gold," *DeBow's Review* 18 (February 1855): 244; *Western Carolinian*, August 29, 1831.

16. *Niles' Weekly Register*, March 26, 1825; *Raleigh Register*, April 5, 1825; U.S. Congress, House, *Assay Offices, Gold Districts, North Carolina, Georgia & c.*, H. Rept 39 to Accompany H.R. 84, 22d Cong., 1st sess., 1831, 12–13 (hereafter cited as *Assay Offices—1831*); Olmsted, "On the Gold Mines of North Carolina," 6; *Free Press* (Tarborough), October 17, 1828; *Catawba Journal* (Charlotte), August 22, 1826; *Western Carolinian*, December 5, 1826.

17. *Western Carolinian*, June 20, 27, July 4, October 10, 1826, February 20, August 17, November 13, 1827, August 18, 1829; *Niles' Weekly Register*, May 16, 1829, *Yadkin and Catawba Journal* (Salisbury), June 16, 1829; Elisha Mitchell, "On the Geology of the Gold Region of North Carolina," *American Journal of Science* 16 (1829), following p. 19.

18. William Blanding, journal, August 6, 1828, William Blanding Papers, Special Collections Department, Duke University Library; *Western Carolinian*, October 27, 1829.

19. *Catawba Journal*, August 29, September 12, 1826; Henry Barnard, "The South Atlantic States in 1833, as seen by a New Englander," ed. Bernard C. Steiner, *Maryland Historical Magazine* 13 (September and December, 1918): 339; Blanding, journal; Daniel A. Tompkins, *History of Mecklenburg County and the City of Charlotte, from 1740 to 1903*, 2 vols. (Charlotte: Observer Printing Company, 1903), 1:130, 2:121; *Mecklenburg Jeffersonian* (Charlotte), March 28, 1845; Joshua Forman to Arthur Bronson, May 1, 1829, Speculation Land Company Records, Southern Historical Collection; Alfred Graham to William A. Graham, March 31, 1829, William A. Graham Papers, Southern Historical Collection.

20. *Greensboro Daily News*, January 11, 1925; Edward W. Phifer, "Champagne at Brindletown: The Story of the Burke County Gold Rush, 1829–1833," *North Carolina Historical Review* 40 (autumn 1963): 489–490; Isaac Avery to William B. Lenoir, July 4, 1829, Lenoir Papers, Southern Historical Collection; *Western Carolinian*, June 13, 1829.

21. *Miners' and Farmers' Journal*, March 10, 17, April 21, May 25, July 13, August 10, 31, September 7, 1831; *Western Carolinian*, June 23, July 7, 1829; *Free Press*, September 22, 1827, July 23, 1830; *Niles' Weekly Register*, April 18, May 16, 1829; *Yadkin and Catawba Journal*, June 16, 1829.

22. *Western Carolinian*, September 6, 1825, January 16, March 7, 1827; *Laws of North Carolina, 1827*, c. 103; Charles Fisher to Henry W. Connor, January 14, 1828, statement, March 8, 1829, memorandum, April 6, 1829, agreement with Edmund DeBerry, April 10, 1829, Fisher Family Papers, Southern Historical Collection.

23. *Laws of North Carolina, 1836–1837*, c. 55; *Greensboro Patriot*, June 13, 1829; *Free Press*, February 22, May 3, 1831; *Western Carolinian*, February 28, 1831; *Miners' and Farmers' Journal*, June 1, 1831; *Niles' Weekly Register*, April 7, 1832.

24. *Free Press*, June 7, 1831; James S. Buckingham, *The Slave States of America*, 2 vols. (London: Fisher, Son, and Company, 1842), 2:220; *North Carolina Spectator and Western Advertiser* (Rutherfordton), May 21, 1830; *Western Carolinian*, May 25, 1830; *Miners' and Farmers' Journal*, November 15, 1830, March 10, 24, 1831; George C. D. Odell, *Annals of the New York Stage*, 15 vols. (New York, 1927–1949; reprint, New York: AMS Press, 1970), 3:691.

25. Bryan Earl, *Cornish Mining: The Techniques of Metal Mining in the West of England, Past and Present* (Truro, England: D. B. Barton, 1968), 9, 17, 24, 35; Arthur C. Todd, *The Cornish Miner in America* (Truro: D. B. Barton, 1967), 16, 17; Alfred K. H. Jenkins, *The Cornish Miner: An Account of his Life Above and Underground from Early Times* (London: Albert and Unwin, 1927), 171, 203–241 passim; *Western Carolinian*, July 25, 1826.

26. Buckingham, *The Slave States of America*, 1:42, 2:220; Green, "Gold Mining," 11; *Hillsborough Recorder*, July 8, 1829; *Niles' Weekly Register*, May 21, 1831; Sixth Census of the United States, 1840: Mecklenburg County, North Carolina, Population Schedule, National Archives, Washington, D.C. (microfilm, State Archives, Division of Archives and History, Raleigh); David H. Strother (Porte Crayon), "North Carolina Illustrated: The Gold Region," *Harper's New Monthly Magazine* 15 (August 1857): 296–299; Barnard, "The South Atlantic States in 1833," 348; Green, "Gold Mining," 17.

27. Franklin L. Smith, "Notices of Some Facts Connected with the Gold of a Portion of North Carolina," *American Journal of Science* 32 (1837): 133; Barnard, "The South Atlantic States in 1833," 347; Green, "Gold Mining," 15; Phifer, "Champagne at Brindletown," 492; Henry B. C. Nitze and George B. Hanna, *Gold Deposits of North Carolina* (Winston: M. I. and J. C. Stewart [North Carolina Geological Survey Bulletin 3], 1896), 153; Willard P. Ward, "The Gold Deposits of the Southern States," *Engineering and Mining Journal* 9 (1870): 392; *Free Press*, January 11, 1831; Alfred Graham to William A. Graham, March 31, 1829, William A. Graham Papers, Southern Historical Collection; Robert S. Starobin, *Industrial Slavery in the Old South* (New York: Oxford University Press, 1970), 215, 219–220, 223–224; *Sun* (New York?),

August 29, 1897; John Hope Franklin, *The Free Negro in North Carolina* (Chapel Hill: University of North Carolina Press, 1943; reprint, 1995), 134–135.

28. William G. Ouseley, *Remarks on the Statistical and Political Institutions of the United States* (Philadelphia: Carey and Lea, 1832), 173; *Niles' Weekly Register*, May 21, 1831.

29. Rothe, "Remarks on the Gold of North Carolina," 36, 38; *Catawba Journal*, October 17, 1826; *Sun*, August 29, 1897; *Niles' Weekly Register*, May 21, 1831; *Western Carolinian*, August 4, 1829, September 30, 1828; *Raleigh Register*, August 10, 1858; Peter D. Swain to Benjamin Elliott, August 29, 1843, Benjamin P. Elliot Papers, Special Collections Department, Duke University Library; Phifer, "Champagne at Brindletown," 493; Barnard, "The South Atlantic States in 1833," 338; Green, "Gold Mining," 16–17.

30. Franklin L. Smith to Daniel M. Barringer, December 25, 1825, Daniel M. Barringer Papers, Southern Historical Collection; *Western Carolinian*, July 30, 1832; *Miners' and Farmers' Journal*, April 30, 1835.

31. *North Carolina Spectator and Western Advertiser*, March 26, June 11, 1830; *Miners' and Farmers' Journal*, April 28, November 16, 1831; expense accounts in volume 6, William Henry Burwell Papers, Southern Historical Collection; *Western Carolinian*, May 12, October 20, 1829; Isaac T. Avery to Willie P. Mangum, February 28, 1834, in Henry T. Shanks, ed., *The Papers of Willie P. Mangum*, 5 vols. (Raleigh: Department of Archives and History, 1950–1956), 2:109; *Free Press*, June 7, 1831; Rothe, "Remarks on the Gold of North Carolina," 38–42.

32. *Miners' and Farmers' Journal*, October 25, 1830, January 25, March 17, 24, 1831, February 8, 1832; Earl, *Cornish Mining*, 17, 83; Charles Singer et al., *A History of Technology*, 5 vols. (London: Oxford University Press, 1955–1958), 4:64; *Western Carolinian*, October 27, 1829; Blanding journal.

33. Stephen P. Leeds, "Gold Ores and Their Working," *Mining Magazine* 7 (July–August 1856): 29–30; *Western Carolinian*, October 27, 1829; Charles Fisher to Samuel P. Carson, March 22, 1830, in *Assay Offices—1831*; *Miners' and Farmers' Journal*, March 2, 24, 1831, March 28, 1832; T. H. Watkins, *Gold and Silver in the West* (Palo Alto, Calif.: American West Publishing Company, 1971), 193–194; Otis Young, *Western Mining* (Norman: University of Oklahoma Press, 1970), 69–72.

34. Partz, "The Reid Mines," 164–165; *George Reid v. George Barnhart and others*, Case 3633, Supreme Court Original Cases, fall term 1845, State Archives, Division of Archives and History (also reported as 54 N.C. [1 Jones Eq.] 142 [1853]); Knapp, *Golden Promise*, 16–18.

35. Barnard, "The South Atlantic States in 1833," 340–342; *Free Press*, January 11, August 23, 1831; *Miners' and Farmers' Journal*, November 15, 1830, March 10, November 30, December 28, 1831, January 25, May 29, 1832; *Charlotte Journal*, October 20, 1837. See also Larry C. Hoffman, "The Rock Drill and Civilization," *American Heritage of Invention and Technology* 15 (summer 1999): 56–60.

36. *Miners' and Farmers' Journal*, February 1, March 24, 1831.

37. Franklin Gold Mining Company, *Charter, with Amendments and Descriptions of their Mines* (New York, W. M. Mercein and Son, 1835), 17; Barnard, "The South Atlantic States in 1833," 341; *Miners' and Farmers' Journal*, March 22, 1830, June 15, 1831, September 4, 1832, May 7, 1835; *Western Carolinian*, February 21, 1831; *North Carolina Spectator and Western Advertiser*, November 26, 1830; John Taylor Jr. and George Francis, "Report upon the Possessions and Properties held on Lease of the North Carolina Gold Mining Company," September 5, 1832, Fisher Family Papers, Southern Historical Collection.

38. *Western Carolinian*, March 31, July 28, 1829; James Martin Jr. to Willie P. Mangum, December 17, 1831, in Shanks, *Mangum Papers*, 1:440–441; William A. Graham to Thomas Ruffin, August 10, 1829, in J. G. de Roulhac Hamilton, ed., *The Papers of Thomas Ruffin*, 2 vols. (Raleigh: Department of Archives and History, 1918–1920), 1:509–510; *Miners' and Farmers' Journal*, November 1, 1830; Green, "Gold Mining," 152–155; Phifer, "Champagne at Brindletown," 497; *Yadkin and Catawba Journal*, June 16, 1829; Charles Fisher to Samuel P. Carson, March 22, 1830, in *Assay Offices—1831*, 17–18.

39. *Western Carolinian*, July 11, 1826, August 4, 1829; *Miners' and Farmers' Journal*, September 21, 1831, April 24, 1832; Barnard, "The South Atlantic States in 1833," 347; Green, "Gold Mining," 16.

40. *Western Carolinian*, February 12, 1828; *Assay Offices—1831*, 1; *Miners' and Farmers' Journal*, February 22, 1835.

41. *Miners' and Farmers' Journal*, March 17, April 28, 1831, October 27, 1832, March 2, 1833, February 8, 1834; *Assay Offices—1831*, 1, 15; John H. Wheeler, "Gold Mines of North Carolina," *American Almanac* (1841): 211–217; *Western Carolinian*, November 16, 1824; Green, "Gold Mining," 138; Olmsted, "On the Gold Mines of North Carolina," 13; Taylor and Francis, "Report upon the Possessions and Properties held on Lease of the North Carolina Gold Mining Company," September 5, 1832, Fisher Family Papers, Southern Historical Collection.

42. *Assay Offices—1831*, 1–2, 12, 15, 19–21; *Free Press*, February 19, 1830; *Western Carolinian*, February 23, 1830; *North Carolina Spectator and Western Advertiser*, March 26, 1830, July 2, August 27, 1831; *Miners' and Farmers' Journal*, July 16, 1831, February 10, 26, March 12, 19, 1835; Dexter C. Seymour, "Templeton Reid, First of the Pioneer Coiners," *American Numismatic Society Museum Notes* 19 (1974): 225–267; Clarence Griffin, *The Bechtlers and Bechtler Coinage; and, Gold Mining in North Carolina, 1814–1830* (Forest City: Forest City Courier, 1929), 4–5; Thomas Featherstonhaugh, "A Private Mint in North Carolina," *Publications of the Southern History Association* 10 (March 1906): 70–71; Rodney Barfield, *The Bechtlers and Their Coinage: North Carolina Mint Masters of Pioneer Gold* (Raleigh: North Carolina Museum of History, Division of Archives and History, Department of Cultural Resources, 1980); "Report on the Establishment of a Mint in the State of North Carolina," Legislative Papers, 1830–1831, Box 452, State Archives [subsequently publication 29 for 1830 (Raleigh: Lawrence and Lemay, 1830)]; Carpenter, *Gold in North Carolina*, 16; Green, "Gold Mining," 141–142; *Charlotte Journal*, July 24, September 11, October 23, 1835.

43. Carpenter, *Gold in North Carolina*, 16; Phifer, "Champagne at Brindletown," 499–500; Henry B. C. Nitze, "Gold Mining in the Southern States," *Engineering Magazine* 10 (February 1896): 822; Census of 1840, quoted in *American Almanac* (1842), 211.

44. *Miners' and Farmers' Journal*, September 27, November 1, 1830; "Mining: Its Embarrassment and Its Results," *Mining Magazine* 2 (June 1854): 636–638; Walter R. Crane, *Gold and Silver, Comprising an Economic History of Mining in the United States* (New York: John Wiley and Sons, 1908), 18–19; *Charlotte Journal*, November 10, 1837, March 9, 1838; Bruce Roberts, *The Carolina Gold Rush* (Charlotte: McNally and Loftin, 1971), 16–18; Leeds, "Gold Ores and Their Working," 353–354; Nitze and Hanna, *Gold Deposits of North Carolina*, 109; *Miners' and Farmers' Journal*, March 14, 28, 1832; Green, "Gold Mining," 151–152; Henry B. C. Nitze and H. A. J. Wilkens, *Gold Mining in North Carolina and Adjacent South Appalachian Regions* (Raleigh: Guy V. Barnes [North Carolina Geological Survey Bulletin 10], 1897), 150–151.

45. *Laws of North Carolina, 1831*, c. 108; *Miners' and Farmers' Journal*, December 6, 20, 1834; *Charlotte Journal*, July 3, 1835; Ebenezer Emmons, *Geological Report on the Midland Counties of North Carolina* (New York: George P. Putnam and Co., 1856), 176; Stephen P. Leeds, "The Rudisel Gold and Copper Mine of North Carolina," *Mining Magazine* 2 (May 1854): 518.

46. Green, "Gold Mining," 143–145; *Charlotte Journal*, December 8, 1837; Charlotte Mint Register of Gold Bullion, Records of the U.S. Mint, Record Group 104, National Archives, Washington, D.C.; Clair M. Birdsall, *United States Branch Mint at Charlotte, North Carolina: Its History and Coinage* (Easley, S.C.: Southern Historical Press, 1988); Daisy W. Bridges, "Carolina Gold and the U.S. Branch Mint in Charlotte," *Antiques* 104 (December 1974): 1041–1046; Anthony M. Stautzenberger, *The Establishment of the Charlotte Branch Mint: A Documented History* (Austin, Tex.: the author, 1976), 145; Tompkins, *History of Mecklenburg County and the City of Charlotte*, 2:131.

Gold Mining in North Carolina, 1840–1915

Brent D. Glass

Dr. Glass is executive director of the Pennsylvania Historical and Museum Commission and the principal scholar of the Gold Hill mining district, North Carolina's most famous gold-mining region. He is a member of the council of the American Association for State and Local History and numerous other historical organizations. Among his publications are A History of the Textile Industry in North Carolina *(1992) and (with Richard F. Knapp)* Gold Mining in North Carolina: A Bicentennial History *(1999), as well as several articles in* Pennsylvania Heritage.

During the seventy-five-year period under consideration in this paper, the gold-mining industry in North Carolina reflected a full range of activity that included impressive growth and spectacular failure. Between those two extremes, a diverse cast of characters moved across the stage—German farmers, Cornish miners, slave laborers, Wall Street investors, engineers, and inventors (including Thomas Edison)—all seeking to sustain an industry that depended upon skillful management, experienced workers, and technological innovation. Those qualities, unfortunately, were in short supply in most of the state's mining districts, as was the supply of gold itself—a shortage in what the miners called "paying quantities." As a result, the mining industry prospered briefly in the 1850s, revived in the 1870s and 1880s, and sputtered to a close by the early decades of the twentieth century.

During most of that period, the Gold Hill mining district served as a case study of the industry at its best and at its worst. The district extended through southeastern Rowan County, northeastern Cabarrus County, and into a narrow strip along part of Montgomery (now Stanly) County. The Yadkin River formed its northeastern boundary for eight miles, from which the district extended in a southwesterly direction for eighteen miles to the village of Mount Pleasant. In the center of the district grew the town of Gold Hill, located about fourteen miles southeast of Salisbury within a two-mile ridge, which formed a watershed with streams flowing northwest to the Yadkin River and southeast to the Rocky River. To the southwest were the Beaver Hills, the only other prominent topographical feature in an otherwise flat valley.[1]

The most important geological feature in the area was the Gold Hill fault, which ran north-south along the west central section of the district. The rocks on the eastern side of the fault line, mostly slates and schists, contained the district's most important mineral deposits. The character of the ore at Gold Hill resembled that of other mineral regions in the Carolina slate belt. Deposits of gold-bearing ores formed at least ten distinct veins running generally in a northern direction. Two types of veins were found in Gold Hill: one that primarily contained gold, the other in which copper ores dominated. Each type of vein also included quantities of silver and lead. All ores found in the district contained some amounts of those minerals, although samples with a higher concentration of gold usually held a minimum of copper, and vice versa. Most geologists believe that the gold and copper veins formed at different geological times during periods of minor faulting that occurred after the formation of the Gold Hill fault.[2]

The Gold Hill district emerged in the early 1840s at a time when the mining industry in North Carolina had rebounded from a series of failures in the previous decade. The establishment of the federal branch mint in Charlotte in 1837 offered a market for local mines, and the general economic recovery from the panic of 1837 further stabilized the financial community. Mining production steadily advanced, with the value of annual deposits at all federal mints averaging nearly three hundred thousand dollars in the decade after 1837 and that of deposits at the Charlotte mint approaching four hundred thousand dollars a year by 1850.[3]

The first discovery of gold at Gold Hill occurred in September 1842 on the farm of Andrew Troutman. Two local farmers were working that surface deposit when they soon struck a rich vein of gold ore and sank a shaft that eventually reached two hundred feet. Gold seekers on neighboring farms identified at least five additional vein deposits over the next year. "We were utterly surprised," wrote a visitor in 1844,

> to find almost the whole surface of a ridge, some one and a half or two miles in extent, torn up by the pick and shovel of the miner. Here was a deep deserted pit with gaping mouth, inviting as it were, man and beast to sudden destruction by a misstep. There was a long trench running through the woods for many yards, which had been made in "searching for the veins." Here again were a company of operators in and about a shaft or pit of some 60 to 75 feet in depth pursuing a vein into the bowels of the earth, and from thence drawing the rich ores. A little way off and there is another company similarly engaged. A little beyond and there is still another, and another, and another, and so the woods are filled.[4]

Many of these miners were veterans of the gold rushes of previous decades in North Carolina's Piedmont region. They were familiar with basic techniques of excavation and extraction. They understood pumping procedures and the use of gunpowder in deep mining. They also introduced hoisting equipment and quicksilver at Gold Hill. Nevertheless, poverty and backwardness dominated Gold Hill's landscape in those early years. The new settlers utilized the most crude methods in their zeal to remove as much ore as possible without investing scarce capital. They relied on "sheer muscle power" and horse-powered mills and hoisting equipment rather than developing more reliable but more expensive steam-powered technology. They lived in small log cabins "daubed with clay . . . of a real primitive order with clapboard roofs and weight poles; with a chimney at one end as wide as the whole house [and] very few that exhibit brass nobs on paneled doors, whilst there are many of rough boards with string and latch." A longtime resident later recalled these first miners at Gold Hill as "poor men with rude machinery."[5]

The mine operators at Gold Hill organized the district through a series of lease agreements with local farmers who charged between one-sixth and one-tenth of all profits as a rent or land toll. The miners also formed partnerships among themselves and worked in small companies consisting of ten to thirty men. Some of those companies owned slaves or rented slave workers from nearby farms. Between 1842 and

1853 at least twenty such companies operated in the Gold Hill district for some extended period of time. The story of two of the most successful of these companies is critical for understanding the development of the Gold Hill district. Both companies reflected the fundamental components of the state's mining industry. Led by experienced men from other mines in the region, both relied on the deep-mining expertise of Cornish miners to exploit what proved to be the most productive veins in the district. Two shafts—the Barnhardt and the Earnhardt—became the core properties in building the reputation of Gold Hill as the state's most successful gold-mining district.

George Barnhardt established a company of miners in 1843 and worked a shaft of ore that eventually would bear his name. A descendant of German immigrants, he was the son-in-law of John Reed and gained his mining experience at the Reed mine in the 1830s. Barnhardt quickly assumed a leading role in business and civic affairs in the community and is credited for selecting "Gold Hill" as the district's name.[6] As the reputation of Gold Hill spread, veterans from other mining districts in the state arrived, among them Ephraim and Valentine Mauney, who joined the Barnhardt company in the mid-1840s. The Mauneys' Pennsylvania German family migrated to North Carolina in the eighteenth century and settled in Lincoln County. Before moving to the Gold Hill district, the Mauney brothers worked at the Carter Gold Mine in Montgomery County and gained some knowledge about deep mining and milling techniques. They eventually became two of the most prominent citizens in the community, the owners of commercial businesses and real estate and principals in the promotion and management of the Gold Hill mines well into the 1870s.[7]

The second major property in Gold Hill was located on the farm of Phillip Earnhardt. Another pair of brothers, Moses and Reuben Holmes, was responsible for the development of that property. Like George Barnhardt and the Mauney brothers, the Holmes brothers had prior experience in the mining business near their home in Davidson County before moving to Gold Hill in 1846. Despite some unfavorable returns in the early years of operations, the Holmes brothers slowly accumulated a cash reserve and began investing in steam-driven equipment.[8] By 1850 they employed thirty workers, including six slaves, and earned a healthy return on their initial investment. They further expanded their company through a partnership with two Cornishmen—David Martin and John Peters—who had likewise arrived in Gold Hill in 1846. The Holmes, Martin, and Peters company immediately upgraded the pumping operations at the Earnhardt mine, which allowed for exploration at greater depths. Within the first year, the new partnership was earning a profit of fifty-six hundred dollars a month, more than seven times greater than the 1850 profits of both companies combined.[9]

Even the most profitable mines in North Carolina did not come close to competing with the spectacular production of California and other western states. By 1850, fifty thousand miners were working the mines of California and producing $41 million worth of gold. Production peaked in 1852 at $81 million and averaged out at $45 million annually by 1857. By comparison, North Carolina's mines employed fewer than eight hundred men (and fifty women) in 1850 and produced a record high of $485,000 in 1849. Nevertheless, there is no evidence to suggest that the discoveries in the West attracted large numbers of Tar Heels from the mining regions. The total number of North Carolinians in California was little more than one thousand in 1850 and only fifteen hundred people in 1860. Although few in number, those miners were valued for their experience by the "Argonauts of '49."[10]

The California gold rush probably spurred a renewed interest in the mines of North Carolina and neighboring states. Investors from New York and other northern cities began shopping for mining properties amid the growing national prosperity of the early 1850s. Lawmakers in Raleigh accommodated that interest by liberalizing the incorporation law for companies engaged in mining and manufacturing. During the decade, nearly fifty mining companies received approval to incorporate. New operations began at Gardner Hill and McCullock in Guilford County, Conrad Hill in Davidson County, and the Phoenix and Reed mines in Cabarrus County. Many of those mining companies also pursued copper production. Most companies, however, existed only through the manipulations of the stock market in New York. Speculation in mining stocks and the desire for quick profits by the owners of those securities led to an insufficient supply of capital. Once a deep mine was allowed to close and fill with water, the initial start-up costs could discourage new investment. One by one, most of North Carolina's mining properties succumbed to the contractions of available capital after only brief periods of operation.

The celebrated Reed mine was a good example of that pattern. John Reed's son-in-law and grandson purchased the mine after Reed's death in 1845. They enjoyed some success using the same techniques employed in the 1830s. However, they failed to pay their debts, and by 1853 a New York-based company acquired their interests. Under the direction of chemist and mining superintendent Louis Posselt, the company invested considerable capital in underground development, an enginehouse, a millhouse, and a dozen other structures. Nevertheless, the company had ceased operations by the end of 1854.[11]

The Gold Hill mining district proved to be the exception to that cycle of boom and bust. By the end of 1853 Moses Holmes had attracted the interest of northern investors in the major mining tracts in the district. For $315,000, the Gold Hill Mining Company purchased those properties and embarked on an ambitious period of development over the next five years, a period during which the district earned its reputation as "the prince of mines"—the richest mining property east of the Mississippi.[12] That aggressive building campaign began in earnest late in 1854 after miners struck a rich deposit of gold ore near the Earnhardt mine. That deposit, later known as the Randolph mine, eventually reached a depth of eight hundred feet and became the single most valuable gold-bearing mine in the South. Geologists, miners, and investors agreed that the Randolph mine "has contributed most to the celebrity of the Gold Hill district and is in reality 'the Gold Hill mine.'"[13]

Propelled by brisk trading in its stock and the discovery of the Randolph shaft, the Gold Hill Mining Company invested more than sixty thousand dollars in less than two years to erect new structures for shafts and millhouses, a reservoir, and elevated trestles; dwelling houses for miners; and extensive timbering underground. A new eighty-horsepower steam engine drove the water pumps, the Chilean mills, and the log rockers lined with quicksilver.[14] "The steam mills," wrote one observer, "seem to partake of the general feeling of confidence and puff, puff as deliberately and grand as if they knew every breath was worth a dollar. . . . The creaking joints of connecting rods, the spatter of small jets of water, the rattle of cogs and the deep base harp of monster crushers unite in a sort of chorus of which 'O take your time Miss Lucy' is the prominent idea. . . . The piston is the only part that seems to care whether anything is done. It is leaping out and in like the tongue of the excited viper; and small steam leaks, here and there, persistently, but vainly hiss it, every motion, for its indecent haste."[15]

As deep mining proceeded, the organization of the work force became more complex. By the late 1840s, miners began working in shifts, and by the mid-1850s the underground crew worked three shifts of eight hours each, while those above ground worked two twelve-hour shifts. Each work shift or "force" included miners and laborers. Miners performed the skilled jobs—drilling, timbering, and blasting underground and mechanical operations above ground. Miners made up approximately half the nearly 130-man work force, and about half of the miners were Cornishmen.

Laborers performed manual chores such as ore cobbing (chipping waste rock from lumps of ore) or carrying ore underground and at the mill. Laborers also assisted miners in drilling and milling operations. Most laborers—either native-born whites or black slaves—came from nearby farms.[16] The presence of slave workers at Gold Hill increased in direct proportion to the development of deep mining. About forty slaves, or nearly one-third of the work force, labored for the Gold Hill Mining Company, and most of those men worked underground. They assisted their Cornish supervisors as drill strikers. The sight of black and white workers laboring cooperatively was not unusual in the gold mines of the South, and one visitor to Gold Hill concluded that "Negroes are . . . among the most efficient laborers."

Illustrator Porte Crayon rendered this drawing of two African American miners working at Gold Hill. Engraving from David H. Strother (Porte Crayon), "North Carolina Illustrated: The Gold Region," *Harper's New Monthly Magazine* 15 (August 1857): 294.

While most mining companies owned slaves outright, the Gold Hill Mining Company and a few other mines rented their slave labor for an average annual fee of $125. In addition, the company provided free room and board for each slave at an average of $55 per year and wages of $18 per month. Therefore, each slave cost the company nearly $400 per year, while the average salary of a native white worker was about $250 per year. Nevertheless, the company absorbed the increased rental cost because slaves were more dependable and less likely to leave the company on short notice.[17]

The Cornish miners stood at the top of Gold Hill's labor hierarchy. They typically earned twice as much as their native white counterparts, and some Cornish foremen brought home a monthly wage of sixty dollars. Their influence extended to all aspects of the mining community, including the establishment of the Methodist Church. Ironically, the very respect accorded those foreign-born miners in deep-mining districts such as Gold Hill placed them in highly hazardous situations. Many of the injuries and deaths reported at Gold Hill occurred underground as a result of explosives, poor footing, and falling objects within the shafts. The *Carolina Watchman*, a Salisbury newspaper, reported a Cornishman's death in 1852 in dramatic detail: "A most estimable man, John Stediford, was killed in Martin and Peter's mining shaft. This pit is 220 or more feet deep descending perpendicularly. It is lined with plank. One of these 14 inches by 12 feet broke from its fastenings and descended end wise. Stediford was at work at the bottom, with his pick, stooping down. The plank made no noise in its descent but came down without touching the sides. The end of the plank struck him fairly on the back of the neck, and he was whirled completely over by the tremendous blow. . . . [H]e never spoke or breathed afterwards."[18]

In addition to the dangers of work in the mines, a major concern during this period was the overall health of the people living in the district. Outbreaks of smallpox and typhoid fever at Gold Hill usually occurred in the spring when disease-bearing insects bred and spawned along the water sources and water-filled excavations. Even before the knowledge of germ theory by modern scientists, many people in the mining region understood the grim consequences of industrial pollution. One of the most remarkable accounts of public health concerns is found in the mortality census of 1860: "The water at Gold Hill is thrown out of the mines at the depth of some 675 feet by means of engines at the rate of 200 gallons per minute, [a] compound of sulphur, copper, and copperas which is of a poisonous character, which has the effect of destroying vegetation, fish, frogs, snakes, and all water quadruples. It also destroys the land over which it passes in time of floods. It also produces chronic diarrhea and Flux and [in] the great mineral belt running from North to South embracing the counties of Guilford, Randolph, Davidson, Stanly, Rowan, Cabarrus, Union and Mecklenburg, I am told the use of the water produces the same Diseases."[19]

Concerns over worker safety and public health were secondary to the instability of the mining industry itself. Even during the peak years of the mid-1850s, reports issued by the Gold Hill Mining Company clearly indicated a precarious financial condition. To keep Wall Street investors satisfied, the company had designed a strategy to repay its debts and also pay regular dividends to the stockholders. At the same time, investments in mining and milling technology had to be made in order to maintain and, preferably, increase the production of gold. Funds for those expenditures could come from the gold deposits at the United States mint, the sale of stock, or modest assessments upon stockholders. The company attempted to sustain the confidence of investors and keep the mines open, but the depression in mining stocks caused by the panic of 1857 restricted the amount of money available for reinvestment.[20]

The lack of cash to maintain and upgrade the physical plant at Gold Hill compounded the already challenging realities of Gold Hill's geology. The gold in the district and throughout much of the North Carolina's gold region occurred in a very fine mixture of minerals that required great care and attention in the milling process. The technology employed at Gold Hill utilized Chilean mills—large stone basins lined with quicksilver, with millstones positioned vertically to grind the ore into a paste. Most experts agreed that such mills were inefficient and ill-suited for the type of gold deposits found in North Carolina. Some estimated the loss of potential gold product resulting from dependence upon that technology at nearly 80 percent.[21] A dramatic decline in productivity in 1859 and 1860 confirmed those operational problems. The Gold Hill Mining Company suspended all dividends and capital improvements by the end of 1860, ceased operations in the summer of 1861, and went into receivership in October 1861.[22] The same pattern of retrenchment occurred throughout the mining region. In 1860 only three hundred men in the entire state worked in gold mines, and nearly half of those workers were at Gold Hill. The state's gold-mining industry, which had produced more than $17.5 million worth of coins and possibly an equal amount in other forms in the antebellum years, had passed its peak and was quickly winding down.

Clearly, the decline of gold mining in North Carolina was under way well before the Civil War began, but the outbreak of hostilities in the spring of 1861 sealed the fate of the industry. "Millions to buy guns but not a dollar to buy a pick or shovel," complained one mine promoter, who lamented the "draft which is depleting our population rapidly."[23] While most of the miners marched off to war and the mines filled with water, the owners of defunct mines such as Gold Hill found new opportunities during the war years as producers of bluestone for use as a building material and copperas, a sulfate of copper, for use as an astringent.

Unlike most mining communities in North Carolina, Gold Hill never became a ghost town. The population fell to less than three hundred by the end of the Civil War, but the village managed to survive as a crossroads marketplace for local farmers and as a source of seasonal work for independent miners who reworked the great piles of ore wastes found throughout the district. At its peak in the 1850s, as many as eight hundred people lived and worked in Gold Hill. During that time, the district had become the most prominent mine in the state, perhaps the best documented industrial community in the antebellum period. Gold Hill was the yardstick by which all other mining properties in the South were measured. Its geology, corporate structure, technology, and labor management were points of reference for entrepreneurs, geologists, and journalists, who visited regularly and who produced a wonderful collection of narratives that embody a very detailed account of life in North Carolina's most famous mine.

Following the Civil War, the philosophy of the New South inspired renewed interest in gold mining in North Carolina. According to that doctrine, industrial and technological development were essential components of economic prosperity and the recovery of the South in the aftermath of a devastating war. While many southerners resisted those ideas and clung to agrarian values and a rural life-style, a small but influential corps of leaders emerged in the postwar era to promote an aggressive campaign of industrialization.[24]

For the last third of the nineteenth century and in the early decades of the twentieth century, gold mining in North Carolina was of marginal importance. Symbolic of the low status of the industry was the downgrading of the Charlotte mint to an assay office, a designation retained largely through political pressure until 1913. Southern gold mines did serve a useful purpose as proving grounds for a variety of inventions and innovations designed to help retrieve the precious metal stubbornly embedded in low-grade vein deposits. During that period, Charlotte's Mecklenburg Iron Works served as a major supplier of machinery for the mining industry, including steam boilers, engines, pumps, and processing equipment. The company's most famous product was its iron stamp mill, a high-quality variation of the California gold mill. Unfortunately, the low-grade quality of North Carolina's ore required mine owners to supplement the stamp mill process with more primitive methods of recovery.

Another response to the disappointing results from stamp mill operations was to experiment with chemical treatments of the ore. Chlorine, a highly toxic and dangerous gas, gained acceptance as a method of extracting gold at many North Carolina mines. In the 1890s, cyanide replaced chlorine as the most promising chemical agent. The cyanide process involved mixing a solution of sodium cyanide with crushed gold ore and then exposing the ore in the open air. The gold dissolved in that mixture and later separated out onto sheets of zinc. Heating the solution in a retort was the final step in the process.[25]

None of the chemical treatments or mechanical processes employed in North Carolina's gold region produced the revenue needed to sustain long-term operations. Nevertheless, the promoters of mining and manufacturing held firm to their belief in progress through technology and invention. That faith appeared to be justified when the living symbol of the age of invention, Thomas Edison, visited the gold region in 1890. Edison's interest in ore-separating technology was well known, but his experiments with the ores of North Carolina were shrouded in secrecy. His arrival in Charlotte in January 1890 produced a wave of intense speculation, and his activities absorbed the local press for several months. Many observers believed that Edison had found a miraculous new method to extract gold from refractory ores through electricity. A great sense of anticipation mounted as he moved from mine to mine to confer with local engineers and send samples back to his laboratory in New Jersey. By the time Edison reached the Gold Hill district on March 4, the local press could not contain their enthusiasm. "Here comes the Wizard of Menlo Park," wrote the editor of the *Carolina Watchman*, "and he says he can, with his wonderful friend and servant electricity, separate the gold, even when mixed with the closest and most pestiferous of its companions. If the electric process can be done at all inexpensively the gold mines of North Carolina will boom, indeed."[26]

In fact, Edison's primary interest was not gold but iron. He had designed a magnetic ore separator to save magnetite found in great quantities in the eastern United States. After he left North Carolina in the summer of 1890, he established an extraordinary milling operation near Ogdensburg, New Jersey. He built a giant machine nicknamed the "Ogden Baby," which was capable of crushing six tons of ore in thirty seconds, and an industrial village named Edison to house the mill's four hundred workers. Before he could bring the mill into full production, however, the discovery of rich deposits of iron ore on Minnesota's Mesabi range dramatically increased the nation's supply. As a result, the price of iron dropped sharply throughout the 1890s, forcing Edison to close "Ogden Baby." The closure was one of the few failures in his otherwise illustrious career.[27]

Although Edison did not speculate in gold mining, investors from Great Britain did assume a large share of ownership of North Carolina's mining properties. The British economy was expanding worldwide at the time, and the development of the limited liability stock company encouraged small investors to participate in the global marketplace. Between 1860 and 1901, more than five hundred British limited mining companies were incorporated, and twenty-three of them operated in North Carolina. In 1881, for example, Gold Hill Mines Ltd. of London purchased the famous mining property for $125,000 from Moses Holmes, the same man who had sold New York investors the identical tract of land for $315,000 some thirty years earlier.[28] Although the company sent a small crew of Cornish miners and invested about $150,000 in new equipment, they never produced an appreciable amount of gold during the first four years of operations. To keep the mines from closing, the British company "reorganized" three times over the next decade by dissolving an existing company and selling discounted shares of a new company to stockholders. Needless to say, owners of those mining stocks received no return on their investment.[29]

The most notable contributions of British miners in North Carolina were made by a young engineer named Egbert Barry Cornwall Hambley, who arrived in 1881 with the Gold Hill Mines Ltd. company.[30] While Hambley was a student at the Royal School of Mines in Kensington, his father died, forcing the young man to leave school and find employment. He joined the company as a mining captain and assistant to the principal engineer. He left the company in 1884 to join one of England's most successful engineering firms. His assignments took him to mining regions throughout the world—India, the Gold Coast and Transvaal in Africa, Mexico, California, Spain, and Norway. During those travels he acquired considerable knowledge of the latest mining techniques, as well as the design of modern power plants.

Hambley returned to Rowan County in 1887 and spent the next eleven years as a consulting engineer to eight British mining companies. In the late 1890s he convinced several northern investors to develop a site on the Yadkin River known as the Narrows for hydroelectric power. He organized the North Carolina Power Company and at the same time engaged the support of George Whitney, a financier from Pittsburgh and a confidant of banker and industrialist Andrew Mellon. The Whitney Development Company, with Hambley as general manager, established its headquarters in the Gold Hill mining district and set up a gold-milling plant.

The facility at Gold Hill was only a small part of the Whitney company's bold and ambitious plan to develop the resources of the region; the scheme included the promotion of mining, manufacturing, real estate, and utilities. The pivotal element in the Whitney-Hambley partnership was the construction of a hydroelectric dam—35 feet high and 1,100 feet in width—at the Narrows of the Yadkin designed to generate 27,000 horsepower. Public response to the Whitney project was overwhelming. "The great master minds at Whitney," exulted one newspaper, " are engaged in carrying out a scheme to grapple with and subdue elemental nature, forcing her with many inventions to lend her untamed energy, wasted for ages, to the direction of human intelligence, that much good may result of the world of man."[31]

Unfortunately, a series of events led to the ultimate failure of the Whitney company within its first decade. The first setback came in 1904 when an accident at the company's Barringer Mine near Gold Hill left eight men dead and the company in debt to families and creditors. At the construction site along the Yadkin River, several typhoid fever epidemics depleted the labor force each summer. One of the victims of typhoid fever was E. B. C. Hambley, who died in August 1906 at the age of forty-four. His death prompted tributes throughout the state, including an editorial in the *Charlotte Observer*. The final blow to the Whitney enterprise came after the panic of 1907 when Whitney was forced to sell most of his business to his friend Andrew Mellon at an enormous discount. The company filed for bankruptcy protection in 1910.[32]

Gold mining in North Carolina came to an end not with a whimper but with a bang in the person of Walter George Newman. A Virginian who had spent his adult life as a sailor, a jockey, a railway express messenger, an advertising salesman, and a self-proclaimed world traveler, Newman was easily the single most memorable character in the state's mining history. Once again, the Gold Hill district provided the stage for the exploits of a man who was more comfortable mining cash in the canyons of Wall Street than attending to the business of mining in North Carolina.[33]

Within a year after his arrival in Gold Hill in 1898, Newman had established two companies—the Union Copper Mining Company and the Gold Hill Copper Company. His interest in copper reflected the growing awareness among geologists and mining experts that at Gold Hill and other mines in the state the cost of processing ore for its gold content had become prohibitive. Newman constructed more than one hundred new buildings at the Union Copper property and renovated dozens of existing buildings at the Gold Hill Copper property. He and his partners ran the Union mine at full capacity for several years, processing hundreds of tons of ore in a smelter on site and shipping hundreds more to northern smelters. Returns on that activity were not sufficient to produce a profit, and the mine closed in 1907 after the price of copper fell during the economic crisis of that year. Two years later, Newman tried to reopen the Union Copper mine, but within a few years he forfeited his ownership in a court-ordered foreclosure sale.

The Gold Hill Copper Company likewise failed under Newman's management. Despite his predictions of success and his lavish spending, the company was operated more frequently by court-appointed receivers than by Newman himself. After 1905 the company altered its operations and concentrated exclusively on gold production, again without success. Newman left Gold Hill in 1909, presumably for good, but made a dramatic and triumphant return in 1913.[34] For two turbulent years, he threw himself into the business of mining with his full energy. By that time, however, his reputation for reckless spending and personal instability severely limited his capacity to raise funds for general operations. His aggressive tactics attracted the attention of the United States Senate when it appeared that he had used the stationery of a Senate committee to promote his company. Although he avoided formal charges and even managed to turn a small profit at the end of 1914, he could not secure long-term financing. The company suspended its operations in April 1915. That June, a mining journal reported that "A news dispatch from North Carolina now announces the appointment by the court of receivers for the Gold Hill property, which has been shut down and idle . . . and adds that Walter George Newman . . . is not at the mines. It did not require a periscope of perspicacity to foresee this denouement."[35]

By the time of his death in a New York hotel room in 1918, Newman had acquired a legendary status based chiefly on his flamboyant and eccentric personality. While his repeated failures in business discredited North Carolina's mining industry among potential investors on Wall Street, his erratic behavior inspired a rich legacy of stories that endured in the memories of residents for several generations. Newman was equally capable of sudden bursts of charity—he frequently threw dollar bills and coins from his carriage—as he was of promoting cruelty and violence. Often in a drunken state, he sponsored cockfights and reveled in "battles royal," especially among black workers, to whom he offered cash prizes and drinks to the survivor of a group melee. With his trademark derby, umbrella, solid-gold scarf pin, and white suit, Newman made a lasting impression even more memorable than the poor performance of his mining enterprises.[36]

During the brief periods in which the mines of Gold Hill did operate, hundreds of workers flocked to the mining district seeking what was called "public work"—nonfarm labor that paid cash wages. As many as 350 miners worked in the district in 1900, and a similar number found employment during periods of renewed mining activity in 1906 and 1913. During those times of peak production, work crews consisted of four groups. A small "executive force" included a general manager, a superintendent, a mill foreman, an assayer, and a bookkeeper. A "top force" performed odd jobs on the surface and operated the power plant. Skilled workers such as blacksmiths, carpenters, and machinists, as well as common laborers, water boys, cooks, and servants, were included in that category. A "mill force" ran the stamp mill, while the fourth and largest unit, the "underground force," consisted of timbermen, miners, machine runners, and laborers.

The ethnic and racial composition of the work force in the early twentieth century differed markedly from the antebellum period. Only a handful of foreign-born miners settled in districts like Gold Hill. Most of the men were from North and South Carolina, with blacks comprising about one-third of the work force and overall population. Jobs and wages reflected the prevailing social practices of racial segregation and discrimination. While skilled white workers at Gold Hill could earn more than $2.00 per day, black underground laborers were paid a daily wage of from $1.00 to $1.60. Regardless of racial background, most miners occupied rental houses provided by the company, often a one-story shack measuring fifteen feet square, in which three to five miners lived.[37]

Ironically, the best year for gold production in almost three decades—1915—was also the last year of extensive mining activity in North Carolina. A renewal of mining at several sites around the state besides Gold Hill had occurred during the first fifteen years of the twentieth century. New methods such as dredging and cyanide processing did produce favorable results at mines along the Uwharrie River and at the Howie mine in Union County and the Phoenix mine in Cabarrus County. Other producing mines included the St. Catherine and Rudisill near Charlotte, the Coggins in Eldorado, and the Iola near Troy, both in Montgomery County. By 1915 a combination of scandalous business practices, high operating costs, and diminishing quantities of gold led to a steady erosion of investor confidence.[38] The outbreak of World War I drew most available capital away from marginal industries such as gold mining in North Carolina.

After 1915, gold mining provided occasional employment at a handful of locations around the state. One company opened the Coggins mine to a depth of 550 feet in 1920 and later erected a fifty-stamp mill operated by electricity. By 1930, however, all operations ceased. The Great Depression inspired a brief flurry of activity in the gold region, with the rising price of gold, low labor costs, and new chemical processes all cited as contributing factors. At the celebrated Gold Hill mine, a mining engineer named C. R. Hays took up residence in the old hotel and office building after World War II and dreamed of producing gold from the Randolph shaft. For two decades, he lived alone with a large brood of dogs and stacks of company records from the nineteenth and early twentieth centuries. In 1969 a fire killed Hays and the dogs and destroyed many of the records, thus providing a final tragic chapter for the Carolina gold rush.[39]

Fortunately, a few vestiges of the gold-mining industry in North Carolina have survived. Highway historical markers at Gold Hill, the Reed mine, the Portis mine, the Charlotte branch mint, and other sites commemorate the industry, and several sites have been listed in the National Register of Historic Places. In 1992, citizens in

Gold Hill formed a new organization dedicated to preserving the district's mining heritage. In Guilford County, near Jamestown, a businessman restored the historic McCullock gold mill as a reception center. Most notably, the Reed Gold Mine in Cabarrus County achieved National Historic Landmark status in 1966 and underwent extensive restoration by the state of North Carolina in the 1970s. Since 1977, the state has operated that historic site, at which visitors are offered exhibits on mining history and technology, guided underground tours, a restored and operating stamp mill, and the opportunity to pan for gold along Little Meadow Creek. Although the returns are small, visitors show many of the same symptoms of "gold fever" as their predecessors two centuries ago.

Two noteworthy landmarks of the mining era are associated with E. B. C. Hambley, the British engineer whose various enterprises included the ill-fated Whitney Development Company. One of them is his mansion in Salisbury, completed three years before his death. Designed by Charlotte architect Charles C. Hook, the three-story stone structure served as both a residence and a lavish showpiece at which to entertain prospective investors. Of greater significance to the process of industrialization is the town of Badin in Stanly County. That small company town, located just a few miles from the Gold Hill mining district, is the home of a plant presently operated by the Aluminum Company of America (Alcoa). Although Hambley did not live long enough to fulfill his plan for industrial development in North Carolina's central Piedmont region, his basic concept of manufacturing facilities driven by hydroelectricity ultimately became a reality. In 1912 a French company purchased the Whitney property and began construction of a dam and power plant at the Narrows of the Yadkin River. It also built for its workers a new town named Badin, which featured distinctive French colonial houses. The outbreak of World War I led to the abrupt end of the French company's efforts. Alcoa purchased the property in 1915 and completed the dam and power plant in 1917. Over the next fifty years, Alcoa constructed three additional dams along the Yadkin River. Alcoa's Badin Works became a major area employer and a part of one of the nation's most successful manufacturing enterprises.[40]

The Narrows Dam at Badin is a lineal descendant of the miners and farmers who created one of North Carolina's first industries. While gold mining brought prosperity to very few and often produced a bleak and barren landscape, there are important connections between the history of that industry and the emergence of what has become known as the Piedmont Industrial Crescent. Promoters of mining introduced new technology, formed companies, raised substantial amounts of capital to finance their operations, assembled and organized large numbers of workers, and earned the support and attention of government and press. They instilled a belief in progress as an inevitable and desirable condition and in technology as the instrument of progress.

That belief in progress through economic development is still a driving force in North Carolina. For more than thirty years, a modern "gold rush" has taken place, with Charlotte once again at its forefront as a center of banking and finance. This latest push toward progress, two centuries after Conrad Reed's remarkable discovery along Little Meadow Creek, reflects the spirit and dreams of the miners and farmers of his generation. Whether the agrarian values, character, and history of the region will survive in harmony with this extraordinary growth and development is a central question at the dawn of a new century.

Notes

1. Francis B. Laney, *The Gold Hill Mining District of North Carolina* (Raleigh: North Carolina Geological and Economic Survey [Bulletin 21], 1910), 13.

2. Laney, *The Gold Hill Mining District*, 83, 116–118; Henry B. C. Nitze and George B. Hanna, *Gold Deposits of North Carolina* (Winston, N.C.: M. I. and J. C. Stewart [North Carolina Geological Survey, Bulletin 3], 1896), 86.

3. P. Albert Carpenter III, *Gold in North Carolina* (Raleigh: North Carolina Department of Environment, Health, and Natural Resources [North Carolina Geological Survey Information Circular 29], 1993 [reprinted 1995, 1999]), 17 [table of North Carolina production].

4. *Carolina Watchman* (Salisbury), May 18, 1844.

5. Thomas L. Clingman, *Gold Hill Mine in Rowan County* (Washington, D.C.: McGill and Witherow, 1875), 5; *Carolina Watchman*, May 18, 1844.

6. John Hill Wheeler, *Historical Sketches of North Carolina from 1584 to 1851* (Philadelphia: Lippincott, Grambo and Company, 1851), 395–397.

7. Bonnie Mauney Somers, *Three Mauney Families* (Kings Mountain, N.C.: the author, 1967), 45.

8. Jerome Dowd, *Sketches of Prominent Living North Carolinians* (Raleigh: Edwards and Broughton, 1888), 256–260.

9. Charles U. Shepard, *Description of the Gold and Copper Mines at Gold Hill, Rowan County, North Carolina* (Charleston, S.C.: n.p., 1853), 7.

10. Rodman Paul, *Mining Frontiers of the Far West* (New York: Holt, Rinehart and Winston, 1963), 18–21; Malcolm J. Rohrbough, *Days of Gold: The California Gold Rush and the American Nation* (Berkeley: University of California Press, 1997), 15, 123; James W. Williams, "Emigration from North Carolina,1789–1860" (master's thesis, University of North Carolina at Chapel Hill, 1939), 78–80.

11. Richard F. Knapp, *Golden Promise in the Piedmont: The Story of John Reed's Mine*, rev. ed. (Raleigh: Division of Archives and History, North Carolina Department of Cultural Resources, 1999), 16–21.

12. Dowd, *Sketches of Prominent Living North Carolinians*, 258–260; Book 40, pp. 85–93, Rowan County Deeds, State Archives, Division of Archives and History, Raleigh.

13. Ebenezer Emmons, *Geological Report of the Midland Counties of North Carolina* (New York: George P. Putnam and Company, 1856), 165.

14. *Report of the Gold Hill Mining Company, January 21, 1856* (New York: Nathan Lane and Company, 1856), 9.

15. *Carolina Watchman*, September 14, 1854.

16. Gold Hill Mining Company Papers (microfilm), Private Collections, Mf.P. 104, State Archives, Division of Archives and History; Seventh Census of the United States, 1850: Rowan County, North Carolina, Population and Industrial Schedules, National Archives, Washington, D.C. (microfilm, State Archives).

17. Robert Starobin, *Industrial Slavery in the Old South* (New York: Oxford University Press, 1970), 11, 48, 138; David H. Strother (Porte Crayon), "North Carolina Illustrated: The Gold Region," *Harper's New Monthly Magazine* 15 (August 1857): 297–298; Gold Hill Mining Company Records (microfilm), Private Collections, State Archives; Gold Hill Mining Company Records, Vol. 5, Southern Historical Collection, University of North Carolina Library, Chapel Hill; Rowan County Miscellaneous Papers, State Archives; Eighth Census of the United States, 1860: Rowan County, North Carolina, Slave Schedules; Charles T. Jackson, *Report on the McCullock Copper and Gold Mining Company* (New York: McSpedon and Baker, 1853), 11.

18. *Carolina Watchman*, September 20, 1850. Gold Hill Mining Company Records, Vol. I, Southern Historical Collection.

19. Eighth Census, 1860: Rowan County, Mortality Schedule (microfilm, State Archives).

20. *Report of the Gold Hill Mining Company, January 21, 1856*, 9–11.

21. Emmons, *Geological Report of the Midland Counties*, 158, 160, 182–183.

22. Receiver's Records, 1862, 1867, Rowan County Miscellaneous Records, State Archives; Book 42, pp. 370–373, Rowan County Deeds, State Archives.

23. Calvin Cowles to Ebenezer Emmons, January 2, 1861, and April 10, 1862, A. L. Young to Calvin Cowles, March 23, 1864, Calvin Cowles to Robert, Holmes and Company, February 2 and December 12, 1863, Calvin J. Cowles Papers, Private Collections, State Archives.

24. C. Vann Woodward, *Origins of the New South* (Baton Rouge: Louisiana State University Press, 1951), 110–120, 124; W. J. Cash, *The Mind of the South* (New York: Alfred A. Knopf, 1941), 177, 188; Paul Gaston, *The New South Creed: A Study in Southern Mythmaking* (New York: Alfred A. Knopf, 1970), 7, 196–198; Richard Knapp and George Stinagle, "A Preliminary Report on Gold Mining in North Carolina with Particular Emphasis on the Reed Gold Mine" (research report, Historic Sites Section, Division of Archives and History, 1976, on file at the University of North Carolina Library), 172.

25. Richard Knapp, "A Preliminary Report on Mining Technology and Machinery at the Reed Gold Mine and Other Gold Mines in North Carolina" (research report, Historic Sites Section, Division of Archives and History, 1973, on file at the University of North Carolina Library), 196; *American Journal of Mining* 72 (1900): 494; *North Carolina Herald* (Salisbury), March 23, 1888; see also Calvin J. Cowles Papers, Southern Historical Collection, and Richard Eames Jr. Papers, Private Collections, State Archives.

26. *Carolina Watchman*, March 6, 13, 1890.

27. Matthew Josephson, *Edison* (New York: McGraw-Hill Company, 1959), 370–378; W. Bernard Carlson, "Edison in the Mountains: The Magnetic Ore Separation Venture, 1879–1900" (paper delivered to the Society for the History of Technology, Newark, New Jersey, 1979).

28. Book 59, pp. 273–275, Book 64, p. 554, Rowan County Deeds, State Archives; Clark Spence, *British Investments and the American Mining Frontier, 1860–1901* (Ithaca: Cornell University Press, 1958), 20; Alfred P. Tischendorf, "North Carolina and the British Investor, 1880–1910," *North Carolina Historical Review* 32 (October 1955): 512.

29. Gold Hill Mining Company Records, vol. 15, Southern Historical Collection; *North Carolina Herald*, September 21, 1887, April 24, 1889.

30. *Dictionary of North Carolina Biography*, s.v. "Hambley, Egbert Barry Cornwall."

31. *Dispatch* (Lexington, N.C.), February 14, 1906; *Concord Times*, April 20, 27, 1906.

32. *Charlotte Observer*, August 14, 1906; Harvey O'Connor, *Mellon's Millions* (New York: John Day Company, 1933), 56, 65 ff, 76.

33. *Baltimore Sun*, September 11, 1932; *Evening Journal* (New York), September 19, 1918.

34. *Salisbury Daily Index*, December 9, 1899; *Engineering and Mining Journal* 68, March 10, December 23, 1899.

35. Brent D. Glass, "King Midas and Old Rip: The Gold Hill Mining District of North Carolina" (Ph.D. diss., University of North Carolina at Chapel Hill, 1980), 368–383.

36. *Baltimore Sun*, September 11, 1932.

37. Receiver's Records, 1900–1907, Rowan County Miscellaneous Records, State Archives; Gold Hill Mining Company Records (microfilm), Private Collections, State Archives.

38. Joseph H. Pratt, *The Mining Industry in North Carolina during 1899* (Raleigh: Edwards and Broughton, 1900). There are numerous subsequent reports in this series, published for the North Carolina Geological Survey.

39. Herman J. Bryson, *Gold Deposits in North Carolina* (Raleigh: North Carolina Department of Conservation and Development [Bulletin 38], 1936), 74; Knapp and Stinagle, "A Preliminary Report on Gold Mining in North Carolina," 187; Bruce Roberts, *The Carolina Gold Rush* (Charlotte: McNally and Loftin, 1971), 54–57.

40. Brent D. Glass, *Badin, A Town at The Narrows* (Albemarle: Stanly County Historic Properties Commission, 1982).

Kernow Comes to Carolina: Cornish Miners in North Carolina's Gold Rush, 1830–1880

Elizabeth Hines

Dr. Hines, associate professor of geography at the University of North Carolina at Wilmington, has studied at the University of North Carolina at Greensboro (B.A.), the University of Kansas (M.A.), and Louisiana State University (Ph.D.). Among her published articles are "McCullough's Rock Engine House: An Antebellum Cornish-style Gold Ore Mill near Jamestown, North Carolina" (in Material Culture*, 1995) and "Cousin Jacks and the Tarheel Gold Boom: Cornish Miners in North Carolina, 1830–1880" (in* North Carolina Geographer*, 1997).*

Cornubia

Like Scotland and Wales, the county of Cornwall is located on England's Celtic periphery—not nestled but stretching seaward, as if away from England and toward the New World. It is a remote peninsula of rugged topography, in which no place is very far from the sea, separated from Devonshire and the rest of England by the broad valley of the Tamar River and bleak Dartmoor (figure 1). The historian A. L. Rowse remarked in his book *Tudor Cornwall* that, yes, Cornwall *was* an English county, although by conquest, but it was also something more: "a homogenous society [of its own] defined by language and having a common history underneath, like Brittany, Wales, Scotland or Ireland, reaching back beyond Normans and Saxons, beyond even Rome [in these islands] to an antiquity of which its people were still dimly conscious."[1] Cornwall is haunted by legends. The ruin of Tintagel, perched high on the cliffs of the rocky North Atlantic coast, is the reputed castle of King Arthur, with "Merlin's cave" conveniently tucked below. The nearby town of Camelford claims descent from Camelot. Another misty memory claims that Malpas, near Truro, is where Tristan landed with Iseult.

Cornwall's Celtic Christian heritage, although rich with legend, is more than myth. Irish missionaries, the "Celtic Saints," brought Christianity to Cornwall in the fourth century A.D., while the rest of the island was still pagan. The Cornish embraced the teachings of the saints, which bulwarked their opposition to the then-heathen Saxon invaders to the east. They managed to retain their religion, and their language, until the Reformation, when powerful English reformers attempted to stamp out every Celtic, as well as Catholic, memory.[2] But nearly a thousand years of Celtic Catholicism would not die quietly. Celtic churches, crosses, holy wells, and other sacred sites remain common landscape features, and many of Cornwall's more than two hundred parishes are named for Celtic saints (figure 2).

The legends of the saints and their works also abound. One tells how an Irish missionary, Saint Perran, was bound to a millstone and flung seaward from an Irish cliff. Miraculously, the stone floated, carrying the saint to Cornwall's north coast,

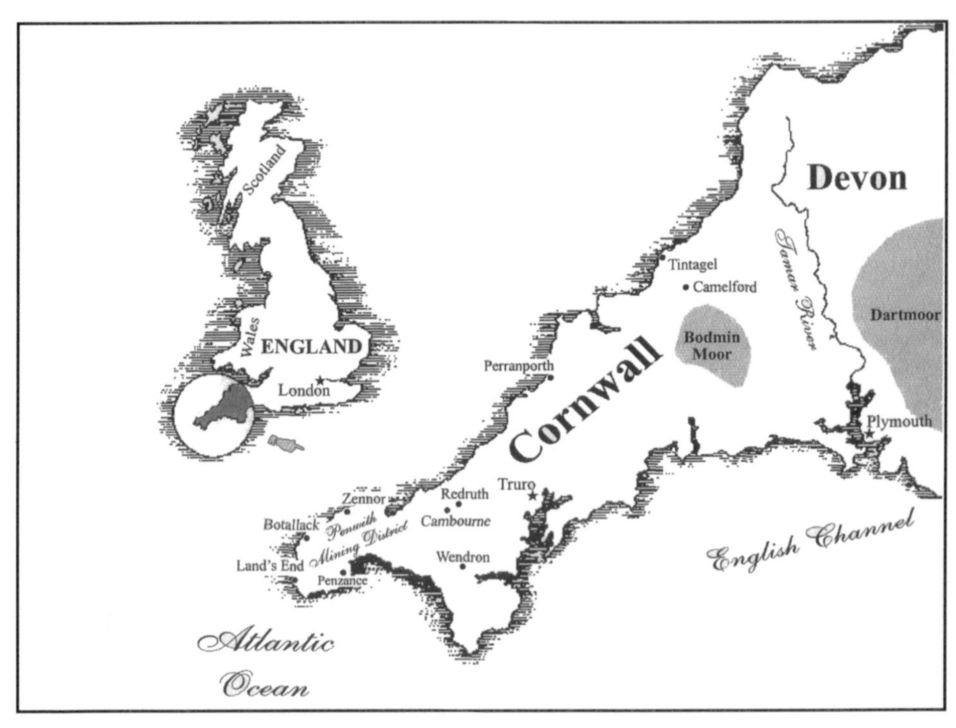

Figure 1. *Map supplied by the author.*

Figure 2. A Celtic cross at the twelfth-century Chapel of Saint Just in Roseland, Cornwall. *Photograph by the author.*

where he landed near Perranporth (so named in the nineteenth century). There he built his oratory on the dunes and, later, discovered tin or, more likely, improved the method of smelting and so became the miners' patron saint. His standard, a black-and-silver cross on a white background, became modern Cornwall's flag. Today a thick but translucent veneer of Anglicanism covers Cornwall's indelible Celtic past.

As late as the mid-1500s, Cornish natives often greeted the increasingly frequent English visitors with suspicion, replying, when spoken to, "*Meea navidna cowzasawzneck*"—I can speak no Saxonage.[3] But the Cornish people were nearly irrevocably Anglicized during the reign of Elizabeth I, when twenty years of war with Spain brought reluctant Cornwall, with its strategic position and excellent ports, into the

English fold.[4] Cornish animosity toward England peaked in 1548 with the Act of Uniformity, which enforced the substitution of the English *Book of Common Prayer* for the Latin/Celtic mass, which the Cornish had known for centuries, and which was accompanied by confiscation of precious church ornaments and artifacts. Resentment toward the English exploded in the Prayer Book Rebellion of 1549, when twenty thousand Cornishmen marched on London, and simmered thereafter.[5] But Anglicization triumphed in the 1600s when Cornish was supplanted by English as the language of commerce and Cornish was stigmatized as the language of the poor. By the nineteenth century only a few Cornish speakers remained.[6] Cornish historian Hamilton Jenkin tells the story of a young nineteenth-century Cornish student's description of Cornwall. The student exclaimed that "he's kidged (joined) to a furren country (England) from the top hand (Tamar River)," to which the entire school responded approvingly.[7] Henry Jenner began the twentieth-century revival of the Cornish language with the 1904 publication of his *Handbook of the Cornish Language*. Shortly thereafter, Jenner formed a Cornish *Gorsedd* (a "coming together of dignitaries known as bards, to celebrate their [Cornish] culture in a very public ceremony"), an organization similar to those in Wales and Brittany, thus initiating the modern revival of the Cornish-language movement.[8] At present there is a small but serious modern movement for home rule in Cornwall (figure 3) and a more moderate proposal to have the Cornish declared a "national minority," like the Welsh and Scots, within the Council of Europe's Framework for the Protection of National Minorities in Strasbourg.[9] Cornwall is alternately called "Kernow," its Cornish name, on highway signs, Web pages, and many other markers around the county.

Figure 3. "Free Cornwall" graffiti scrawled in Cornish at the Botallack mine site near Penwith in Cornwall. *Photograph by the author.*

Until the twentieth century, most Cornish emigrants were pushed and pulled abroad—to the United States, South Africa, Australia, and Brazil—rather than "up the country" into England.[10] Most of the emigrants left the failing mining economy of Cornwall's West Penwith, including the towns of Camborne and Redruth, and other mining districts in the nineteenth and early twentieth centuries for the new mining areas of the English post-colonial world. The Cornish could provide the mining expertise to tap the mineral wealth of the Americas, South Africa's Rand, and New South Wales. At the present time, there is more immigration *to* Cornwall, because of its booming tourist industry. It is a so-called "twee" (upscale) destination for vacationers from "up the country," as well as from the European mainland.

The history of Cornish mining must be examined to understand Cornish emigration abroad. Elizabethan author Richard Carew wrote of the richness and importance of the Cornish mineral deposits in his 1602 *Survey of Cornwall*: "Why seek we in corners for petty commodities, when as the only mineral of Cornish tin openeth so large a field to the country's benefit?—this is in working so pliant, for sight so fair and in use so necessary, as thereby the inhabitants gain wealth, the merchants traffic, and the whole realm a reputation: and with such plenty thereof hath God stuffed the bowels of this little Angle, that it overfloweth England, watereth Christendom, and is derived to a greater part of the world besides. . . ."[11]

Historically the Cornish have been occupied in three basic endeavors: farming, fishing, and mining—two of which are common throughout the British Isles and the rest of Europe. One of the few English impositions to which the Cornish did *not* object was the (fourteenth-century) Enclosure Act, because, like the *bocage* landscape of Brittany, and unlike the Saxon commons, Cornwall's lands had been subdivided beginning in Neolithic times as the stony fields were cleared and the hedgerows erected.[12] The moderate, humid climate and pockets of fertile soils supported adequate to excellent farming in areas with gentle slopes. However, along the rocky north coast and along the cairns of the peninsular backbone down to the Lizard in the south, the steep and stony slopes, lashed by the gales of the North Atlantic in winter, are amenable to neither farming nor husbandry.

Many Cornishmen have taken to the sea, primarily to fish but at certain times for trade, smuggling, and defense in this sea-surrounded peninsula scalloped with harbors. In Cornwall one is never far from the sea; there, from a few promontories in the west and southwest, the Atlantic and the English Channel can be seen on a clear day just by glancing from left to right, and the salt air often penetrates miles inland. The lure of the sea is a fact of life in Cornwall. People farm and fish everywhere in Europe, but Cornwall, unlike much of the rest of northern Europe, provided the possibility for a third primary occupation in the extraction of minerals, especially copper and tin, although some gold was also mined. Cornish copper and tin deposits have been mined at and near the surface since ancient times (figure 4). Many of Cornwall's rivers are silted up from centuries of "tin streaming." Bronze-age Greeks likely traded with Cornish tin merchants.[13]

Figure 4. An abandoned tin mine in a hillside at the Botallack mine site near Penwith in Cornwall. *Photograph by the author.*

By the late seventeenth century, large numbers of Cornish men and their families (figure 5) had devoted themselves to procuring mining wages as the Industrial Revolution gathered steam. They earned poor wages while performing dangerous work for the "mineral lords," who controlled all of Cornwall's mineral rights. That those lords were of Norman-English, rather than Cornish, nobility exacerbated the miners' discontent. Upon the death of mineral lord Charles Bassett in 1788, a monumental Celtic cross was erected on Carn Brea in Penwith, a long-mined granite outcrop on which are found several Neolithic sites. The Bassett family claims that Charles was so loved by the thousands of miners in his employ that after they marched thirty thousand strong at his funeral, they quarried the granite and raised the impressive memorial (figure 6). Conversely, the miners' descendants have a tradition that their ancestors'

Figure 5. These nineteenth-century "bal maidens" of Cornwall (*bal* means mine in Cornish) were needed to break ore with mallets "at grass" or at the surface. *Photograph of Tincroft mine, Illogan, Cornwall, ca. 1890; reproduced courtesy Cornish Studies Library, Redruth, Cornwall.*

pay was held until proof of attendance at the funeral was confirmed, then contributions were accepted to finance the cross. Those who failed to attend and/or contribute were fired.[14]

Figure 6. Bassett Monument at Carn Brea in Cornwall. *Photograph by the author.*

Although unconfirmed, the varying accounts of the origins of the Bassett memorial intimate the attitudes of one class toward the other and offer some insight into one of the "push" factors that resulted in the Cornish diaspora of the nineteenth century. Of course, the most powerful "push" factor was the deterioration of the global minerals market and the exhaustion of Cornish tin reserves after centuries of moderate exploitation and decades of rapid depletion following industrialization. Ironically, it was not tin depletion that initially devastated the Cornish economy but rather a short-lived boom in copper exports beginning about 1750, which accompanied Cornwall's, and indeed England's, rapid industrialization.[15] In addition to a vast increase in technological investment on the part of the mineral lords, there was so much money to be made during the boom that farmers abandoned their fields for the mines, damaging the locally produced food supply. The number of mining laborers grew from six thousand in 1800 to twenty-eight thousand in 1838 during the boom, but the number of farmers declined.[16]

Deep mines always flooded. For centuries the water table was the limit of the miners' efforts. The steam-powered technology of the early Newcomen water-pumping engines and the later Bolton and Watt engines, especially after extensive improvements to the design by Cornish engineers after Watt's patent expired in 1800, allowed deeper shafts and more efficient retrieval and processing of ores, albeit at greater expenditures by newly organized investment companies.[17] Although more workers were needed during the boom, they required new skills to operate the machinery. It became more difficult to get work, and, once employed, conditions were more dangerous and the labor pool was more tightly controlled by managers with their eyes on the bottom line.

Given Cornwall's difficult agricultural environment, the reduced agricultural force soon found it nearly impossible to adequately supply the increasing industrial force with barley and potatoes, staples in western Cornwall.[18] In addition, some local farmers hoarded small lots of grain, which might have gone to local markets at lower prices, to accumulate large and profitable bulk sales for export. Food riots erupted in Penwith and other mining districts at the beginning of the nineteenth century. The 1812 crop failure and concomitant famine struck the final blow, and emigration ensued.[19]

Although the copper boom continued until the early 1850s, and tin was mined into the twentieth century, Cornish miners began arriving in small numbers in the Carolinas in the 1820s, about twenty years after the discovery of gold in the Piedmont. Soon their numbers increased in the southeastern Piedmont from Georgia to North Carolina as gold discoveries continued.[20] By the 1850s they were employed in the lead mines of the Great Lakes region and in gold and silver mines throughout the West. These opportunities were coincidental with the crash in global markets for Cornish copper and tin when huge Chilean copper reserves and Bolivian tin first flooded and then devastated the market.[21]

Tarheelia

The accidental discovery of a seventeen-pound gold nugget in 1799 in Cabarrus County, North Carolina, was followed by the first gold rush in the United States. Conrad Reed, a boy of twelve, found the flatiron-sized nugget in Little Meadow Creek on his father's farm. He showed it to his father, John Reed, who thought it would make a dandy doorstop, and there it rested for several years until it was suggested that it be taken to Fayetteville for identification. The assayer swindled the elder Reed, who later realized that he had been cheated, complained, and apparently was ultimately compensated. When word got around, the rush was on. Local farmers began searching the stream beds and digging up the hillsides. Until the California gold rush of 1849, John Reed's placer works in time accounted for the largest number of gold nuggets—those weighing a pound or more (figure 7). When Reed and his partners, and later other Piedmont gold miners, had exhausted many of the surface deposits, they began to dig up the hillsides, following veins. Often such pits collapsed, injuring the diggers. When the miners shored up the pits and hit hard rock, they used explosives to drive deeper. Deaths and maimings resulted. Clearly, professional expertise was needed.

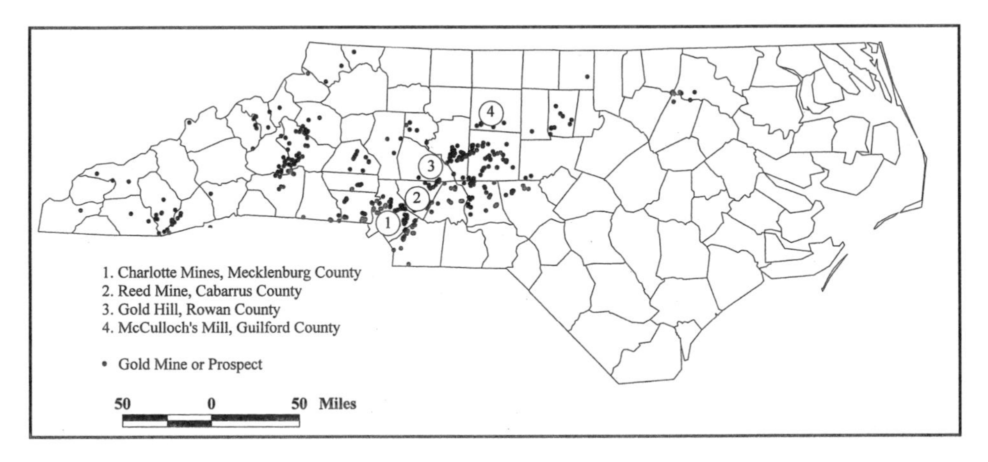

Figure 7. Map: Gold mines and prospects in North Carolina. *Source: U.S. Geological Survey minerals management survey.*

The boom attracted the attention of northern and foreign investors, especially the English, and operations grew with the application of outside capital. To protect the investment, companies brought in from throughout the world skilled miners, engineers, and masons, many of whom were Cornish.[22] In the 1830s rich veins of gold were discovered beneath Charlotte, in Mecklenburg County. A London mining company, represented by Count, or Chevalier, Vincent de Rivifanoli, brought to the area Cornish miners and engineers, who drilled deep shafts at the Capps and Rudisill mines beneath the streets of Charlotte. Rubble raised to the surface was used for some of the roadbeds in the young town. As a result, Charlotte can with some degree of veracity claim that its streets were paved with gold.[23]

In 1831 a South Carolina planter named Charles McCulloch had a Cornish-style dry-laid stone mill building built by a Cornishman, Elizier Kersey. Very little else is known about McCulloch, but the resemblance of his building to hundreds of mining whims in Cornwall is remarkable. That he would hire a Cornish engineer to build it and then fit it, most likely, with a Harvey's (of Hayle, Cornwall) walking-beam steam engine further exemplifies his knowledge of and perhaps affinity for Cornish mining know-how. The mill building, which housed a steam engine—not to pump water, as in Cornwall, but to drive Chilean or arrastra crusher stones (figures 8 and 9)—was centrally located near several profitable mines. McCulloch sold the mill in 1848 to John Gluyas, a Cornish mining engineer.[24]

Figure 8. Chilean millstones at the Reed mine in Cabarrus County. **Figure 9**. Gold ore-crushing beds carved in the granite bedrock at McCulloch's Mill in Guilford County, North Carolina. *Both photographs by the author.*

Gluyas, whose name was most likely taken from St. Gluvius in Cornwall, was born in the Cornish parish of Wendron in 1796. He brought his family to New York City in 1834, then moved them the following year to Charlotte, North Carolina, where he first worked as a steam engineer and machinist at the Capps mine.[25] His mining career in the Piedmont included the supervision of operations at several deep gold mines in Mecklenburg, Cabarrus, Davidson, and Montgomery Counties from 1835 to 1847. During the latter year he moved again to Gold Hill in Rowan County, where he joined with William Trealoar, another Cornishman, in leasing land and mining and processing loads of gold, silver, and copper.[26] In 1848 he purchased McCulloch's mill, but it is not certain that he ever operated it. He died in Gold Hill in 1858.

John Gluyas was unusual in that he left at least a limited record of his life in North Carolina and had a distinguished career as a mining engineer and entrepreneur. Two generations later another Cornish immigrant made important contributions to the industrial development of the state. He was the world-traveling Cornish mining engineer Egbert Barry Cornwall Hambley. Hambley was born in Penzance, Cornwall, in 1862 and attended the Royal School of Mines at Kensington as a young man. An early employer, J. J. Truran, head of several British companies doing business overseas, sent him to Gold Hill in 1881. There Hambley held the position of assistant superintendent of mines for three years. In 1884 he returned to England, where he secured a post as a special engineer in the Indian Gold Mining Company of Glasgow Bank's South Indian gold fields, which he held for two years. In 1886 he visited "in a professional way" the gold fields of Africa's Gold Coast, then went to the mining regions of Spain, Norway, South Africa's Transvaal, Mexico, South America, and California. In 1887 he returned to and settled in North Carolina's Rowan County, where he organized several London-based companies to develop gold and other mines in the Piedmont. He had a pumping station, complete with a Cornish pumping engine, built on the Yadkin River near Gold Hill to supply the mines with water. Subsequently he managed newly industrializing cotton mills and budding utility companies and developed the Rowan Granite Quarry, which no doubt employed Cornish miners as well. He and his family were still living in Salisbury in Rowan County—in a mansion he had built with granite from his quarry—when he died from typhoid fever at the age of forty-four in 1906.[27]

Few of the other Cornish people who came to North Carolina left such an enduring record of their lives—although a 1952 article in the *Salisbury Sunday Post* cited Matthew Moyle and Nicky Trevethan (Moyle and Trevethan being two unmistakably Cornish names) as having supervised crews of Cornish and white and black American miners at Gold Hill in the 1850s. That feature story was taken from another article written considerably earlier by David Hunter Strother, who wrote and illustrated for *Harper's New Monthly Magazine* in the 1850s under the pen name "Porte Crayon."[28] The drawings that accompanied Porte Crayon's article depict typical Cornish miners working at Gold Hill during that decade. Specifically, the drawings of Moyle and Trevethan and the Cornish equipment, technology, and activities have become familiar to students of the Reed mine in Cabarrus County (figure 10) and Gold Hill. A Cornish kibble—a bucket used for lowering and raising men, ore, and equipment—(figure 11) remains at the Gold Hill museum, along with other distinctive Cornish mining artifacts depicted in Porte Crayon's 1857 *Harper's* article.

Many Cornishmen came with their families to the copper and gold areas of the Appalachian Mountains in the mid-nineteenth century.[29] They often remained in the

Figure 10. Porte Crayon's drawing of Cornish miners Mat Moyle and Nicky Trevethan at Gold Hill, from David H. Strother (Porte Crayon), "North Carolina Illustrated," 292.
Figure 11. Cornish kibble at Gold Hill. *Photograph by the author.*

East and Southeast, even after the 1849 California rush, occasionally turning to agriculture or commerce, as was the case in Gold Hill, where several Cornishmen who had come in the 1840s and 1850s to mine were listed in the 1860 census as farmers, grocers, and tavern keepers.[30] Did the Piedmont landscape remind them of Cornwall? Alternatively, the East's adjacency to the Atlantic placed only an ocean, rather than a continent and an ocean, had they migrated west, between them and home. As with other economic immigrants, many did return to Cornwall, despite increasingly limited mining activity and economic opportunity there. In fact, Cornwall has many American place-names, among them Dakota Farm, Nevada Terrace, Michigan House, and a section of Redruth known as "Cally."[31] Obviously, none are from North Carolina, with perhaps the exception of several Gold Hills, although there are Gold Hills in many other gold-bearing locations worldwide.

Cornish miners often moved in family or community groups and thus differed from the more mobile (and often single) mining immigrants. They were generally known to be family oriented and strongly affiliated with their beloved Methodist Church, another symbol of their repudiation of the English as a result of their rejection of Anglicanism following John Wesley's proselytizing trips to Cornwall in the 1740s.[32] By the 1850s, when northern investment companies employed Cornish miners at the Reed mine and at Gold Hill,[33] nearly half of all the miners at Gold Hill had been born in England.[34]

Cornish skill and technology was invaluable in the mines of North Carolina, in and around Auraria and Dahlonega in north Georgia, and later throughout the West

Figure 12. The Botallack tin mine, Penwith, Cornwall. *Photograph by the author.*

in California in the 1850s, Utah in the 1860s, New Mexico in the 1870s, and Arizona, Colorado, and Montana in the 1880s.[35] Cornish miners had also reached the mining areas of the Great Lakes in Wisconsin by the 1830s, enjoying a boom in the 1850s, and in Michigan's Upper Peninsula by the 1870s.[36] They had solved the practical problems of blasting and supporting shafts; hoisting men, equipment, and debris; and pumping dry and ventilating the hundreds of tin and copper mines in Cornwall—many of which reached depths of thousands of feet. In fact, the Crown Mines of Botallack on Cornwall's north coast (figure 12) featured deep vertical shafts that were connected to horizontal tunnels, known as levels, which reached thousands of feet beneath the North Atlantic. The descendants of nineteenth- and early-twentieth-century miners in Zennor and other Penwith mining towns still recall family stories of the eerie rumble of the boulders as they rolled on the sea floor above their heads during storms.

Cornish miners and engineers made a significant contribution not only to North Carolina's now nearly forgotten gold industry; their presence helped to form the early industrial landscape of the state. The hard-rock men who arrived after the large and small surficial nuggets had been taken demonstrated how the subsurface ores, which were bound to hard bull quartz and granite at great depth, could be profitably extracted. Engineers such as John Gluyas and Egbert Hambley and the mason and builder of McCulloch's grinding mill, Elizier Kersey, were among the vanguard in the design and use of steam technology in the Piedmont, which was rapidly adopted for non-mining uses in mills and utilities.

North Carolina's mining and industrial history is richer for the contribution of a few remembered Cornish engineers and numerous anonymous Cousin Jacks (and Jinnies)—so called because all Cornish miners claimed to have a cousin in Cornwall who would immigrate to work the mines. Indeed, the Cornish diaspora, which began in earnest in the nineteenth century and continued into the twentieth, deposited Cornish miners around the globe. As an 1893 visitor to Cornwall noticed, "the Cornish miners have mostly emigrated." To find a Cornish miner engaged in his ancient craft, one would have to look in North or South America, South Africa, Asia, or Australia.[37] Among many old Cornish adages is this: "Wherever in the world a hole is sunk in the ground, you will be sure to find a Cousin Jack at the bottom of it, searching for metal."[38] The mines of North Carolina's gold rush were no exception.

Unfortunately, those men and their Cornish contemporaries left very little of Cornwall's flavor—a culture that predates Roman Britain—in North Carolina's Piedmont. Only McCulloch's enginehouse; the old Methodist church and cemetery (figure 13) at Gold Hill; Hambley's granite quarry; and the mining landscape and a few mining artifacts at the Reed mine, Gold Hill, and a few other sites remain. In California's Grass Valley and Michigan's Upper Peninsula, one can at least get a "pasty" or a saffron cake in a bakery and hear Celtic music played at festivals.[39]

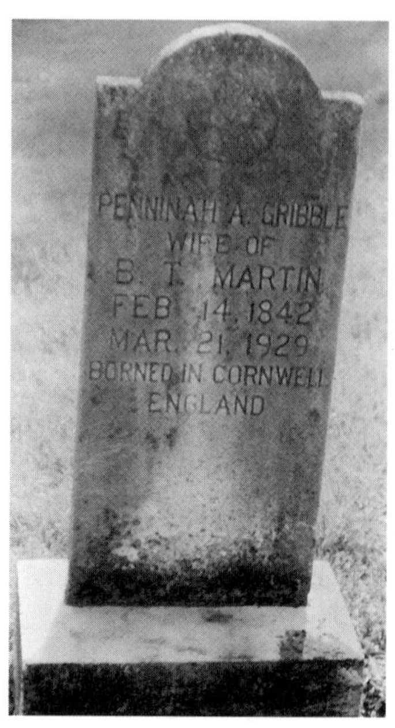

Figure 13. Cornish headstone in the Methodist church cemetery in Gold Hill. *Photograph by the author.*

When a people lose their language, they lose the most important element of their culture—the primary means by which culture is transmitted to succeeding generations. When the English-speaking Cornish emigrated to other English-speaking lands, the first arrivals were seen as essentially English rather than Cornish. Their descendants, generally, thought of themselves as having come from England rather than Cornwall. In America they became Americans within a generation and have now largely forgotten their ancient heritage.

Notes

1. A. L. Rowse, *Tudor Cornwall: Portrait of a Society* (London: Jonathan Cape, 1941), 20.

2. John Jenkin, *A First History of Cornwall* (Cornwall, England: Dyllansow Truran, 1984), 21.

3. Rowse, *Tudor Cornwall*, 20–21.

4. Rowse, *Tudor Cornwall*, 23.

5. Phillip Payton, *Cornwall* (Fowey, Cornwall: Alexander Associates, 1996), 136.

6. *www.clas.demon.co.uk* [the Cornish-language Web page]. The myth that the Cornish language died in 1777 with the purported last native speaker Dolly Pentreath is not accurate. Although Ms. Pentreath's house, featuring a commemorative plaque, still stands in the tiny Cornish channel port of Mousehole (pronounced "muzzle"), Cornish as a native language did not die out entirely until the nineteenth century, when the last speaker died in Penwith.

7. A. K. Hamilton Jenkin, *The Cornish Miner* (1927; reprint, Truro, Cornwall: David and Charles Newton Abbot, 1972), 274.

8. Matthew Spriggs, letter to author, 1999.

9. *www.coe.fr/eng/legaltxt/157e.htm*. Member states of the Council of Europe convened at the Framework Convention for the Protection of National Minorities, which met in Strasbourg, France, in 1995. They resolved in Section I, Article 3, part 1, that "Every person belonging to a national minority shall have the right freely to choose to be treated or not to be treated as such . . ."; and in Article 6, part 2, to "take appropriate measures to protect persons who may be subject to threats or acts of discrimination, hostility or violence as a result of their ethnic, cultural, linguistic or religious identity."

10. John Jenkin, *A First History of Cornwall* (Cornwall: Dyllansow Truran, 1984), introduction.

11. Carew also published *The Excellecie of the English Tongue* in 1614. His *Survey of Cornwall* was edited with an introduction by F. E. Halliday; along with maps by John Norden, it was reprinted in 1969 by Adams and Dart of London.

12. Rowse, *Tudor Cornwall*, 33.

13. Payton, *Cornwall*, 40.

14. Laurence Holmes, letter to author, 1996.

15. John Graham Rule, "The Labouring Miner in Cornwall, c. 1740–1870: A Study in Social History" (Ph.D. diss., University of Warwick [England], 1971), 1–2.

16. Rule, "The Labouring Miner in Cornwall," 9, 2.

17. Payton, *Cornwall*, 203.

18. Rule, "The Labouring Miner in Cornwall," 121–122.

19. Rule, "The Labouring Miner in Cornwall," 126.

20. Elizabeth Hines and Michael Smith, "The Rush Started Here" (manuscript in progress, 1999).

21. Cecil Todd, *The Cornish Miner in America* (Truro, Cornwall: Barton, 1967; reprint, Spokane, Wash.: Arthur H. Clark Co., 1995), 19.

22. Elizabeth Hines, "Cousin Jacks and the Tarheel Gold Boom: Cornish Miners in North Carolina, 1830–1880," *North Carolina Geographer* 5 (winter 1996): 4.

23. Charlotte, North Carolina, Public Works Department/Engineering Division, *A Historical and Engineering Report on the Rudisill Gold Mine (1826–1938)* (Charlotte: Public Library of Charlotte and Mecklenburg County, 1976).

24. Elizabeth Hines, "McCulloch's Rock Engine House: An Antebellum Cornish-Style Gold Ore Mill Near Jamestown, North Carolina," *Material Culture* 27 (winter 1995): 8, 23.

25. John Gluyas Papers, Southern Historical Collection, University of North Carolina Library, Chapel Hill.

26. Brent D. Glass, "'Poor Men with Rude Machinery': The Formative Years of the Gold Hill Mining District, 1842–1853," *North Carolina Historical Review* 61 (January 1984): 1.

27. A. L. Rowse, *The Cornish in America* (Redruth, Cornwall: Dyllansow Truran, 1967), 129; Samuel A. Ashe, Stephen B. Weeks, and Charles L. Van Noppen, eds., *Biographical History of North Carolina*, 8 vols. (Greensboro: Charles L. Van Noppen, 1905–1917), 2:133–135; Richard F. Knapp and Brent D. Glass, *Gold Mining in North Carolina: A Bicentennial History* (Raleigh: Division of Archives and History, North Carolina Department of Cultural Resources, 1999), 117–118, 149.

28. Brent D. Glass, "King Midas and Old Rip: The Gold Hill Mining District of North Carolina" (Ph.D. diss., University of North Carolina at Chapel Hill, 1980), 407; David H. Strother (Porte Crayon), "North Carolina Illustrated: The Gold Region," *Harper's New Monthly Magazine* 15 (August 1857): 289–300.

29. Colamer Abbott, "Cornish Miners in Appalachian Copper Camps," *International Review of the History of Banking* 7 (1973): 199–219.

30. Glass, "King Midas and Old Rip," 159.

31. Todd, *The Cornish Miner in America*, 15.

32. Thomas Shaw, *A History of Cornish Methodism* (Truro, Cornwall: D. Bradford Barton, Ltd., 1967), 12.

33. John Rowe, *The Hard Rock Men: Cornish Immigrants and the North American Mining Frontier* (Liverpool, England: Liverpool University Press, 1974), 142.

34. Seventh Census of the United States, 1850: Rowan County, North Carolina, Population Schedule, National Archives, Washington, D.C. (microfilm, State Archives, Division of Archives and History, Raleigh).

35. Payton, *Cornwall*, 237–238.

36. Todd, *The Cornish Miner in America*, 28–29, 114–138.

37. Payton, *Cornwall*, 222.

38. Payton, *Cornwall*, 242.

39. Jean Jollife, personal communication with author, 1996.

The Bechtler Mint

Rodney D. Barfield

Mr. Barfield is president of Virginia Museum Consultants and is currently consulting on exhibits connected with restoration of the Whalehead Club in Corolla. He previously served as chief of exhibits at the North Carolina Museum of History in Raleigh, supervisory curator of the Museum of the Cape Fear in Fayetteville, and director of the North Carolina Maritime Museum in Beaufort. His publications include The Bechtlers and Their Coinage: North Carolina's Mint Masters of Pioneer Gold *(1980) and* Seasoned by Salt: A Pictorial History of the Outer Banks *(1995).*

The story of the Bechtler mint has been mined for generations. The facility's history runs much deeper than the shafts put down in the western Piedmont of North Carolina. Its telling not only chronicles an important era in the state's history but also embodies a part of the American tapestry of the first half of the nineteenth century. The Bechtler mint was the only significant private gold manufactory in America until the California gold rush commenced in 1848. The little mint in the foothills of North Carolina handled more than three million dollars' worth of the precious metal during the 1830s and 1840s, a production of domestic gold that rivaled that of the federal mint at Philadelphia. Circulation of Bechtler coins in the Southeast during those years exceeded even that of United States coinage.

The unprecedented success of the Bechtler issues was fueled by early gold strikes in the southeastern United States, particularly North Carolina, South Carolina, Virginia, and Georgia. Those finds, initially placer mines staked by farmers armed with little more than picks and shovels, were sufficiently rich to put several minters in business and to attract national and international attention to the area.[1] The California strikes at mid-century dwarfed the earlier efforts in the Southeast. As the enormous rush developed in the West and untold wealth began to pour from the barren hills a continent away, the North Carolina gold story and the contributions of the Southeast to mining lore and technology were overwhelmed and quickly faded from the public's interest. But prior to 1848, North Carolina was the gold field of the nation. Gold from the North Carolina fields found its way to the United States mint as early as 1804. For more than twenty years thereafter, North Carolina was the only state in the Union that produced gold for the mint. Virginia and South Carolina did not send bullion to Philadelphia until 1829, and Georgia did not send its first shipment until 1830.[2]

Dr. William Thornton, clerk at the United States Patent Office and noted architect, was one of the first prominent outsiders to form a joint-stock company to exploit the North Carolina mines. The 1806 venture included Albert Gallatin, treasurer of the United States, as a member of its board of trustees. By the 1830s more than fifty mines were operating in North Carolina alone, and the following year the Old North State sent $250,000 worth of gold to the federal mint. It remained the nation's largest producer of domestic gold until 1848.[3]

The early gold industry in North Carolina created a scattering of boom towns across the western Piedmont as miners, and the peripheral humanity that attended their efforts, poured into the sparse outback in search of a rich strike. Mining veterans from Europe and South America joined the rush. New capital financed sluice and vein mining as the alluvial mines played out. The mining industry at times employed as many as twenty-five thousand people and for a short period ranked as the state's second major economic pursuit.[4]

It is not possible to calculate the amount of gold taken from the soil of North Carolina. There was no local mint to accept the metal and render it into coin. Much of the metal was bought up by local merchants, jewelers, brokers, and speculators. A part of it found its way to New York and to Europe, where prices were higher. A congressional investigation of the industry in 1834 concluded that the national mint received no more that one-third of the state's total production, the remainder being exported, "consumed in the arts, and . . . circulating under private stamps in at least two of the States in which gold is found."[5] Most of the gold mined in North Carolina left the mining region. One of the state's merchants calculated that his bullion accounts totaled nearly three hundred thousand dollars from 1828 to 1839 and that three-fourths of it went to New York, where the metal sold at higher prices. Much other gold went to Europe, which likewise paid a higher premium for the valuable metal.[6]

Resentment over the loss of the precious resource to northern and foreign markets grew during the 1820s. Local politicians, newspaper editors, and influential citizens began to campaign for a government branch mint or assay office in the gold region. They argued that the volume of production alone warranted local branches, not to mention the loss of the resource to the "rusting coffers" of Europe and the inconvenience of hauling the metal to the national mint at Philadelphia. The lack of a government assay office in the region forced miners to sell their output to local jewelers, merchants, or brokers at a large discount or travel to Philadelphia to have it assayed. The expertise and integrity of the local brokers was suspect, often with good reason. The trip to Philadelphia was a journey of several hundred miles fraught with danger and discomfort.[7] The absence of any means for converting gold into a convenient and credible economic commodity, coupled with the government's lack of interest in the problem, or at least its slowness to deal with the issue, prepared the way for a local solution.

In 1830 North Carolina's gold production was worth half a million dollars, according to a state legislative inquiry of that year. The actual figure was, of course, much higher. One contemporary authority estimated a figure of ten million dollars in gold by 1840.[8] Yet, in the midst of all that natural wealth, in the heart of what was then the nation's "gold country," there was an extreme scarcity of specie. The money situation was so severe by 1831 that a local legislator asked the state to postpone tax collections in the region for a year because the population did not have the currency with which to pay. Barter was the common practice of economic exchange. The sight of a man dropping a nugget on the counter at the general store or pouring a little dust from a quill for a bottle of spirits was not uncommon. Miners and merchants usually carried a set of pocket scales for weighing gold, so common was the metal as a trade commodity.[9] Given the currency problems in the gold region, the lack of a recognized office to determine gold standards, and the frequent removal of the precious metal from the area, it is surprising that private minting and assaying were not attempted sooner and by more people. That a few attempts were made is evidenced by newspaper and mint reports such as the following example: "Several skilled assayists have established themselves in the gold regions, and have acquired so much reputation for accuracy, that their pieces of Gold marked 'five dollars' pass every where as half eagles."[10] (Note that the writer did not refer to the stamped pieces as money or coinage. They were simply gold pieces stamped with the worth of their assayed value. The notion of private coinage was still a novel idea in the gold region.)

Government action on the gold issue, even after acknowledgment of the need for branch mints, was hampered by other political considerations: the question of hard versus soft currency , the efficacy of assay offices over regional mints, and the location of government facilities. Sectional differences pitted northeastern congressmen against those from the South and West who together formed a coalition. Not until 1835 did Congress finally resolve the issues and vote to construct branch mints in North Carolina, Georgia, and Louisiana. It was 1837 before the first branches accepted bullion and struck the first coins. By that time, millions of dollars worth of gold had been bought, sold, and traded throughout the Southeast, and North Carolina boasted a private mint that rivaled the United States mint in its production of gold coins.[11]

The local solution in North Carolina to the problems of miners and merchants—how to establish a fair standard, a fair value, and a means of convenient transport of the gold ore—appeared in the person of Christopher Bechtler, a German immigrant who opened a private mint in Rutherfordton in 1831. Bechtler (1782–1842) emigrated to the United States from Pfortzheim in the Grand Duchy of Baden in 1829. He was already a noted gunsmith and goldsmith in his native Germany in a town famous for its metallurgy.[12] Bechtler came to America in the company of two sons and another relative initially identified by local writers as his son-in-law but who was in fact his nephew. The latter's wife was also in the group. The exact relationship and kinship between Christopher Bechtler and that relative was unknown for years and has been problematic for historians and others who have tried to piece together the family history. The story is complicated by the fact that the relative carried the senior Bechtler's Christian name and was known as "Christopher Bechtler Junior." According to the scant records available, it was Bechtler's original intent to settle in Philadelphia. In his application for citizenship, he stated that he intended to make Philadelphia his home.[13] He went to some trouble to open a business in the city and, presumably, to set up a household. He is listed in the Philadelphia city directory of 1830 with the designation "Jeweler."[14]

Philadelphia did not hold the Bechtlers for long. In April 1830, some seven months after disembarking at New York, Christopher Bechtler Sr. bought the first of several parcels of land in Rutherford County. He purchased the acreage through a fellow German resident of the village, Martin Kibler, for $350.[15] Bechtler could not purchase the property outright because he was still an alien. Over the years, he acquired additional tracts of land adjoining the original lot until his property totaled 246 acres.[16] It is not known if Bechtler moved to North Carolina with the intention of opening a private mint. A writer who interviewed him in 1837 suggested that he moved to Rutherfordton to be near the nation's major domestic supply of gold.[17] If that was his intent, the craftsman could not have located himself more conveniently. Rutherfordton lay in the heart of the North Carolina gold region and was eminently accessible to the recently opened mines of South Carolina and Georgia. It was also, incidentally, an area heavily populated by Germans, perhaps another factor in Bechtler's decision.

Numerous jewelers followed the mining industry in North Carolina, and jewelry and watches were the mainstay of the Bechtler family business, as it had been in their native Baden. The family continued in the business for some twenty-five years in North Carolina, and Christopher Jr. followed the trade when he relocated to South Carolina. The family continued to identify with jewelry and watches even after the elder Bechtler had built a large and renowned business in private coinage.

When Bechtler appeared in the foothills of North Carolina in 1830, the land was awash with speculators, prospectors, scientists, government agents, tradesmen, and merchants who had invaded the western Piedmont as its reputation as the "Golden State" spread throughout the country. During the late 1820s, newspapers from Georgia to New York and even in Europe carried accounts of gold finds in North Carolina. The state press reported gold-related events daily. New mines opened monthly, and each new mine brought fresh reports of untold wealth.[18] Bechtler wasted no time in addressing the important need of miners and businessmen to have their gold assayed and stamped. Freshly arrived in the village of Rutherfordton, he placed the following advertisement in a Rutherfordton newspaper in July 1831:

NOTICE TO GOLD MINERS & OTHERS. C. BECHTLER informs all interested in the gold mines and in assaying and bringing the gold of the mines into ingots or pieces of a standard value, that he is now prepared to assay and stamp gold, to any amount, to a standard amount of 20 carats, making it into pieces of $2.50 and $5.00 value, at his establishment 3 1/2 miles north of Rutherfordton, on the road leading from Rutherfordton to Jeanstown. The following are his prices:

For simply fluxing rough gold, 3/4 per cent.
For fluxing gold—to be stamped, 1/2 per cent.
For assaying gold; any quantity less than 3 lbs., $1.00
For stamping, 2 1/2 per cent.
July 2, 1831.[19]

Bechtler's enterprise in minting began with and was encouraged by the local community. A brief comment by Roswell Elmer Jr., editor of the newspaper in which Bechtler's advertisement appeared, informed readers of the paper that Bechtler had "undertaken this enterprise at the suggestion of several gentleman [*sic*] of the highest standing among our miners, for the purpose of putting into use the actual resources of this region as a circulating medium in the transaction of business."[20] Bechtler was, Roswell assured his readers, a man of "strict honesty," and the editor expressed the hope that the assayer would not disappoint public confidence, "as it has been with some others who have attempted the assaying and stamping of gold . . . ," a comment that perhaps referred to Templeton Reid, a Milledgeville, Georgia, minter of German descent who had closed his short-lived enterprise the previous year.[21] Roswell went on to explain that the Bechtler mint was not the result of personal whim but the invention of necessity, then enumerated a litany of miners' woes: the lack of circulating specie, the flight of gold from the region, and the lack of a government assay office.[22]

Bechtler was not the first private assayer or even minter in the gold region. His best-known predecessor was Reid, who like Bechtler was an artisan in the best nineteenth-century sense of the word: an inventor, manufacturer, merchant, jeweler, watchmaker, and gunsmith. He struck his first run of gold coins in 1830—about fifteen hundred dollars' worth of quarters, half eagles, and eagles.[23] Unlike Christopher Bechtler, whose coining enterprise had the support of the local community before he commenced his enterprise, Reid was immediately attacked for his presumption. An individual who signed himself "No Assayer" launched a campaign against Reid in a Georgia newspaper, charging him with producing inferior coins and with violation of the Constitution, the greater offense, in the estimate of the writer. The critic had tested the German's coins at the federal mint, where they had been shown to be worth less than their nominal value.[24] The newspaper campaign undermined the public's confidence in Reid's integrity, and his business closed after about three months of operation. Reid had stamped about seven thousand dollars' worth of gold.

The Reid mint is important to the Bechtler story because it provided a standard of sorts for comparison. The Reid operation prepared the way for Bechtler by sensitizing Bechtler and his backers to the potential for dishonest business practices inherent in operating a private mint. When Bechtler opened his shop less than a year after Reid had closed, he made lengthy statements in the newspaper concerning the process of arriving at fair gold values, the percentages to be charged for the work, and his intention of returning to each customer a sample of his assay, in order that the customer might have it "assayed elsewhere to find its value, that no deception or fraud may be practiced. . . ."[25]

It was relatively simple for a merchant and especially an assayer to swindle his customers, and the private assayer had always been regarded with some suspicion by the public. Gold, of course, mixes readily with silver, copper, and other metals to form an impure alloy. The same amount of gold by weight may differ considerably in value, depending on the ratio of gold to other metals present in the sample. That ratio is expressed in terms of fineness and carats. An assayer was able to profit handsomely by simply misrepresenting the fineness of a given amount of gold ore. It has been suggested that Templeton Reid fell victim to the mysteries of alchemy not by design but from his inability to flux the metal to determine its purity. Christopher Bechtler went to some effort to ensure his customers that he would bring their gold "to a standard" and that his assays of gold would be conducted "by a fire ordeal [a very refined process of smelting] for the purpose of ascertaining its exact fineness, and he will be held accountable for the amount of the value of the whole ascertained. . . ."[26] As will be discussed below, however, Bechtler too might have been plagued with some of Reid's problems in producing coinage of uniformly high purity.

Christopher Bechtler Sr. worked in Rutherfordton for about twelve years. During that time, by his own accounting, his mint coined a quantity of gold worth about $2,250,000 and fluxed into bars and ingots almost two million pennyweights of gold valued at $1,384,000. To put those figures in perspective, Bechtler's mint processed, during a period of just over a decade, more North Carolina gold than either the Philadelphia mint or the Charlotte branch mint.[27] John Wheeler, superintendent of the Charlotte mint, confirmed that output when he offered the following accounting of North Carolina gold production to 1840:

Coined at the United States mints, to 31st Dec., 1839, $3,000,000
Amount of bullion passed through Mr. Bechtler's hands, $3,625,000
Bullion sold to manufacturers, sent to Europe, carried in bars to the west, Etc., Etc., $3,375,000.[28]

Measured from another perspective, Christopher Bechtler's mint generated a larger percentage of North Carolina's gold production than did any of the private mints that processed California gold.[29] Bechtler coinage was the strongest currency in western North Carolina as late as the Civil War. At times it enjoyed a greater circulation than United States coins. It also circulated widely in South Carolina, Georgia,

Representative coins of various values made at the Bechtler mint. *Photographs courtesy North Carolina Travel and Tourism Division.*

Kentucky, and Tennessee. Bechtler's reputation for integrity was so strong that even at mid-century there were business contracts that specified payment in Bechtler currency.[30]

Bechtler achieved that high public regard in only twelve years and became something of an institution in western North Carolina and particularly in the Rutherfordton community. His mint came to be viewed as a local resource that was visited by anyone of note who passed through the region. George W. Featherstonhaugh was one such visitor. A scientist and geologist who traveled widely in the United States, Featherstonhaugh made a tour of the gold region in the late 1830s and called on Bechtler at Rutherfordton. He likewise found the mintmaster to be a man of high character and integrity and described his work as "singular."[31] Bechtler's production of $2,250,000 worth of small-denomination hard currency in just nine years brought a significant degree of fiscal relief not only to Rutherfordton but also to western North Carolina and to some extent a larger geographic area. The figures are all the more remarkable in light of the fact that the state's total tax revenues for 1835 amounted to only $71,749[32] and that its largest construction projects at that time, the Wilmington and Raleigh (161 miles long) and Raleigh and Gaston Railroads (86 miles long), both completed in 1840, cost $1,300,000.[33] Christopher Bechtler remained an integral part of the Rutherford community until his death in 1842. He was mentioned in newspapers around the state, and his work was scrutinized and criticized by the federal government. He was praised and defended by the citizens of Rutherford County as his fortunes and reputation and those of the county became increasingly interdependent.

But Bechtler's reputation did not rest solely upon his coinage, inasmuch as he was not only a minter but also a jeweler, a watchmaker, and a gunsmith. Working with his son, Augustus, Bechtler offered an array of repair services and merchandise. The two were enamored of firearms and expended much effort experimenting with unconventional designs and rapid-fire systems. George Featherstonhaugh commented on their experiments and proclaimed the two "pre-eminent for their ingenuity."[34] Featherstonhaugh was particularly impressed with the speed and accuracy of the Bechtler designs. He described a rifle tested by Augustus that would fire eight bullets in a minute and another with a chain load that held sixty caps and revolved by the catch of the trigger. The weapon was accurate at a distance of 165 yards, and Augustus struck a target at that range "with great success."[35] Featherstonhaugh was so impressed with the Bechtler guns that he purchased a "pistol rifle" that "kills at one hundred yards" and had his name inlaid with gold on the weapon.[36] Other notables commented on the Bechtler gunsmithing business with the same superlatives used by Featherstonhaugh. Judge William Battle visited the then elderly minter in 1841 and was struck by the unique designs of the weapons and by their beauty. He wrote to his wife that he had seen "a great many beautiful rifles and pistols, and among others, a snuff box pistol and a walking stick rifle."[37]

There are many unanswered questions concerning the Bechtler households, family members, and the history of the Bechtler businesses. The simultaneous existence of two Christopher Bechtlers in the same town in the same trades creates much of the confusion. There is no doubt, however, that the household of Christopher Bechtler Sr. included two sons—Augustus and Charles. Census records and deeds bear this out.[38] The entry in the 1840 census for Christopher Sr. is written "Chrst." A separate entry indicates that a "Chrs Bechtler"—a reference to Chistopher Bechtler's nephew—was between the ages of twenty and thirty and was head of a household that included two females between the ages of twenty and thirty. Bechtler sleuths have tended to impose a neat chronological order to the Bechtler fortunes, starting with Christopher Sr., who started the business with his sons and nephew, was succeeded by Augustus, who, lacking his father's skills, allowed the business to decline and was in turn succeeded by Christopher Jr., who oversaw the demise of the business. That neat sequential order of events results in part by regarding the Bechtler newspaper advertisements as originating at the same source—that is, that Bechtler Sr. was the source of all the ads. There are problems with that reading, however, one being location.

Many writers have assumed that the Bechtler enterprise was located in the village of Rutherfordton. While the 1830 ads for watchmaking and jewelry locate the business in Rutherfordton "(opposite Mr. Wm. Twitty's tavern)," the 1831 ads locate the assaying and stamping services "3 1/2 miles north of Rutherfordton on the road leading from Rutherfordton to Jeanstown."[39] Other sources, among them George Featherstonhaugh, who claimed to have visited the Bechtler "farm," place the coining business outside the village. The geologist never mentioned an intown business, even though that is where he kept his lodgings.[40] Judge Battle likewise made a trip to the farm outside Rutherfordton and, like Featherstonhaugh, made no mention of a business in town.[41] Finally, the will of the elder Bechtler bequeathed to his son Augustus his entire estate, including "all the Tools and Instruments belonging to & used about the Shop or establishment," and located the establishment and house outside the town.[42]

The explanation for the discrepancies in the location of the Bechtler business is that Christopher Jr., the nephew, operated a separate and distinct business from the elder Bechtler, and that business was situated in Rutherfordton, across from Mr. Twitty's tavern. Christopher Jr. was listed in the census records by the abbreviation "Chrs"; in similar fashion, his watchmaking and jewelry ads list him as "Mr. Chr. Bechtler," whereas the assaying ads from the same newspaper identify Christopher Sr. as "Mr. C. Bechtler."[43] Thus Christopher Sr. and Augustus were engaged in the coining business and the watchmaking and jewelry trade at the house outside Rutherfordton (at which the elder Bechtler also mined for gold on a limited basis), while Christopher Jr. practiced the trade of watchmaker and jeweler at his own shop in the town.

Christopher Bechtler Jr. has not received very good press from the writers who have looked back at the Bechtler story. Local writers have made him something of a scapegoat for the decline of the Bechtler fortunes while giving the elder Bechtler all of the credit for the impressive little mint that put Rutherfordton on the national map. Rutherford County local historian Clarence Griffin is the source of much of the negative perception of Christopher Jr. In his 1929 article on the Bechtlers, Griffin states categorically that Christopher Jr. was responsible for the collapse of the family fortunes as a direct result of his dubious integrity and his hard drinking.[44] Relying upon an old newspaper article as his authority, Griffin wrote that Christopher Jr. would "drink from two to four glasses of lager beer before breakfast, and as he went from his shop at lunch time he would go by the saloon, stopping to consume from two to four more glasses of beer, smoking as he did so. This performance would be repeated again at night, after quitting work, and frequently between morning and luncheon periods."[45] Unfortunately, Griffin's "reliable authority" appears to have been an article in the Shelby (N.C.) *Highlander* that quoted one R. J. Daniel of Mooresboro, who appears to have been more interested in the amount of beer consumed by Bechtler than in the quality of his metallurgy.[46] W. E. Hidden, the noted geologist and discoverer of the mineral Hiddenite, weighed into the fray in a response to the Daniel article. Hidden asserted that Christopher Jr. never prospered in the coining business, never made an accounting of the estate he inherited to the family, and died in poverty in Charlotte, "blind and decrepit." "After embezzeling money from his uncle's estate," Hidden wrote, "he became a general failure, and trouble and disaster followed him everywhere."[47]

In fact, the nephew counts for far more in the family fortunes than he has received credit for, and there is no documentary evidence to support the charges of theft or poor management against him. Rutherford County records reveal that the younger Bechtler was appointed executor of the estates of both Christopher Sr. and his son Augustus, who died a year after his father. Indeed, Christopher Sr. himself named his nephew to serve as executor of his will in the event of Augustus's death.[48] There is no indication that Christopher Jr. was anything but responsible in settling the estates of his cousin and uncle. A court-appointed committee investigated his administration of the two estates and reported that debts had been made good and that the estate of the elder Bechtler owed Christopher Jr. "a small balance."[49]

Perhaps Clarence Griffin and W. E. Hidden found the younger Bechtler a convenient foil for the occasional criticism concerning the quality of Bechtler coinage. Or they may have used the nephew as an easy explanation for the demise of the once notable mint near Rutherfordton and thus as a diversionary target for those who might be inclined to direct barbs at the local legend, Christopher Sr. It is instructive to remember that Hidden offered his critique of the Bechtlers more than half a century after their exit from Rutherfordton and that Griffin wrote almost a century after the death of the elder Bechtler.

Christopher Jr. did have the misfortune to preside over the demise of the Bechtler coining business, but the destiny of the mint was well out of his control by the time he succeeded to the position of mintmaster. He inherited the business following the death of Augustus in 1843—shortly after the death of Christopher Sr. The opening of the Charlotte branch mint in 1837 had already cut heavily into the Bechtler business, and the California finds in 1848 would empty the once-booming western Piedmont of its prospectors and brokers. The following table shows the rapid decline in Bechtler coinage following the opening of the Charlotte branch mint, even while the elder Bechtler was still in control of his business.

	Coined	Fluxed
Jan. 1831–Dec. 1834	$109,732.50	396,804 dwts.
Dec. 1834–Dec. 1835	695,896.00	711,583 dwts.
Dec. 1835–Aug. 1836	471,322.50	397,410 dwts.
Aug. 1836–May 1838	770,239.50	201,141 dwts.
May 1838–Feb. 1840	194,560.00	24,060 dwts.
	$2,241,840.50	1,729,998 dwts.[50]

The Charlotte branch mint was a mere seventy miles away from Rutherfordton and, as was anticipated by government officials, who frowned on private mints, the federal facility lured away much of Christopher Bechtler Sr.'s trade. Production of Bechtler coinage dropped by half a million dollars in value from 1838 to 1840, and his fluxing business practically came to a halt. After Christopher Jr. inherited the mint, in much-reduced circumstances, he attempted for several years to maintain the unique coining business. In a postscript to an ad he ran for his jewelry and watch-making business in 1847, he added: "The Watch-Making & Gold-Coining business attended to as heretofore. . . ."[51] Christopher Jr. maintained his trade in Rutherfordton until 1857 or 1858, when he moved to Spartanburg, South Carolina. He was listed as an elder in the Rutherfordton Presbyterian Church in 1857 and sold his house there in 1858. Even after he moved out of North Carolina, Christopher held property in Rutherfordton and advertised his business there as late as 1869.

A century later, in 1963, a New York auction house offered for sale a private collection of currency that included a number of Bechtler coins. Preparations for the sale revealed new information about the Bechtlers while at the same time adding to the historical uncertainty concerning the family. In the sales catalogue were six tin-types of people collectively identified as "The Bechtler Family." The individuals were identified as Christopher Sr., his son Augustus, two daughters, and the son and wife of Christopher Jr.[52] In fact, only the last two identifications are accurate. The other subjects are also members of the family of Christopher Jr., as the following table, gleaned from the population schedule of the 1850 census, suggests.[53]

Name of Person	Age and Sex	Occupation	Place of Birth
Christopher Bechtler	50M	Watchmaker	Germany
Sophia	43F		Germany
Edward	19M		N.C.
Louisa	17F		Ga.
Augustus	16M		N.C.
Christopher	14M		N.C.
Frederic	5M		N.C.
Anneliou	1M		N.C.
Annelia	1F		N.C.

Comparing the photographs with the census records and estimating the ages of the people portrayed, it is likely that in addition to Christopher Jr. and his wife Sophia, the other named individuals in the photographs are three of the couple's four sons and their two daughters. Furthermore, the tintypes bearing the images were not produced until the 1850s, more than a decade after the deaths of Christopher Sr. and his son Augustus. The clothing of the figures appears to fit the 1860s.

The precise composition of the Bechtler family is not the only historical conundrum in this saga. The integrity of Bechtler coinage has been debated without resolution since the first coins were struck. The debate has been pressed with more emotion than science. Prejudice has often blunted the objectivity of otherwise unbiased opinion. Local historian Clarence Griffin, whose writings have influenced most deliberation on the subject for the past fifty years, accepted no criticism of the elder Bechtler and his son Augustus and found nothing positive about Christopher Jr. Griffin's unwavering position undermined his own arguments when more temperate judgment would have won him a more sympathetic hearing. With his unyielding positive opinion on the elder Bechtler's coinage, Griffin wrote that "exactly the proper amount of gold was in every coin. Never was there the slightest hint of dishonesty on the part of Christopher and Augustus Bechtler. They died poor but respected."[54] Griffin obviously could not have known whether the Bechtler coinage actually achieved the totality of accuracy and quality he ascribed to it. His authoritarian stance was needlessly defensive: his characterization of father and son as "poor but respected" was gratuitous and does not reflect the actual value of the Bechtler estate.

Local contemporaries of Christopher Sr. were no less fervent in their faith in the mintmaster. When a merchant from nearby Salisbury had some of his Bechtler coins discounted by the United States mint, he complained in a local newspaper about the quality of the money. The newspaper printed the complaint but appended a defense of Bechtler that concluded: "May it not be that the error is in the Mint itself?"[55] Travelers through the gold region offered the same blanket assessment of Bechtler and his operation. As previously mentioned, George Featherstonhaugh was very taken with Bechtler and his operations and wrote with some enthusiasm about the "perfect confidence" that prevailed between Bechtler and his customers. The traveler accepted Bechtler and his business on a large degree of faith and repeated the local wisdom that his product had been assayed at the federal mint and proven accurate, though such was not the case.[56]

Bechtler had a fair amount of critics, many of whom were no better informed in their opinions than were his partisans. United States mint officials were uniformly negative in their assessments of the quality of Bechtler coinage and in their attitudes concerning the very existence of the Bechtler mint. The official position was stated by mint superintendent John H. Landis, who declared: "There can hardly have been any reason of necessity for either of these enterprises [the Reid and Bechtler mints], since neither community was beyond the reach of assay offices where gold could have been disposed of."[57] That sort of insensitivity to local problems in the gold region was partially responsible for the creation and success of the private mints. In an 1841 report, R. M. Patterson, director of the federal mint, exhibited somewhat more understanding of the local situation. Though opposed in principle to private mints, he admitted to their convenience to miners and merchants in the mining regions. While also noting the inferior quality of Bechtler coins, he acknowledged that they successfully competed with government-produced coinage.[58]

Mint officials were not merely venting spleen against Bechtler coinage, despite their bias against private mints. The mint reports are too consistent to explain away. United States assayer George B. Hanna of the Charlotte assay office returned the same negative verdict on the Bechtler coinage: "The character of the stamping varied greatly, that of the dollar pieces being very poor, and these were extensively counterfeited."[59] More comprehensive assessments of Bechtler coins by other qualified assayers are also negative. In 1842 United States assayers Jacob Eckfeldt and William E. DuBois made numerous examinations of the coins at the federal mint and concluded: "There is not much variation in the weight, but the fineness . . . is exceedingly irregular and inferior, causing an average loss of 2 1/2 per cent on the nominal value. A safe estimate of the value of single five dollar pieces, taken 'as they come,' would be $4.84."[60] As the two assayers pointed out, that percentage is not much of a loss on a single coin but could amount to a considerable amount on a large shipment.

Interestingly enough, especially in light of local opinions regarding Bechtler coinage, the output of coinage by Augustus and Christopher Jr. was found to be of a higher quality than that turned out by Christopher Sr. Assayers Eckfeldt and DuBois reported that the coins they tested up to 1842 were deficient by as much as 6 percent (for an average loss of 3 percent), while those struck in 1843—presumably by Augustus—averaged only 1.5 percent below nominal value. Five-dollar pieces tested in the late 1840s—presumably those struck by Christopher Jr.—averaged a very high $4.94 per coin.[61]

The North Carolina Museum of History mounted its first exhibit of Bechtler coins in 1978, and I had the pleasure of working with the collection that formed the exhibit. The persistence of the debate over the integrity of the Bechtler stamps emerged as a nettlesome point that needed to be addressed in the exhibit. The problem, of course, was how to determine the quality of the coins without damaging them. We interviewed jewelers, mint officials, and assayers about ways to assay the museum collection without debasing the coins, only to find that anything less than melting the pieces would result in inaccurate readings. In response to that information, we took a rather extreme measure. We heard of a nuclear-activation analysis program at North Carolina State University that was used in criminal cases to identify minute samples of organic material by comparative analysis, using a computer. We contacted the university's nuclear engineering department and explained our dilemma. The director of the program was intrigued and offered to "assay" our coins without damaging them in any way, so we submitted them to "nuclear activation analysis."

Nuclear analysis is dull stuff now, but twenty years ago it could still raise an eyebrow. It is a technique that measures or "counts" gamma and X-rays emitted from materials exposed to neutrons. Radioactive isotopes resulting from neutron exposure emit gamma rays that are measured on a solid-state detector connected to a computer. The computer makes identification and produces an analysis. Nuclear activation analysis measures a sample of material, such as a Bechtler coin, against a known quantity, such as a United States coin of determined fineness and weight by exposing both to neutrons and comparing the radioactivity of the known sample to the unknown. The nuclear engineer who conducted the tests referred to the technique as "nuclear fingerprinting."

The museum collection of Bechtler coins, 148 pieces, was irradiated in a nuclear reactor over several weeks, along with a United States gold coin and a sample of 24-carat gold. By making the atoms "talk" to us, the coins were "assayed" without damage. The analysis confirmed the 1842 and 1849 findings of the United States mint assayers: that the fineness of Bechtler gold coins is irregular and inferior. It further belied the traditional bias that favored coins produced by Bechtler Sr. over those turned out by his son and nephew; indeed, our analysis revealed that the coinage of Augustus and Christopher Jr. was superior to that of the elder Bechtler.[62] Those findings in no way diminish the stature of the old mintmaster nor detract from his historical importance. His business operations tell us much about the private entrepreneur of the early 1800s. The Bechtler family exemplifies the creative and industrious individuals who settled North Carolina's western counties and used their skills to bring order to the rough-and-tumble ways of the frontier.

Christopher Bechtler's economic impact on North Carolina was significant; his contributions to the nation's numismatic history seem remarkable almost 160 years after his death. The survival of his coins and firearms has enabled us to reach across the decades and, through the use of modern technology, reconstruct an important part of his and his family's experiences. That is the proper use of artifacts and a proper alliance of the humanities and science.

Notes

1. Fletcher M. Green, "Gold Mining: A Forgotten Industry of Antebellum North Carolina," *North Carolina Historical Review* 14 (January, April 1937): 8.

2. J. T. Pardee and C. F. Park Jr., *Gold Deposits of the Southern Piedmont*. United States Department of the Interior Geological Survey Professional Paper 213. (Washington: Government Printing Office, 1948), 27.

3. "Assay Offices—Gold Region, South," United States Congress, House Reports, 1st sess., 23d Cong., Vol. 3, Nos. 314–445, Report No. 391 (April 4, 1834), 1.

4. "Assay Offices—Gold Region, South," 2.

5. "Assay Offices—Gold Region, South," 1; "The Gold Mines of North Carolina," *Merchants' Magazine* 11 (July 1844): 63.

6. "The Gold Mines of North Carolina," 63.

7. "Assay Offices—Gold Region South," 2.

8. "The Gold Mines of North Carolina," 64.

9. Green, "Gold Mining," 137.

10. *Tarboro Free Press*, October 10, 1834.

11. "The Gold Mines of North Carolina," 64; Green, "Gold Mining," 140–142.

12. Bechtler naturalization papers in "Miscellaneous Notes," typed copy of Rutherford County Court Minute Docket, Charles Lee Coon Papers, Special Collections Department, Duke University Library, Durham.

13. Bechtler naturalization papers.

14. Maurice Brix, *List of Philadelphia Silversmiths and Allied Artificers, 1682–1850* (Philadelphia: privately printed, 1920), 8.

15. Martin Kibler to Christopher Bechtler, April 18, 1836, Book 43, p. 250 (microfilm), Rutherford County Deeds, State Archives, Division of Archives and History, Raleigh. The deed refers to a prior transaction that took place on April 25, 1830—obviously the date on which Kibler originally acquired the property on Bechtler's behalf.

16. Index (microfilm) to Rutherford County Deeds, State Archives.

17. George W. Featherstonhaugh, *A Canoe Voyage Up the Minnay Sotor*, 2 vols. (London: R. Bentley, 1847; reprinted, St. Paul: Minnesota Historical Society, 2 vols., 1970), 1: 328.

18. Green, "Gold Mining," 9–10.

19. *North Carolina Spectator and Western Advertiser* (Rutherfordton), July 2, 1831.

20. *North Carolina Spectator and Western Advertiser*, July 2, 1831.

21. *North Carolina Spectator and Western Advertiser*, July 2, 1831.

22. *North Carolina Spectator and Western Advertiser*, July 2, 1831.

23. Dexter C. Seymour, "Templeton Reid, First of the Pioneer Coiners," in *American Numismatic Society Museum Notes* 19 (1974): 225–266.

24. Seymour, "Templeton Reid," 233–241.

25. *North Carolina Spectator and Western Advertiser*, August 27, 1831.

26. *North Carolina Spectator and Western Advertiser*, August 27, 1831.

27. "The Gold Mines of North Carolina," 63–64.

28. "The Gold Mines of North Carolina,"64–65.

29. B. W. Barnard, "The Private Coinage of Gold Tokens in the South and West," *South Atlantic Quarterly* 16 (April 1917): 163.

30. Clarence Griffin, *The Bechtlers and Bechtler Coinage; and, Gold Mining in North Carolina, 1814–1830* (Forest City: Forest City Courier, 1929), 9; Clarence Griffin, *History of Old Tryon and Rutherford Counties, North Carolina, 1730–1936* (Asheville: Miller Printing Company, 1937), 202; Book 44, p. 234, Rutherford County Deeds, State Archives.

31. Featherstonhaugh, *A Canoe Voyage Up the Minnay Sotor*, 330.

32. Hugh Talmage Lefler and Albert Ray Newsome, *North Carolina: The History of a Southern State*, rev. ed. (Chapel Hill: University of North Carolina Press, 1963), 355.

33. Lefler and Newsome, *North Carolina: The History of a Southern State*, 347.

34. Featherstonhaugh, *A Canoe Voyage Up the Minnay Sotor*, 330.

35. Featherstonhaugh, *A Canoe Voyage Up the Minnay Sotor*, 331.

36. Featherstonhaugh, *A Canoe Voyage Up the Minnay Sotor*, 331.

37. William H. Battle to Lucy M. Battle, May 3, 1841, Battle Family Papers, Southern Historical Collection, University of North Carolina Library, Chapel Hill.

38. Sixth Census of the United States, 1840: Rutherford County, North Carolina, Population Schedule, entry 1112, National Archives, Washington, D.C. (microfilm, State Archives).

39. *North Carolina Spectator and Western Advertiser*, July 28, 1830, July 2, 1831.

40. Featherstonhaugh, *A Canoe Voyage Up the Minnay Sotor*, 327, 330.

41. William H. Battle to Lucy M. Battle, May 3, 1841, Battle Family Papers.

42. Book 44, p. 76, Rutherford County Deeds, State Archives.

43. *North Carolina Spectator and Western Advertiser*, July 28, 1830, July 2, 1831.

44. Griffin, *The Bechtlers and Bechtler Coinage*, 4–5.

45. Griffin, *The Bechtlers and Bechtler Coinage*, 4–5.

46. "The Bechtler Coinage," reprint from the *Shelby* (N.C.) *Highlander* in *Numismatist* 25 (April 1912): 122.

47. *Numismatist* 25 (May 1912): 164.

48. Will of Christopher Bechtler Sr., November 28, 1842 (probated February 1844), Book E, p. 122 (microfilm), Rutherford County Wills, State Archives.

49. Estate of Christopher Bechtler Sr., Book B, p. 657 (microfilm), Rutherford County Estates Records, 1831–1854, State Archives.

50. Jacob R. Eckfeldt and William E. DuBois, *A Manual of Gold and Silver Coins of All Nations, Struck within the Past Century* (Philadelphia: Assay Office of the Mint, 1842), 160.

51. *Western North Carolina Republican* (Rutherfordton), July 29, 1847.

52. *George O. Walton Fabulous Numismatic Collection of United States Gold, Silver & Copper Coins, Paper Money, Foreign Gold Coins* (public auction sale catalog) (New York: Stack's, 1963), 157.

53. Seventh Census, 1850: Rutherford County, Population Schedule (microfilm), State Archives.

54. Griffin, *The Bechtlers and Bechtler Coinage*, 6.

55. *Charlotte Journal*, July 24, 1835, quoting a Salisbury newspaper.

56. Featherstonhaugh, *A Canoe Voyage Up the Minnay Sotor*, 328–330.

57. Barnard, "The Private Coinage of Gold Tokens," 161.

58. House, *Operations of the Mint, 1840—and Medals*, 26th Cong., 2d sess., 1841, H. Doc. 75, vol. 2, 2–3.

59. Henry B. C. Nitze and H. A. J. Wilkens, *Gold Mining in North Carolina and Adjacent South Appalachian Regions* (Raleigh: Guy V. Barnes [North Carolina Geological Survey Bulletin 10], 1897), 42.

60. Jacob R. Eckfeldt and William E. DuBois, *A Manual of Gold and Silver Coins of All Nations* . . . (Philadelphia: Assay Office of the Mint, 1842), 161.

61. Jacob R. Eckfeldt and William E. DuBois, *Supplement to the Manual of Coins and Bullion* (Philadelphia: Assay Office of the Mint, 1849), 227.

62. Jack Weaver, letter to author, October 25, 1976, Vertical File, North Carolina Museum of History, Raleigh.

A Mint for Mecklenburg County: Being the Life and Times of the Charlotte Mint

Richard G. Doty

Richard G. Doty is curator of numismatics at the Smithsonian Institution's National Museum of American History and a former curator with the American Numismatic Society. He is responsible for the care of more than a million items in the National Numismatic Collection. His publications include more than one hundred articles, mostly on numismatics and many in coin magazines, and a half dozen books, among them Coins of the World *(1976),* Paper Money of the World *(1977), and* America's Money: America's Story *(1998).*

Ask an average American to think of the names of places where the nation's coins have been made, and he or she could probably come up with Philadelphia, San Francisco (the latter because of its connection with the California gold rush), and, possibly Denver. But our citizen would likely express surprise upon learning that there was once a mint at New Orleans (and a powerhouse of a mint at that, on occasion); and if you added that there were also mints at Dahlonega, Georgia, and Charlotte, North Carolina, you would probably not be believed. But there were mints in those two places all the same; you may know one of them as the Mint Museum of Art.

If you're wondering why the Republic went to the trouble of building a federal mint in Mecklenburg County, you must keep in mind two considerations. First, there was enough gold in these parts to inspire two *private* coiners—Templeton Reid in Georgia and a German family named Bechtler here in North Carolina. The federal government was naturally interested in what the two coiners were doing and in the competition their products posed to its own. Second, while not precisely in the back of beyond, the interior of the Carolinas and Georgia was difficult to keep supplied with an adequate amount of federal money, especially as the sole mint at the time was located in Philadelphia, hundreds of miles away over wretched roads, and had difficulty supplying even its local customers. A national branch mint at Dahlonega, or at Charlotte, then, made a certain amount of sense. The metal for coining was available, and, once put into operation, the Philadelphia mint's offspring would remove some of the pressure from its parent.

Today I want to examine the fortunes of the Charlotte mint, and I shall be relying on three types of information. Two writers have produced extended monographs on the facility. Anthony J. Stautzenberger's *The Establishment of the Charlotte Branch Mint: A Documented History* appeared in 1976. It is primarily useful for the steps leading up to the opening of the mint in 1838 but gives little information on subsequent events. That gap has been nicely filled by Clair M. Birdsall's *The United States Branch Mint at Charlotte, North Carolina: Its History and Coinage*, published in 1988. Neither of these books is easy to find, but each is well worth the effort. My second source for the Charlotte story is the National Archives of the United States, whose Record Group 104 contains a wealth of information on doings at the Philadelphia mint and its offspring, including Charlotte. There are many gaps in the archival record, especially as time goes on, but without RG104 neither Stautzenberger nor Birdsall could have written their works on the mint. And I could not have written an article on one of the engineers who worked on the Charlotte and Dahlonega projects, Benjamin Franklin Peale.[1] But more about him later.

I mentioned a third source of information, and in some ways it is the most important of the three. I refer to the coins that the new mint struck. We can tell a great deal about the career, the ups and downs, the weaknesses and strengths of any mint, including this one, if we know how to look at the products it made. More about this, too, later.

One essential fact that should be kept in mind about the Charlotte mint is that it was always part of a much more ambitious plan. For the federal government had determined to open not one branch mint but *three*—at Charlotte, North Carolina; Dahlonega, Georgia; and New Orleans, Louisiana. And the choice of the third location reveals something about the Crescent City and about why mints are set up in some places but not in others.

The choices of Charlotte and Dahlonega had at least a superficial logic on their side: they were reasonably close to the gold fields. But there were no significant gold—or silver—mines within half a thousand miles of New Orleans; why would anyone want to put a mint *there*? The answer is that there was more than one way to acquire precious metal for coinage. You could dig it out of the ground, which was what the Carolina miners were doing. Or you could sit in a center of great economic importance—New Orleans, for instance—and take the plenteous coinage coming into your coffers (Mexican silver coins, Peruvian gold ones, and a dozen other types of specie from across Latin America and beyond) and turn some of it into new, American money. In other words, the Charlotte and Dahlonega mints would be coining—while that of New Orleans would be *recoining*. Regardless of the origin of the metal, it would be turned into a new product, more useful for commerce than in its previous state, and that was the main point.

Talk about branch mints had gone on for a number of years, but it was only in the 1830s that conversation led to concrete action. And the action would be two-pronged. The Philadelphia mint would be modernized so that coins could be made by steam power rather than by machinery powered by human beings or by animals. The Treasury had been somewhat tardy in adopting the new ways; it would now attempt to make up for lost time. And the bureaucrats would see to it that the three "daughters"—the Charlotte, Dahlonega, and New Orleans mints—would be equipped to be worked by steam *from the very beginning*. This zeal for modernization was commendable. It would make possible better coins, more consistent coins. And it would make possible the creation of *more* coins—at least, once the minters learned their way around the new machinery. But it could create problems as well as possibilities in places such as Charlotte and Dahlonega. Aspiring moneyers in both locales learned this the hard way, once they began trying to turn idle speculation into concrete accomplishment.

This was because the Charlotte and Dahlonega areas (and if I seem to be lumping them together, this is deliberate: the two locales shared many of the same problems and characteristics) were heavily dependent upon mining, and the two mints to be established were likewise dependent on that extractive process. Unfortunately, the act of coining was an *industrial* process—and a new and anxiety-ridden one just now, given the introduction of the motive force of steam into what had already been a complex and demanding set of practices. And while sufficient technological expertise might be found in or lured to a city of the size and wealth of New Orleans or Philadelphia, getting it to Charlotte or Dahlonega might be something else again. The mining economy, even abetted by agriculture, would be unlikely to create the type of technician needed to build and maintain a state-of-the-art coining facility. And the two sites would offer few inducements for immigrants from technologically advanced areas. The results might have been foreseen: all three branch mints were provided for by an act of Congress passed on March 3, 1835, but only the construction of the New Orleans facility was carried out in a timely fashion. Work at the other two proceeded so slowly and haphazardly that the "Mother Mint" was forced to send a special emissary to the southern Piedmont to prod her lazy offspring to redoubled efforts. The emissary in question was Benjamin Franklin Peale. He was in his early forties when history thrust upon him the role of healer of malingering mints.

Peale deserves a brief introduction, because he was an interesting personality, brilliant in his way. He was a member of the "Philadelphia Peales," a family that included several artists (most of whom bore names of long-departed painters, guaranteed to get them into trouble with their grammar-school classmates). The family also kept the Philadelphia Museum, one of the nation's first such institutions. Benjamin Franklin Peale (or simply Franklin Peale, as he was generally known and signed his correspondence) was of a decidedly mechanical rather than artistic bent. By the time he was assigned to breathe life into the Charlotte and Dahlonega projects, he had had deep experience with the new European steam-powered coining technology, which centered on a coining press that employed a toggle action, which was far more durable and gave far more satisfactory results than the steam-powered screw presses at work in England—or the *hand-operated* screw presses still at work in Philadelphia. Peale gained his expertise in the new moneying between 1833 and 1835, when he was sent to Europe on a fact-finding expedition for the United States Mint. That institution wanted him to concentrate on learning all he could about improved methods of metal refining, but the engineer seems to have taken it upon himself to investigate advanced methods of coining as well, particularly the toggle-action press invented by Diedrich Uhlhorn around 1815 and by then at work in an improved version in Karlsruhe and, most dramatically, in Paris.

Peale took copious notes, as they say, and upon his return to America in the middle of 1835 he was ready to pour all he had learned on the Continent into a series of new machines for the United States Mint. His coining press was to borrow heavily from the Uhlhorn model, as improved by Thonnelier (a gifted inventor who had been associated with Uhlhorn), leading to persistent rumors that he had engaged in industrial espionage.[2] But Peale developed an improved edge-milling machine that resembled nothing on the Continent, and this and his improved coining press (which, married to steam, first went into action in Philadelphia in February 1836, striking medals, and had graduated to coinage by the following November) drove his career forward: by March 1836 he had been named melter-refiner, and he became chief coiner of the Philadelphia mint in 1839. This was after his trouble-shooting expedition to Charlotte and Dahlonega. Considering the difficulties he had encountered in the southern Piedmont, and what he had accomplished there, he had probably earned the promotion.

Reports of incompetence and delay had been spreading north for some months. Of the two facilities, Dahlonega appeared to be encountering the most problems. The local contractor, Benjamin Towns, was dilatory in performing his duties. Moreover, much of what he actually did was done poorly. Joseph J. Singleton, superintendent of the Dahlonega mint, reported that the building's walls had been constructed so cheaply that one of them actually buckled under the thrust of a lateral arch, which formed an integral part of the original building plan.[3]

The woes of the Dahlonega mint were not confined to shoddy masonry. Housing in that backwoods area was in short supply, and so were skilled workmen, whether obtained locally or sent from Philadelphia. And the one person whose advice could have been most valuable proved to be a raging incompetent once he had made his appearance on site. This was the commissioner for the project—a sometime lawyer, Methodist minister, and military man named Ignatius A. Few. Few was pure in heart and great in mind, but he was also constitutionally incapable of carrying projects to their conclusion. Moreover, he appears to have taken a dislike to spending time in Dahlonega, which was unfortunate, considering that he was *supposed* to be there, studiously breathing life into his fledgling mint. A succession of bureaucrats, including Franklin Peale, would have problems with the colonel.

Thus Dahlonega, whose early growing pains foreshadowed its later coining career. What about Charlotte? There were no obstacles of the magnitude of Colonel Few, but there were obstacles enough. There were great difficulties in getting machinery and parts for the mint from their points of manufacture to their place of intended use. When reviewing the histories of the Charlotte and Dahlonega mints, it must be remembered that those who erected the machinery and kept it in productive motion were isolated people on their own. Accordingly, it was essential for them to possess a battery of technological skills adequate for dealing with what was, for those days, some of the most advanced machinery ever seen on these shores. If something broke, they had better be able to repair it—or shunt around it in the manufacturing process. But if something happened elsewhere, away from the premises under their direct and immediate responsibility, there was very little they could do about it except watch the project in their charge abruptly grind to a halt.

And this is precisely what happened at Charlotte. There were then no convenient rail lines between Philadelphia and Charlotte—there had never been the need, and railroads themselves were in their infancy—so the machinery and heavy castings required for the mint would have to come by water. This could not be accomplished at the time, because a severe drought rendered waterborne shipment difficult if not impossible. Low water had emerged as a concern by the spring of 1837; Franklin Peale would still be encountering its effects half a year later. As at Dahlonega, there were also problems with personnel and building construction, and it was the concern with the latter that led to the decision to send Peale on his southern mint-expedition. That decision was taken in August 1837, on the heels of John J. Singleton's account of the collapse of construction at Dahlonega. Mint director Robert M. Patterson made a suggestion to his superior, Treasury secretary Levi Woodbury: "It seems to me . . . that it will be proper to send, to *both* the [new] Gold mints, a confidential & skillful person, who may present us a true account of the execution of the work, and give instructions as to correcting the errors, that have been committed, and completing the work that remains, —such as the erection of the furnaces, &c. I know of no one competent to this task, except our Melter & Refiner, Mr. Peale."[4] Patterson added that Peale could perform that necessary work at no expense to the government beyond carfare. From my experience working in the public sector, I can tell you that this holds a powerful allure for the bureaucratic mind.

What Patterson had in mind for Peale was the role of sophisticated troubleshooter, as good with prideful mint personnel as he was with advanced machinery (much of which Peale had improved or invented in the first place). Such an expert would oversee the final stages of construction and preparation for coinage, ensuring that the United States got its money's worth, so to speak. Persuaded that Patterson was right and that Peale was indeed the man for the job, Woodbury gave his assent three days later. Peale headed south shortly thereafter.

Accompanied by his daughter Anna, Peale left Philadelphia about the middle of September 1837, arriving in Charlotte on the twenty-third. Charlotte was the first mint site he visited, and he immediately ran into trouble. A load of castings and mint stores was several days late, and the work could not proceed until it arrived. Peale could not simply go on to Dahlonega, because that site was in an even less advanced stage of completion than was Charlotte. Nor could he give up and return to Philadelphia: that would render his entire journey meaningless and probably prove malign to the future career of a very ambitious man. For the first time, Peale appears to have lost his nerve. But it was also for the last time: within a few days, he had regained his composure, working out with Col. John H. Wheeler (lawyer, Democratic politician, and future historian), the new superintendent of the North Carolina facility, a plan of action to retrieve the missing parts. But the troubleshooter continued to dwell on the technological backwardness of the area in which he was supposed to set a modern mint in motion. He had a constant, nagging fear that for once he might fail to master the occasion. To be frank, he also had something of the attitude of a Philadelphia snob, confronted with his backward and unpredictable country cousins and they with him, neither quite knowing what to make of the other.[5]

The first month was the worst. By October 11 the castings had still not arrived, and Peale was desperate. He was nearly ready to send his people on to Dahlonega, where they at least could perform useful work. As for himself, he would spend his time in a visit of inspection to the nearby mines that had inspired the two branch mints—and his miserable mission to rescue them. But an excursion to the mining sites helped restore Peale's resolve to persevere: the thing was worth doing after all. He was buoyed as well by a highly supportive letter from Robert M. Patterson, which he found waiting for him upon his return from the mines. And later in the month of October, matters finally turned the corner.

They found the errant castings—in Columbia, South Carolina, of all places. Peale sent two wagons after them as soon as he heard the news. By the first of November, all of the missing castings and mint stores had reached Charlotte, and at long last the aspiring coiners could get on with their tasks. Within a few days, Peale's work at Charlotte was done, or as nearly done as it would ever be, and he began planning for the next leg of the trip. He left for Dahlonega on Saturday, November 10, arriving there the following Wednesday. He would have more frustrating adventures in Georgia than he had had in North Carolina, but before leaving Charlotte he dispatched to mint director Patterson two parting shots on southern mints in general and, in particular, the one he was preparing to depart.

Peale's pessimism resurfaced as he considered the future of Charlotte and Dahlonega. He worried about their overall prospects, for the people chosen to oversee them lacked the necessary technological expertise for the job. While the mints would indeed be set to work one day, "as to the *time* or *how*[,] that is for futurity to shew."[6] And a week later, he related his discovery of the reason for the shoddy workmanship that had gone into the Charlotte mint building. One of the contractors told him "that he secured the contract by a low offer in the expectation of making his profits on the extra work" that would have to be done to make the mint suitable for use.[7]

Peale spent the next few weeks in Dahlonega, whose mint he found in an even worse state than had been reported. For reasons not clear, Philadelphia had ordered the facility constructed out of brick—this in an area in which there was no proper clay but a wealth of granite. The person theoretically responsible for overseeing the construction was Colonel Few, who Peale was now certain was the evil genius behind all of the problems being encountered. Peale did what he could to shore up construction and goad the colonel into action, but he finally concluded that the building, while of shoddy manufacture and dubious safety, would likely have to be accepted as it was, "or there will be no branch [mint] at Dahlonega, a large amount of the appropriation having been spent."[8] On that note of chastened realism, he prepared to return home with his daughter.

Franklin Peale had one more adventure, which must have convinced him that ill fortune may have given up pursuing him south, but only because it was now chasing him back north. He and his daughter left Dahlonega at the end of November, making a circuitous journey back to Philadelphia. While passing through Virginia, the train on which the two were passengers met with "the perils of the most horrible accident . . . that has occurred for years. A fine new engine . . . was thrown off the track by an iron plate rail which was raised up at one end. The engine was brought up against the side of the ditch, where a fearful crash took place."[9] Miraculously, no one was killed, although twenty were hurt, some seriously. The Peales escaped relatively unscathed, the mint technician crediting their good fortune to the fact that they had been riding in the middle passenger car, which, instead of being crushed by the force of the accident, was simply raised off the tracks and thrown to one side. Shaken by the adventure, Peale and his daughter slowly made their way back to Philadelphia and safety. By the twenty-third of December, he was back at work, zealously trying to forget the frustrations of the past three months.

A brief mention of Peale's subsequent career is in order. He never again served as a troubleshooter in the southern Piedmont, his increasing responsibilities at the "Mother Mint" keeping him close to home. He eventually ran afoul of the authorities there by striking medals for sale on the mint's time and with the mint's materials and machinery—but for his personal profit. Having, as he saw it, brought the coiner's apparatus to a state of perfection, he probably felt that he was acting within his rights. But he was dismissed all the same in December 1854, left the mint, and never returned. He died in the early 1870s. His devoted wife presented a marble bust of her late husband to then mint director Henry Linderman on the last day of April 1873. The bust's current whereabouts are unknown.

Peale's charges opened for business in the spring of 1838. The Charlotte facility had received his approval in a report to the mint director, written on the day he returned to work. Much of Peale's negativity concerning the North Carolina branch mint had vanished (one suspects that this was a direct result of his experiences at Dahlonega, compared to which Charlotte's situation looked positively rosy): as he now saw it, the Charlotte commissioner had performed an admirable job under difficult circumstances, choosing the best, most durable materials for the building's construction and extending to Peale a free hand in the installation of its machinery—which of course was of superior quality, because Peale had invented it. The Carolina mint was pronounced ready for coining and had already received bullion for the purpose.

This engraving of the Charlotte branch mint appears on a four-dollar bill issued by the Bank of Charlotte in the 1850s.

The picture was far bleaker at Dahlonega, where Colonel Few still stood as a barrier to progress, and where irate mint superintendent Joseph J. Singleton told contractor Benjamin Towns that the mint would commence operations on February 1, 1838, whether it was finished or not. Singleton's ultimatum impressed Robert M. Patterson back in Philadelphia, who decided to remove Colonel Few and replace him with the forceful superintendent, who at least promised action. This proved easier said than done: the colonel hung on to his position like a leech, and the squabble was still unresolved in the middle of 1840. Dahlonega indeed began coining in the spring of 1838; it is hard to see how it was able to do so.

Both it and its sister at Charlotte were in business for twenty-three years, commencing operations in the spring of 1838 and concluding them in the spring of 1861 as early casualties of much larger events. The New Orleans branch was active during those same years, but, alone among the three, it was reopened as a mint in 1879, remaining active for precisely thirty years. And alone among the southern branches, New Orleans struck silver coinage as well as gold, a reflection of the Latin American raw materials that came through its doors. Charlotte and Dahlonega struck only gold, and only in certain denominations. And to Charlotte's portion of the coinage we now turn.

A contemporary view of the building in which the coinage was made graces a note from the Bank of Charlotte, issued on the Fourth of July, 1853. The note was the equivalent of four gold dollars, struck at the town's mint that very year. The bill's denomination appears strange at first, but it makes sense from a couple of directions. First, various areas harbored traditional affinities for particular denominations—people in parts of Georgia liked three-dollar bills, for instance. Second, bank notes in a wide range of denominations were an absolute necessity in a nation still short of coinage—which the United States had always been and would continue to be for a decade more. You needed a four-dollar bill just in case somebody handed you a ten for a six-dollar purchase.

My impression of the Charlotte mint is that it began coining on a reasonably sound basis, that its people knew how to put the new industrial moneying processes to work with satisfactory results, and that, in general, its record was more impressive than that of its Georgia counterpart. It certainly coined in greater quantity than did Dahlonega—at least during most years. While it is unwise to generalize, Charlotte tended to produce about a fifth to a quarter more of a given denomination in a typical year than did Dahlonega. But Dahlonega did strike one denomination never seen at the Carolina mint: a three-dollar gold piece (in 1854), which met no real demand but was nonetheless struck at the parent mint in Philadelphia all the way to 1889—an unknown genius having figured out that you could buy a sheet of one hundred three-cent stamps with the coin.

The gold coins of Charlotte (and Dahlonega and New Orleans, and eventually San Francisco, Carson City, and Denver) bore designs determined in Philadelphia. How could it be otherwise? The machinery used to strike them was perfected in Philadelphia, and the dies to strike them were made in Philadelphia, at the senior facility. But those dies, first to last, were always given a special mark, so that everyone could discern the origins of the gold and the place of coinage: Charlotte got a *C* as a mint mark, while Dahlonega received a *D*. At first, those mint marks were found on the obverses of the coins, appearing just above the date. Half eagles ($5.00 gold pieces) and quarter eagles ($2.50 gold pieces) were the only denominations struck at Charlotte and Dahlonega down to 1849, but the designs, holdovers from an earlier era, were soon modernized.

Obverse and reverse of half eagle (five-dollar gold coin) struck at the Charlotte mint during 1838, the first year the facility was in operation. Note *C* mint mark above date on obverse. *Photograph courtesy Smithsonian Institution, National Numismatic Collection, Douglas Mudd.*

An artist named Christian Gobrecht had recently redesigned United States silver coinage. He now did the same for the nation's gold. His left-facing Liberty head broke no new ground, but it was somewhat neater in conception than the previous, "classic head" design by William Kneass. Nor was Gobrecht's eagle a work of art, although it functioned fairly well in the demanding confines of a small gold coin. Artistic or not, the nation would use the Gobrecht Liberty and eagle for the next seventy years. The Charlotte and Dahlonega mints began striking coins with the new designs in 1839. In that year alone, the mint mark remained above the date on the obverse. The following year, it migrated to the reverse, where it stayed.

There is something about the gold coins of Charlotte that suggests to me that the coiners there were more in control of their craft than were their cousins in Dahlonega. The Charlotte coins are more carefully struck, more sharply struck, than their Georgia equivalents. In the case of two 1852 half eagles, for example, even though the Charlotte coin has *more* wear from circulation, its devices and denticles are actually sharper, more completely finished off, that they are on the Dahlonega coin. I am a dedicated student of the history of coining technology, and when I see differences such as these, I conclude that the industrialization of money was working reasonably well at one venue and not at all well at the other.

But there was one period during which the mint at Dahlonega won the contest—by default. On July 28, 1844, the Charlotte mint's then superintendent, a gentleman named Green Washington Caldwell (a Charlotte lawyer and another Democratic politician), had the melancholy duty of advising his superiors that there had been a fire at the facility. The fire had occurred on the previous morning. At first glance, the situation might have been far worse: Caldwell reported that the bullion and books had been saved, and, while some of the machinery had been damaged, he was cautiously optimistic. But over the next few weeks his optimism vanished.

Caldwell had not been in Charlotte at the time. He was in poor health and was recuperating at his mother's house a dozen miles or so out of town. He had left his apartment at the mint (he lived on the premises) in the care of a friend, who promised to sleep there every night during his absence. But the friend had reneged, and the wrong people had found out: "I am now satisfied that my rooms were robbed previous to the fire, and that the Mint was set on fire to conceal the robbery," Caldwell wrote. A suspect had been rounded up: "In mine of the *7th* Inst [August 7, 1844] I informed you that Col: [*sic*] Gaither's Boy had been arrested on suspicion of being the incendiary, and promised to give you the result of the investigation—Upon investigation before the Judge the State's Witnesses who were principally Slaves, told a very contradictory and unsatisfactory story, so much so that the Judge was compelled to discharge him. The evidence on that investigation satisfied the Judge that the origin of the fire was the effect of design and not of accident."[10] Caldwell assured Philadelphia that he had not ceased his efforts to ferret out the real perpetrator or perpetrators. But meanwhile, there were the results of a fire to contend with, and to set right. How much had been damaged or lost? How would that affect coinage at the Charlotte mint?

The answer was that a good deal had been damaged and that it would affect the mint's production more than the most pessimistic observer could have anticipated. An estimate by the coiner, a skilled mechanic named John R. Bolton, concluded that while the steam engine that conducted power to the machinery could indeed be repaired and made ready for work in a short time, and that while some parts of the mint apparatus (including the rolling mill, which was of crucial importance, especially in the coining of gold) had indeed escaped relatively unscathed, other parts of the mint had suffered far greater damage. Chief among them were the two coining presses. With luck, enough parts could be cannibalized to put one of them back in service, but the other would be lost. The refinery was nearly a total loss as well. In sum, the Charlotte mint had sustained a major setback. And the scarcity of technological capital in the immediate vicinity would retard the facility's recovery.[11]

Debate over the mint's future went on through the summer of 1844. At one point, it was suggested that when the building was reconstructed it should only be one-and-one-half stories high: all of the moneying operations should take place on the first floor and in the basement, thereby eliminating, in the words of one correspondent, the "Superintendent[']s Family Apartments . . . upon the ground of economy, but also upon those of comfort and convenience in the processes of Minting, besides the protection it would afford in case of accident."[12] The correspondent added that Superintendent Caldwell wanted no establishment of his own in the building except for his offices and a sleeping room. Caldwell's experience with the robbery had probably persuaded him that less was more—or at least more secure.

An estimate of repairs was duly drawn up and sent to Philadelphia. It amounted to $22,550, of which $8,100 would go toward stone and brickwork; $10,450 for carpentry and materials; and the remaining $4,000 for painting, plastering, and glazing. The Charlotte estimate did not include amounts for machinery or the skilled technicians whose services would be required, although Congress eventually earmarked an additional $10,000 for new machinery and tools.

While the mint awaited help, the fire had a most dramatic and deleterious effect on its coinage production: from the summer of 1844 until the autumn of 1846, there *was* none.[13] But the facility was at last refurbished and was once again ready for work. By October 1846 it was again striking coins, and it quickly made up for lost time. As one might expect, Dahlonega overshadowed it that first year, having had a full twelve months in which to coin. By 1847, however, the roles had been reversed, with Charlotte producing a total of 107,377 coins to Dahlonega's 80,109. And the Carolina mint maintained its lead or at least rough parity through the majority of the years remaining to it.

In 1849 two new denominations were added to the constellation of American coinage, both of them a reflection of events far to the west. On the top of the scale, a double eagle, or twenty-dollar gold piece, would utilize much of the gold from the California fields. On a lesser level, there would be a new gold dollar, a tiny coin weighing less than one-twentieth of an ounce. Charlotte and Dahlonega never participated in the coining of the former, but they did join in the manufacture of the latter.

The gold dollar went through no fewer than three major design changes during its lifetime. In its first incarnation, its obverse resembled that of the double eagle (no mystery here: the two coins were designed by the same artist, James B. Longacre), while a simple wreath with value and date was employed for the reverse. The mint mark appeared beneath the tie of the wreath if the coin was struck at a branch mint: following a long-standing tradition, coins bore no mint marks if they were struck at the main facility in Philadelphia. In 1854 the coin's diameter was broadened (but the coin was made thinner, so that it contained the same amount of gold as previously), and Longacre placed the head of a Native American princess on the obverse in place of his Liberty head. A cereal wreath replaced the simpler arrangement of previous years. (The artist was fond of the cereal concept, and sooner or later most American coins in which he had a hand displayed the motif.) Two years later the design was changed once again, a different princess now adorning the obverse; the reverse was left as it was. Thus amended, the gold dollar was struck all the way down to 1889—although the southern mints would have no role in its production after the outbreak of the Civil War.

Obverse and reverse of $1.00, $2.50, and $5.00 gold coins manufactured late in the production history of the Charlotte mint. Note that each coin bears the distinctive *C* mint mark on the reverse. *Photograph courtesy Smithsonian Institution, National Numismatic Collection, Douglas Mudd.*

The Charlotte mint participated in the gold dollar's production through 1859, striking coins of all three design types, although none of them in large quantity. That practice paralleled practice at most other mints, which struck the coin sparingly. The exception was Philadelphia, which ground out between two and four million of the tiny pieces per year during the early 1850s and hundreds of thousands each year during the remainder of the decade, despite the fact that the coins' small size and thinness made them notoriously difficult to work with and notably hard on coining dies. The other new coin of the era, the three-dollar gold piece, was never struck at Charlotte, although Dahlonega coined eleven hundred or so of the pieces in their inaugural year of 1854 before promptly terminating production when the coins proved to be unpopular.

The years were closing in on both Charlotte and Dahlonega. The mints, and indeed the nation, were increasingly held hostage to sectional tensions, which had always been present but which no one had the skill or the will to resolve. Both mints saw a tailing-off of coinage in the years between 1858 and 1861. For half eagles, for example, Charlotte's mintage stood at 38,856 in 1858, 31,847 in 1859, 14,813 in 1860, and only 6,879 in 1861. A similar downward spiral is seen for quarter eagles, which Charlotte stopped minting in 1860, and for gold dollars, whose production there ceased in 1859.

Once North Carolina and Georgia left the Union in 1861, a debate as to the future of their branch mints ensued. It was eventually decided that at least a temporary closing of the two as coin-production facilities would be in order. After all, there were more important things to do just now than strike gold coinage. Both mints *were* closed for coining. As it turned out, closed they remained, for the duration of the war and ever afterward.

Each mint had had many adventures during the past twenty-three years. But Charlotte's days as a federal mint ended on an odd, almost elegiac, note. I hope you won't mind if I share the story with you. Eight days after Fort Sumter, the assayer of the Charlotte mint wrote to the then director of the United States Mint, James Ross Snowden. The Charlotte functionary was a long-term employee named Dr. John H. Gibbon, who took great pride in the fact that he had been there as long as the mint itself. Nonetheless, he, and his mint, were overshadowed by larger events. North Carolina governor John W. Ellis had sent what Gibbon felt was a "peremptory" demand that the military commander of the district take possession of the mint property, as well as any funds that might be found on the premises. Dr. Gibbon searched his soul and decided to go out with his state: his letter to Snowden was one of resignation. And then, in an aside, in a few laconic words, the assayer announced the end of a mint, and the end of an era: "While I write[,] the People are hoisting a Confederate flag over the mint building—and on going out—I find an armed guard from a volunteer company at the door[.]"[14]

The recipient of this letter passed on its contents to his superior, Treasury secretary Salmon P. Chase, adding that the New Orleans and Dahlonega mints had likewise been suborned by the insurgents—on January 31 and April 8 respectively.

Snowdon suggested that the products of the three mints be declared inadmissible as legal tender for the payment of debts, observing that it would be easy to recognize them because of their mint marks. But Salmon P. Chase, and the new president who stood behind him, had more pressing matters to consider, and Snowdon's advice went unheeded. The products of the "rebel mints" would remain coins of the realm, even as the reasons and the bases for their creation (and the fact that there had once been no fewer than three national branch mints in the South) would fade from public awareness. I have welcomed the opportunity to share one of their stories with you today, and I can think of no more appropriate venue in which to have done so.

Notes

1. Richard G. Doty, "'An onerous & delicate task': Franklin Peale's Mission South, 1837," in *America's Gold Coinage*, ed. William E. Metcalf (New York: American Numismatic Society, 1990), 67–82.

2. Peale apparently obtained European permission for the drawings that he made; in any case, he introduced a number of improvements to the European version of the toggle-action press, rendering it more durable in a demanding American environment.

3. Joseph J. Singleton to Robert M. Patterson, August 6, 1837, Record Group 104, National Archives, Washington, D.C. (hereafter cited as National Archives).

4. Robert M. Patterson to Levi Woodbury, August 18, 1837, National Archives; emphasis in original.

5. Something of Peale's attitude is captured in a letter to Robert M. Patterson dated October 4, 1837, wherein he referred to Charlotte as "this fag end of creation . . . where the only active beings are the hogs" (National Archives). The long-suffering mint director took this opinion with a pinch of salt (as he did most other missives from his troubleshooter), kept Peale at his task, and sent him on to the even more challenging work at Dahlonega once his labors at Charlotte had been successfully concluded.

6. Franklin Peale to Robert M. Patterson, November 1, 1837, National Archives. The further and frustrating adventures of Franklin Peale are related in my own article, cited in note 1, above.

7. Franklin Peale to Robert M. Patterson, November 8, 1837, National Archives. There's nothing new under the sun, is there? The inventor eventually modified his opinion of Charlotte's prospects—but not until he had had to contend with conditions at Dahlonega!

8. Franklin Peale to Robert M. Patterson, November 25, 1837, National Archives.

9. Franklin Peale to Robert M. Patterson, December 11, 1837, National Archives.

10. Green Washington Caldwell to Robert M. Patterson, August 16, 1844, National Archives.

11. Clair M. Birdsall's reading of the situation was far more pessimistic. Based on surviving testimony, he expressed the opinion that the main building had been destroyed except for the foundations (and some construction materials, which could be reused) and that *both* coining presses had been damaged beyond repair. He also observed that the fact that 1844 was an election year may have been a factor in the lengthy amount of time it took to rebuild the Charlotte mint. While Charlotte was a Whig stronghold, and while the Whigs were still in power, the party was split into two wings: an ineffective one that supported the lame-duck president John Tyler and a very ambitious one that supported the nationalist candidate, Henry Clay. Clay had been against the idea of branch mints in the first place: any rebuilding would have to wait until he had been defeated and James K. Polk (who was a hard-money man born near Charlotte and sympathetic to the mint) had won. Clair M. Birdsall, *The United States Branch Mint at Charlotte, North Carolina: Its History and Coinage* (Easley, S.C.: Southern Historical Press, 1988), 29–30.

12. Unnamed correspondent (probably John R. Bolton) to Robert M. Patterson, September 21, 1844, National Archives.

13. I date its resumption to the fall of 1846 because of the existence of a rough draft of a letter in the National Archives under date of October 30. It was addressed to Robert M. Patterson and was probably written by G. W. Caldwell. The correspondent takes pleasure in telling Patterson that "I have received nearly $200.000 [*sic*] worth of Bullion during this month, that we have made [into] a Coinage of $12,920– [and] that the machinery with the exception of the drawbench works beautifully." The implication is that the mint was just back in production, following a long period of inactivity. In point of fact, it would strike 12,995 half eagles and 4,808 quarter eagles, its first coins since the debacle of 1844.

14. J. H. Gibbon to James Ross Snowden, April 20, 1861, National Archives.

Gold and the Development of Charlotte

David Goldfield

Dr. Goldfield, a historian of the urban South, is Robert Lee Bailey Professor of History at the University of North Carolina at Charlotte. He has written or edited twelve books, including Cotton Fields and Skyscrapers: Southern City and Region, 1607–1980 *(1982),* Black, White, and Southern: Race Relations and Southern Culture, 1940 to the Present *(1990), and* Race, Region, and Cities: Interpreting the Urban South *(1997).*

Let's face it. John Reed was lucky, moving from a Hessian deserter to the owner of a very valuable piece of land in Cabarrus County, North Carolina. In the outline of his rags-to-riches story, he was not unlike many who have come to the Charlotte area to make their fortunes—sturdy German and Scotch-Irish folk who believed hard work was its own reward. Let the lowlanders sip their cocktails and watch the paint peel from their expensive mansions; up here in the Piedmont we'll stick a plow in the ground, a building on a slab, or a cart down a shaft and expect it all to grow. Charlotte is what it is today because people have come from elsewhere, preferably outside North Carolina, bringing their brawn, their brains, and, yes, their luck to the red-clay region. Reed's odyssey from Savannah to Charleston and ultimately to Cabarrus County demonstrated one common route of movement to this area. We were always closer to South Carolina than we were to the North Carolina that existed east of us—which is fortunate because in Charleston, at least, you had wealth, sophistication, culture, good food, and direct ties with New York and Philadelphia. In eastern North Carolina—well, you had turpentine, and today it's pickles and hog farms. It's hard to imagine a book such as *Midnight in the Garden of Good and Evil* being written about Rocky Mount, if for no other reason than everyone there is in bed by that time.

My friend Rick Knapp notes correctly that very few historians have chronicled North Carolina's gold-mining saga during the first half of the nineteenth century. I think the reason for that neglect is that no one expects much of North Carolina in that particular era. The state developed a reputation as the "Rip Van Winkle" state. But it was a poor analogy. At least old Rip woke up.

Now, I don't mean to suggest that John Reed's mine was unimportant in the scheme of things. Reed's mine, along with several other facilities in North Carolina, supplied all the domestic gold coined by the United States Mint between 1804 and 1827, and by the time of the California gold rush in 1849, about $2.6 million worth of gold had been extracted from Charlotte-area mines. In 1869 sixteen gold mines operated in Mecklenburg County; in 1891 the number of mines had jumped to sixty. But to use the phrase "gold rush" to describe any particular time period in the Charlotte area is an exaggeration. At various times there were quite a number of folks, some from as far away as Europe, digging their way through what we now call the Charlotte metropolitan area. Early in the nineteenth century there were even reports of hopefuls digging under the streets of Charlotte, evidently taking literally that old saw about streets being paved with gold in America. Activity was sporadic, however, and the volume, while impressive over a long period of time, paled before the justly more famous California gold rush. San Francisco was a clear beneficiary of that event, much more than Charlotte was of the modest gold-mining operation in its midst. By 1850, for example, Charlotte included scarcely more than one thousand citizens within its borders, compared with almost forty thousand in the city by the bay.

Still, it was a coup for a small town such as Charlotte to receive a branch of the U.S. Mint in 1837. The federal government also opened branches in Dahlonega, Georgia, and New Orleans at the same time. The coins struck at the Charlotte facility between 1838 and 1861, when it closed, amounted to a total of five million dollars, quite a sum in those days. In 1936 the mint building was moved to Randolph Road because, sooner or later, everybody who's anybody winds up in southeast Charlotte. Today one can see fifty of the Charlotte mint coins on display. And today, of course, the mint is no longer the mint but a fine art museum whose contribution to the city and region is greater than that of its predecessor. Money is a fleeting thing, but culture is forever, and the Mint Museum of Art may be the greatest legacy of our gold bicentennial.

In truth, it was not gold but Gov. John Motley Morehead who laid the foundation for Charlotte's growth in the nineteenth century. Morehead was one of those rare politicians who was ahead of his time rather than abreast of the polls. Among his interesting ideas was his proposal to allow free blacks to vote at a time when free blacks voted nowhere else in the United States. Morehead also established North Carolina's first public school system. But most of all, he recognized that the Old North State would remain a slumbering relic unless its wretched transportation was improved—particularly its east-west connections. Such improvement was great news to Mecklenburg County farmers, who led the state in the cultivation of every major crop except tobacco. By the time of the Civil War, rail connections were completed between Goldsboro and Charlotte, but the economic impact was minimal. Gov. Zebulon Vance had another idea—that it would be more advantageous if Charlotte were connected to northern markets through Virginia—and the Queen City was drawn away from eastern North Carolina. By 1860 four rail lines radiated from Charlotte, and the town's population more than doubled to 2,265 people. The railroad-building activity of the 1850s helped Charlotte grow during that decade as much as it had during its first century.

But Charlotte was still a very modest town. In 1860 it could not even qualify for the U.S. census definition of an urban place. The Census Bureau set a population threshold of 2,500 people. By 1860 the nexus of American urbanization lay far from Charlotte. The South's great cities—New Orleans, Savannah, Charleston, and Norfolk—were either just barely holding their own or already in decline. The railroad had begin to divert patterns of trade away from the traditional coastal ports of the South. Southern states built railroads, but northern states built railroad *systems*.

A complex maze of gauges hampered railroad connections in the South, and the region's greatest city, New Orleans, insisted that the Mississippi River would remain the nation's major commercial highway. But as great railroads connected Chicago to Philadelphia and New York, trade took east-to-west and west-to-east directions, and the South became increasingly isolated from the booming regions of the North.

Interior cities of the South were mere scratches in the wilderness, smaller than Charlotte. Atlanta was founded in 1837 and went through various names such as Terminus and Marthasville until a local entrepreneur thought that he should give the name of his hotel to the town in which it was located, so Atlanta became Atlanta in 1847. It's the only city I know of that was named after a hotel. But then folks seem to move into and out of Atlanta like it's one great hotel, so perhaps that's appropriate. Dallas was founded in 1836, and the log cabin of its founder, John Bryan, still exists, though spoil-sport historians have since declared it a fraud. But this *is* Texas, you know. The town, like Atlanta, was little more than a railroad depot until after the Civil War. Nashville was a sleepy place, though considerably older, that came barely alive when the Tennessee legislature was in session. And Birmingham did not exist. Richmond and Louisville were about the only significant interior cities in the South worthy of the name. Both were industrial cities, and their success portended a new trend in southern urbanization that became evident after the Civil War. Still, Mark Twain's characterization of Louisville as "a good place to embalm a dog" summarizes that city's attraction.

Despite the fact that the Charlotte branch mint closed in 1861, that Charlotte's Civil War fame resulted from its designation as a Confederate naval yard, and that at the end of the conflict the Confederate cabinet held its final meeting in the town, much greater things lay ahead for the future Queen City. Gold mines continued to operate in the Charlotte area, and they provided local entrepreneurs with investment opportunities. It was the custom in those days for investors to dabble in an array of economic activities: a diverse portfolio then as now obviously offered a hedge against an economic downturn. The town was fortunate in escaping the physical and psychological destruction of the Civil War that hampered the rapid recovery of cities such as Charleston, Columbia, and Richmond (though not Atlanta). And Charlotte attracted investments from the North and abroad as well. The mines were one attractive source for investment; railroads were another.

When depression struck the nation in 1873, southern-owned railroads faltered, beginning a two-decade process by which northern investors, particularly J. P. Morgan, snapped up southern properties and created a national rail system. The change in ownership proved fortunate for Charlotte because the new system placed the city astride the "Main Street of the South"—the Southern Railway's main line between Atlanta and Washington, D.C. By the 1880s Charlotte was so well connected by rail that a recent South Carolina transplant by way of New York, industrialist Daniel A. Tompkins, could dream of the Queen City as the center of a vast textile empire. With the launching of his "cotton mill campaign," Tompkins proved that his vision was remarkably prescient. By 1905 more than half the looms in the South were

Tryon Street, Charlotte, looking southwest, circa 1887. Engraving from *Harper's Weekly,* April 9, 1887.

located within a one-hundred-mile radius of Charlotte, and by the 1920s the South had surpassed New England as the nation's premier textile region. Charlotte's reputation as a transportation hub also spurred activity in the wholesale-distribution sector, which, after textiles, became the city's leading economic activity and remains an important segment of its economy at the present time.

Charlotte's growing reputation as a place of vibrant economic activity attracted other enterprises, such as James B. Duke's Southern Power Company (presently Duke Energy), which thrived on the textile industry up and down the Carolina Piedmont. It was in one of those factories, in Belmont, in 1906 that a New Yorker by the name of Willis Carrier tried out his new invention, which he called "air conditioning"—a device without which the modern South could scarcely be contemplated, let alone inhabited. William Belk, after first applying his retailing acumen in Monroe, arrived in Charlotte and eventually built a commercial empire. And there were firms directly related to textiles. An example is the Southern Cotton Seed Oil Company, which manufactured Snowdrift Hogless Cooking Fat, a vegetable shortening refined from cotton seed. Even the leftover cotton lint attracted entrepreneurial firms such as Barnhardt Manufacturing and South Atlantic Cotton Waste, which turned the chaff into mattress and upholstery stuffing. Members of the Barnhardt family, incidentally, have been involved in gold mining as well.

In fact, even though gold mining and minting were no longer robust economic activities in the late nineteenth and early twentieth centuries, their presence in the local economy continued to affect other growth sectors. In 1890 Thomas Edison arrived in the Queen City to experiment secretly with ores from North Carolina mines. Many observers felt that he had a new invention for refining gold ore, but his primary interest was in the separation of *iron* ore. Edison's process proved less than stellar, but the great inventor took the opportunity to regale his host, South Carolina émigré Edward Dilworth Latta, about the wonders of electricity and more specifically its application to mass transit. Edison convinced his host of the trolley's potential, and the following year Latta installed Charlotte's first electric trolley line—well in advance of most American cities. Not incidentally, a goodly portion of Latta's trolley line ran through his own holdings, including one of Charlotte's earliest suburbs, the planned community of Dilworth.

All of these activities required money, of course. Before the Civil War the South had banks, but they were insignificant in comparison with the much larger institutions in major northern cities. Southern entrepreneurs either borrowed money among themselves or obtained loans from banks in New York at disadvantageous interest rates. The war destroyed the southern banking system, such as it was, and the region's banks took a long time to recover. Initially, northern and foreign investments played relatively minor roles in the southern economic recovery. As had been the case prior to 1861, local civic leaders pooled resources and helped each other.

In 1880, banks in Massachusetts held five times the capital of all the banks in the South combined. But urban and industrial growth in the South after 1880, especially among the region's interior cities—Charlotte, Atlanta, Birmingham, Nashville, and Dallas—generated working capital sufficient to establish new banks. Men who had invested wisely and diversely in real estate, gold mining, industrial activities, and commercial ventures turned to banking in an effort to put their money to work in another potentially lucrative enterprise. For example, Charlotte real estate developer George Stephens, who developed the still fashionable Myers Park residential neighborhood, in 1901 created Southern States Trust, which he soon reorganized in partnership with textile entrepreneur A. J. Draper. The new banking concern, called American Trust, subsequently merged with the older Commercial National Bank, which was controlled by the Holt family, prominent in textiles. The merged institution eventually called itself North Carolina National Bank, or NCNB, then NationsBank, and presently Bank of America. Its major Charlotte rival, First Union, began as the Union National Bank in 1908. The gold connection in Charlotte banking is evident in the career of Robert N. Miller, who, along with several partners, purchased the Rudisill gold mine in 1878. Among those who held an interest in the company over the next decade was textile magnate Eli B. Springs. Miller went on to become one of the original members of the board of directors of the Commercial National Bank.

The banking industry, fueled by capital from textiles, gold, real estate, and commerce, began to make a distinct visual impact on Charlotte. By 1920 a financial corridor had emerged along South Tryon Street as material evidence of the city's growing financial clout. The corridor included the city's first skyscraper, the Independence Building, erected in 1909. But the banking industry's influence on the city was far more than cosmetic. Local banks increasingly became lenders of first resort to a diverse array of business enterprises. By 1927 Charlotte seemed poised for its greatest financial coup since being selected as a location for a U.S. branch mint nearly a century earlier: the establishment in the city of a branch of the Federal Reserve Bank—and this despite the fact that the "smart money" had predicted that the branch was instead destined for Columbia, South Carolina. As Charlotte historian Tom Hanchett notes in his fine book *Sorting Out the New South City: Race, Class, and Urban Development in Charlotte, 1875-1975*, the branch "maintained the cash reserves of area banks and made loans to them, moved currency and coins in and out of circulation, and provided swift interbank check clearing, which gave Charlotte an additional financial edge over other cities in the region."

The "Fed" gave Charlotte's budding banking industry a big boost and in turn exponentially strengthened the Charlotte-area economy as well. It was not a coincidence, for example, that in 1950 Charlotte's wholesale sales ranked twenty-ninth in the nation, even though the city stood only sixty-ninth in total population. Such overachievement has become commonplace for the city. At about the same time, Charlotte banking institutions began to appear on lists of the nation's largest banks. In 1951 *Business Week* magazine profiled Charlotte as "a paper town—because most of its business is done on paper." Especially green paper.

By 1970 NCNB ranked as the forty-seventh largest bank in the nation, while First Union stood at fifty-ninth. Moreover, by the end of that decade Charlotte banks held more deposits than banks in any other city between Philadelphia and Dallas (within another decade, New York and San Francisco; presently only New York). With all that banking mania, it is easy to lose sight of the fact that banking, like other endeavors, is very much dependent on every other industry and economic activity. For if the loans don't work out and the deposits fall, so do banks. Just ask the folks in Dallas or Houston or almost anywhere in Florida.

In that regard, the federal government has played a crucial role in complementing the efforts of local entrepreneurs in building the city's economy. The decisions to locate the U.S. branch mint here in the nineteenth century and the Federal Reserve branch bank here in the twentieth were merely previews to federal activity from the 1930s onward. During the depression decade, for example, Federal Emergency Relief Administration crews dismantled the old branch mint building brick by brick, rescuing it from the wrecking ball, then moved it to its current location. In the mid-1950s the Federal Highway Defense Act established a nationwide interstate highway system, and few localities benefited more than Charlotte. The opening of Interstate 85 in 1962 and I-77 a decade later reinforced the city's role as a transportation hub and solidified its national prominence as a wholesale distribution center. I-77 provided the city its long-coveted swift access to the Midwest.

The feds also built Memorial Stadium and Memorial Hospital and helped with the construction of Independence Boulevard, a major early crosstown artery. Perhaps more important, federal policies embodied in the Federal Housing Administration and the Veterans Administration provided low-interest home loans to a generation of home buyers, as well as to developers such as C. D. Spangler Sr.

and John Crosland. Those federal initiatives stoked Charlotte's postwar building boom and enabled the city to expand its borders. Most important, it gave tens of thousands of city residents and newcomers an opportunity to own their own homes.

Federal banking policies, especially the advent of interstate banking and legislation that grew out of loan debacles in Florida and Texas during the 1980s and early 1990s, played important roles in the growth of NationsBank into Bank of America, as well as the expansion of First Union and Wachovia. There is a tendency in Charlotte to focus on the individual entrepreneur, and there is ample precedent for doing so; but the Lone Ranger is not a good model for building a diverse economy in a global setting. Governments are key partners in (or obstacles to) economic development. Those who bash the federal government not only don't know what they're talking about; they would be the first to scream bloody murder if the feds suddenly decided to withdraw their programs, which are rife with farm subsidies and a vast array of entitlements for the middle class.

In fact, during the late 1970s a number of northeastern and midwestern political leaders expressed concern that the federal government was perhaps being too generous to the South. Southern lawmakers dominated key congressional committees (and still do). The complaints focused on the fact that the South allegedly sent less in tax revenues to Washington and received more in return than did other regions. Subsequent accounting cast doubt on that assertion, but the feeling that too much federal funding wound up in Dixie prevailed. Part of that perspective derived from the very real plight of the old industrial urban core that spans the Northeast and Midwest, as well as the beginning of a vast migration to the so-called Sunbelt, of which North Carolina is a part. A century and a half earlier, the Old North State lost one-third of its population to other states and territories; out-migration persisted for another 125 years until the 1970s, when the pattern began to reverse itself and more people came here than left.

Not every federal policy worked out well. Urban renewal, despite good intentions, created craters in Charlotte's geography and destroyed a thriving African American neighborhood. What replaced it was not particularly aesthetic. And federal highway policies had a downside in encouraging private over public transport and the paving of the countryside.

Still, one of the constants of Charlotte's history, especially in the twentieth century, has been the effective partnership with the federal government. Another constant has been the entrepreneurial skills possessed by scores of newcomers. When John Reed turned the straw of his life into a golden opportunity, he merely set the pattern that has persisted in the Charlotte area now for two centuries: of men and women, black, white, and brown, coming together to build a city and a region.

Metals and Manacles: A Brief Review of the Role of Slavery in the History of Gold Mining

Robert J. Moye

Robert J. (Joe) Moye, a consulting geologist in Salt Lake City, earned his B.A. and M.A. degrees at North Carolina State University. His career as a geologist in the field moved from development of base-metal and gold deposits in the Southeast to large exploration studies in the American West in search of the same metals. Before becoming an independent consultant, he spent many years with Kennecott Exploration. One of his specialties is evaluating Carlin-type gold systems. Mr. Moye has produced several geological studies about North Carolina.

Introduction

The institution of slavery has been a part of human culture from the earliest written and oral traditions to the present day, despite the current ban under international law. The history of slavery is the history of human civilization. Slavery has no racial, geographic, or historical limitations. Every race on Earth has been both slave and slaver. Societies have justified slavery on economic, moral, philosophical, legal, scientific, and even religious grounds. Few of the world's major religions condemned or even discouraged the institution of slavery until relatively recently in human history.

The principal role of the slave has always been to provide cheap labor, increasing the material wealth and leisure time of the master. The vast majority of slaves throughout history have been used in agriculture, supplying the intensive labor requirements of pre-industrial societies. Slave labor has also played an important role in the extractive industries, especially the mining industry. From Egypt to Greece, Carthage to Rome, and across the sea to the New World, the use of slaves in the mining industry has played an important, and in some cases perhaps even a pivotal, role in shaping the growth of human society and the course of history.

Most history texts ignore both the significance of mining in ancient civilizations and the importance of location and control of mineral resources in general. Moreover, studies of the ancient history of mining and smelting seldom emphasize the role of slave labor. Yet such labor was an important, sometimes critical, economic factor in the success of extractive mineral industries for at least five thousand years and continued to be employed on a societal scale through the twentieth century. This paper is not intended to document the entire history of slavery, to examine the socioeconomic origins of the institution, or to debate the ethical or moral issues of the practice. Rather, it is a brief introduction to the role of slavery in the history of the mining industry, especially the mining of precious metals.

Much of the current focus on the history of slavery centers on the African slave trade of the fifteenth through the nineteenth centuries and on slavery in North America. However, slavery did not begin with the European colonization of the

New World and did not end with the American Civil War. The origins of slavery appear to date to the development of structured societies in human culture. The earliest recorded use of slaves was for agricultural labor, beginning around 3000 B.C. in Mesopotamia. The use of slaves freed slave owners from subsistence labor, provided profit in the form of surplus goods or services, and helped make possible specialized skills and industries. Slave labor provided much of the surplus that built Mesopotamia and Egypt, Greece and Rome. The greater the output of the slaves, the more material wealth and leisure were available to those who profited from that output. The greater the dependence of a society on slavery, the more the elite were free to explore arts, the sciences, philosophy, and government. It can be argued that Western civilization, passed from Greece to Rome, held in trust by Islam, rediscovered by Western Europe, and finally brought to the United States, is strongly indebted to the institution of slavery. Conversely, the development, evolution, and spread of democracy that emerged from that heritage ultimately led to the abolition of slavery.

The role of slavery in societies has varied widely. Historically, slaves have been acquired largely through military conflicts, as a commercial enterprise to make available cheap labor, and through natural increase within existing slave populations. Slavery has also been a punishment for crime, an alternative for debtors and the destitute, and a repository for orphaned or unwanted children. Male slaves have generally been laborers in agriculture and industry, with women and children chiefly employed as domestic servants. Female slaves, and some children and men, have historically been used for sexual gratification, a practice that tragically remains widespread.

Prior to the Middle Ages, slavery often carried little or no social stigma, especially when the enslaved were of the same race and culture as the enslaver. Treatment of slaves was generally humane, with the enslaved often possessing significant legal rights and societal privileges. Greater stigma with fewer rights and privileges was sometimes attached to slaves of foreign race or culture. However, many slaves, particularly domestic servants, were generally incorporated into owning families and cultures, and most slavery, from Mesopotamia to Rome and even in more modern Islamic societies, included legal mechanisms for manumission.

After the fall of Rome, European slavery departed from those earlier traditions and in western Europe was largely superseded by a feudal tenancy system. Peasants, although not technically slaves, were bound to the land under the authority and protection of a lord, much as they had been in Egypt under the Pharaoh. The form of slavery that survived was often harsh, and slaves were commonly regarded as animals or tools, without rights or privileges.

A Roman style of slavery continued in eastern Europe during the Middle Ages and in the Byzantine Empire. The Roman model was also adopted by Muslims in the Middle East. Increasing demand led to a trans-Saharan slave trade by A.D. 650. Saharan traders were already supplying African slaves to Carthage by the fifth century B.C. The trade increased dramatically with the domestication of the camel in the fourth century A.D. The trans-Saharan and transatlantic slave trade of the fifteenth through the nineteenth centuries was particularly brutal and racist when compared to ancient slavery and increasingly hypocritical in the context of the espoused tenets of Western civilization. Manumission was limited, and penalties for defiance or attempted escape were severe. Trans-Saharan slave traders sold nearly five million Africans in North Africa and the Middle East from the tenth through the sixteenth centuries and as many as fifteen million by 1900. Even black Muslims were enslaved. Combined with the transatlantic slave trade, this removal of people depopulated large areas of Africa, weakened existing societies, and facilitated the formation of European colonial empires.

Global enslavement or genocide of indigenous peoples and the relocation of African slaves to colonial frontiers were rationalized as permissible, even desirable, under the growing ethnocentricity and avarice of Renaissance Europe. By the seventeenth century, agricultural slaves generated enormous profits for landowners. The scale and cruelty of agricultural slavery eclipsed all previous slavery and was racist at its foundation, a new paradigm in the history of the institution. Extremes of ethnocentrism had existed throughout history, but there was no precedent for the deliberate effort to subjugate and exploit African peoples and eradicate their culture and identity as human beings.

Legal slavery ended in most of the world in the late nineteenth and early twentieth centuries. It was eliminated in the British Empire by 1838 and ended in French colonies in 1848. Slavery was abolished in the United States in 1865 but continued in Asia and Africa into the twentieth century. Korea officially abolished slavery in 1894 (although it survived there unofficially until 1930); China followed suit in 1906. Abolition in Africa was piecemeal. European bans on the export of slaves did not end widespread domestic slavery, and growing European colonial influence actually encouraged slavery until 1890. Domestic slavery in Africa finally ended between 1924 and 1936 through the actions of the League of Nations.

True chattel slavery—the ownership of human beings—still exists in the Sudan and in Mauritania, and Middle Eastern slave traders continue to capture and sell women and children from Africa. Other forms of slavery—involving children, bonded labor, prostitution, and servitude of indigenous peoples—continue unabated, often with the knowledge and even participation of national governments. According to Anti-Slavery International, there are still more than two hundred million slaves in the world.

The Utility of Slave Labor in Early Mining

The earliest form of mining was the recovery of surficial concentrations of minerals or metals freed from rock by weathering and concentrated by erosion. It is no coincidence that the first metals to become the focus of human industry were those most readily concentrated in this manner, especially gold and to a lesser extent copper and, later, tin. Although not honored in name by history, a Gold Age almost certainly preceded the Chalcolithic. Placer and surficial mining ultimately led to the discovery and exploitation of lode deposits. Underground mining began with shallow excavations that followed outcrops and deepened with demand, experience, and innovation.

The beginnings and earliest evolution of lode mining of metals are lost in antiquity. Excavating in solid rock was very slow, tedious, dangerous work, using primitive stone hammers and deer-antler picks. The process of fire setting and quenching to shatter rock was introduced, and more durable metal tools were developed to increase the speed and efficiency of excavation. Nonetheless, mining in the ancient world remained a dangerous, labor intensive industry. The methods and technology of mining changed little until the introduction of gunpowder in Saxony in A.D. 1627 and the mechanization brought by the Industrial Revolution in the eighteenth and nineteenth centuries.

Mining, especially lode mining with the primitive technology that preceded the Industrial Revolution, was not a desirable career for most people, nor was it generally cost effective to pay fair wages for the work of free men. In addition, mineral deposits were often located in remote, inhospitable regions with little or no local population to provide a suitable labor force. Throughout much of human history, the labor demands of mining operations frequently have been filled by slaves. The role of slave labor in mining can be illustrated by citing various examples throughout history.

The Salt Mines of Africa and the Near East

The importance of salt to early civilizations has long been underappreciated, but it is probably no coincidence that so many centers of human activity developed in arid climates near natural sources of salt. The mining and trade of salt is one of the oldest human industries and has been inextricably linked with slavery since ancient times through the use of slave labor in mining, the purchase of slaves with salt as currency, or the purchase of salt with slaves. Demand for salt, a dietary necessity, has remained high throughout history, and salt has been largely recovered through mining of rock salt and solar evaporation of seawater. The word *salt* comes from the town of Es-Salt in Jordan near the Dead Sea, where salt production may predate organized society and the ancient settlement of Jericho (7800 B.C.).

The acquisition of salt was a major impetus to trade in the ancient world and often the motivation for military conflict and conquest. The Roman occupation of Judea during the time of Herod may have been primarily to control salt production from the Dead Sea and the caravan routes prevalent in the salt trade (Bloch, 1998). In the ancient world, salt was so precious that it was often more valuable than gold, and it is still used as a form of payment in Ethiopia. Roman soldiers were once paid in salt, from which the word *salary* and the phrase *worth his salt* are derived.

Salt mining and evaporative recovery are very labor intensive. The use of slave versus wage labor often depended on conditions at a particular site and the availability of a local labor force. Information on the use of slaves in the salt industry is sparse, except in regard to the African salt trade. Salt production often took place in extreme environments in which willing wage laborers were not available. Slaves mined salt in the Sahara and Sahel regions of Africa and at Teghaza and Taodeni in northern Mali. The salt was loaded on camel caravans and transported to great trade centers in West Africa such as Timbuktu. That process has changed little from the mid-fourteenth century to the present, with clientage replacing true chattel slavery. Elsewhere in the ancient world, slave labor was used in the salt industry at various times and places, and it continues at present in some locales in the form of forced labor by prisoners.

Egypt and Its Empire

Slavery was an institution of Egyptian culture throughout its history. Slaves were largely from Africa and the Near East and included captives taken during military campaigns, conquered peoples, criminals, debtors, and political prisoners. Slaves served in a wide range of industries and as domestic servants but enjoyed certain rights and privileges. The Pharaoh, the temples, and the elite of Egyptian society owned large numbers of slaves, and the households of merchants and tradesmen averaged one to two slaves. Although many slaves worked in the fields, agricultural labor in ancient Egypt was accomplished largely by peasants, who were serfs bound to the land.

The Egyptian civilization mined and quarried extensively throughout northern Africa and the Sinai. Some of the activity may have been done by common people as an obligation to the Pharaoh and the state, especially in areas closer to population centers. This labor force was not available during times of planting and harvesting, however, and much of the mining and quarrying took place in areas too remote, too harsh, and too dangerous for ordinary citizens. Thus, the balance, perhaps the majority, of such work in Egypt was accomplished by legions of slaves.

Stone of superior beauty was sought throughout the empire for the grand temples, monuments, tombs, and palaces of each dynasty. The royal alabaster quarry at Hatnub, near El Amarna, supplied the beautiful "onyx marble" of Egyptian carvings. Under Khufu and later Khafre, the site was worked by criminals and slaves taken in war. The quarry is one hundred feet deep and three hundred feet wide, and the tiny stone huts that housed the slaves can still be found above the workings.

Copper was mined for more than two thousand years in the harsh, forbidding desert of the Sinai—from as early as 3400 B.C. until after the reign of Rameses III. Heavily armed military expeditions escorted prospectors and probably gangs of slaves into this land of equally hostile environment and inhabitants. These campaigns took place in winter, when the heat was less severe, and recovered copper from deposits of malachite, as well as turquoise for decoration. The ancient mining camps in the arid wadis are still marked by heaps of black slag, the stone huts of the miners, and shrines to the goddess Hathor, sometimes known as the "Lady of Turquoise" (Poss, 1975).

Gold was particularly valued in ancient Egypt as the "body of the gods." One of the richest sources was in Nubia to the south, too distant and harsh for free laborers. Thousands of enslaved war captives, criminals, and political prisoners, some with their families, labored in more than one hundred mines in Nubia (later known as "Egypt's Siberia"). With slave labor, no vein was too small or too low-grade to be worked. The brutality of that mining was documented in detail by the second century B.C. geographer Agatharchides, as quoted by the first-century B.C. Greek historian

Diodorus Siculus. Men mined by the fire-setting method in stooped, narrow passages, and children carried the ore to the surface in sacks or baskets. Elderly men and women processed the ore, which was crushed by hand in stone mortars, powdered in querns, and washed across wooden tables to separate the gold. Numerous human bones found in the old mine workings, some crushed by rockfalls, testify to the brutality of the operations (Poss, 1975).

The Silver of Laurium

Slavery is recorded from Mycenae (1500 to 1200 B.C.) and persisted in Greece for centuries. The rise of large agricultural estates during the sixth century B.C. displaced family farms and created a growing demand for slaves, the primary source of labor for those estates by 350 B.C. Slavery was widespread during the Classical period (450-330 B.C.), and slaves may have composed one-third of the population of Athens. Most households had one to three domestic slaves, and it has been suggested that the leisure time available to slave owners made possible advances in the arts, philosophy, science, and politics, even the development of democracy, that were the legacy of Greek culture. The institution of slavery was rationalized by the theory of "natural slavery"—that all foreigners were barbarians and inferior. Aristotle defended natural slavery, and the concept was recognized by Plato, Aristophanes, and Hippocrates. The Greek city-states were the first true slave societies, completely dependent upon slavery.

The largest numbers of slaves in Athens in the fifth century B.C. worked in the silver mines of the Laurium district, located forty miles to the southeast of Athens (Finley, 1981). Thousands of them worked the limestone-hosted, silver-bearing lead-zinc deposits, mining, crushing, separating, and smelting the ore. Mine workings followed the narrow lodes to depths of 350 feet, and excavations were often only two to three feet high (Poss, 1975). Slaves worked underground in ten-hour shifts by the light of small oil lamps. The ore was hand sorted and crushed at the surface, then washed across stone slabs called *laveries* to separate the galena. Rainwater for the washing process was captured, stored, and reused in an elaborate system of large cisterns. The tunneling, ventilation, and lighting in the mines and the crushing, washing, and smelting of the ores were as efficient as any mining operation over the next one thousand years, and more efficient than most of them (Lauffer, 1955).

The mines of Laurium were owned by the state but leased to citizens for a 4 percent royalty (Poss, 1975). Lessors rented slaves to work the concessions, and both lessors and slave owners grew rich. Greek historian and soldier Xenophon reported that the Athenian general Nicias owned one thousand slaves who were leased to mining concessions. It has been estimated that the mines of Laurium produced two million tons of lead and silver worth about a billion dollars, a significant portion of the wealth of ancient Athens. As many as thirty thousand slaves may have been employed in the mines and processing mills at the peak of activity (Lauffer, 1955). Most mine slaves were men, many from areas of Asia Minor with a history of mining, including Thrace and Paphlagonia.

A portion of the royalty from the mines at Laurium was distributed yearly among free citizens of Athens. In the early fifth century B.C., Athens and all of Greece were twice threatened with invasion and conquest by the Persian Empire. The Persians under Darius were defeated on the plain at Marathon in 490 B.C., but the threat remained. The great Athenian statesman Themistocles convinced the citizens of Athens in 483 B.C. to donate their yearly profits from the mines of Laurium to build a naval fleet of one hundred triremes. The discovery of the richest bonanza in the district that same year added significantly to that effort.

The Persian army under Xerxes invaded Greece just three years later. Although Xerxes won the land war and sacked Athens, the outnumbered Greek fleet broke the Persian fleet of about eight hundred vessels at Salamis, and Xerxes was forced to retire, ending the Persian threat. The Greek fleet totaled 317 ships, including 200 from Athens, of which 130 were built with profits from Laurium (Poss, 1975). A Persian victory might have completely rewritten the course of history. It can be argued that the wealth from the mines of Laurium, amassed from the labor of countless thousands of slaves, preserved the culture that would become the foundation of Western civilization and modern democracy.

The Wealth of Macedonia

Philip II (382-336 B.C.), king of Macedonia and father of Alexander the Great, controlled all of the mineral resources in his kingdom, including the vital Mount Pangaeus mining region in southern Thrace. Although legend has it that the auriferous deposits of Mount Pangaeus were discovered by Cadmus the Phoenician, rich silver veins had been worked there as early as 1300 B.C. (Shean, 1997), and the inhabitants of Thasos may have already been mining gold there in the seventh or sixth centuries B.C. (Pareti, 1965). The mineral riches of the area were a prize sought by many conquerors, including Darius during the First Persian War.

In 437 B.C. Athens founded the city of Amphipolis to control the area, which fell to Sparta in 424 B.C. The deposits were being mined as Thracian and Greek concessions, probably by slave labor, at the time Philip II annexed the area in 358 to 357 B.C. (Shean, 1997). Philip raised productivity from the mines to provide an annual revenue of one thousand talents, equivalent to about six million dollars. He built the city of Philippi north of Mount Pangaeus to guard both the region and the gold deposits of nearby Chalcidice. Significantly, there was no formal basis for slavery in Macedonia, except in the mining areas of the Chalcidice Peninsula. The revenue from those mining districts helped to finance Philip's, and later Alexander's, military conquests.

The Glory that Was Rome

The institution of slavery is perhaps more strongly identified with Rome and its empire than with any other ancient society. Large-scale agricultural slavery began with Carthaginian slaves taken during the Punic Wars (264 to 146 B.C.). Those slaves were skilled in agriculture, and their removal to Italy led to the development of full-scale plantation slavery during the second century B.C., producing enormous profits for Roman landowners. Slavery spread throughout the Mediterranean and Europe during this period, following trade routes and the desire for power and material wealth. The institution appears to have been adopted from Rome by Celtic societies and was recorded in Ireland and Wales by the early medieval period.

The labor force involved in mining and quarrying throughout the Roman empire varied according to local conditions. All mineral resources were owned by the empire and were either mined directly by imperial authorities or contracted as concessions. Both the authorities and the contractors used either slaves or free workers or a combination of the two. Conditions in slave mines varied remarkably. In some, slaves were treated relatively well, optimizing per capita productivity, but in others slaves died by the thousands.

The Dolaucothi Gold Mine in Wales was a Roman slave mine for almost eighty years—from around A.D. 75 to 150. Surviving Roman workings there include opencast trenches and pits, the largest measuring 100 by 150 meters and up to 12.5 meters deep. Hand-driven, square-cut adits extend up to 60 meters, and drifts and stopes were worked to a depth of 45 meters (Burnham, 1986). A series of surface tanks and leats (channels) was constructed to store and direct water for mining and gold recovery. The largest tank, situated on a hill, held about 250,000 gallons and was connected to a river eleven kilometers away. Water was released in controlled floods to strip overburden and as small streams for the recovery process. Ore was crushed by hammers and washed across sheep skins. The fleece was then dried and burned to recover adhering gold particles.

Roman slaves mined some five hundred thousand tons of rock at Dolaucothi and recovered perhaps twenty-four thousand ounces of gold. Sixty thousand slaves may have died in those operations, a production of about 0.4 ounce of gold per human life. Rebellious slaves were crucified, and a local summit has long been known as "Crucifixion Hill." The identity of the slaves is uncertain, but a Welsh tribe, the Silures, fiercely resisted the Roman advance. Enslaved Silures may have been miners, although most probably came from mainland Europe, especially from Gaul.

The most richly mineralized region in the ancient world, and the treasure house of the Roman Empire, was Spain. Known to early Greek miners as *Iberia* and to the Phoenicians as *Spania*, Spain became the primary source of mineral wealth for Carthage in the third century B.C. Rome gained control of Spain in the Second Punic War and extensively developed its mineral resources. Spanish silver was the basis for Roman coinage, and the mines provided enormous wealth in gold, copper, lead, and tin. Slaves made up much of the work force. About forty thousand of them are said to have been employed in the silver mines near Cartagena alone (Cunliffe, 1994), and the Romans operated hundreds of mines throughout the Iberian Peninsula. Slaves directed by Roman mining engineers worked shafts to a depth of six hundred feet, drove adits up to two thousand feet long, and excavated thousands of feet of drifts by fire quenching and hammer and chisel (Poss, 1975).

Julius Caesar, at the age of thirty, served as quaestor, custodian of the treasury, and assistant to the provincial governor in Farther Spain in 69 B.C. The vast wealth in slave-mined metals that flowed through his hands in one year enabled him to amass a fortune, ensure his popularity in Rome, and launch a career that shaped subsequent Roman history.

Europe after Rome

Lode mining diminished in Europe after the fall of Rome but again became important as Europe emerged from the Dark Ages with the rise of the Carolingian dynasty. Charlemagne unified a kingdom that included present-day France and Germany. He expanded his empire east of the Rhine and captured the world-class Erzgebirge mining district in present-day Germany and the Czech Republic. Enslaved Saxons began working the rich silver- and gold-bearing lead deposits in A.D. 770 (Poss, 1975), and the concessions were granted to Charlemagne's sons in 786. Production from those mines enabled Charlemagne to issue new silver currency, which replaced gold as the principal coinage of western Europe. Charlemagne was crowned Carolus Augustus, Emperor of the Romans, by Pope Leo III in A.D. 800. The mining expertise developed in that area would ultimately provide the technical foundation for the development of the modern mining industry, as documented in *De Re Metallica*, by Georgius Agricola, in 1556.

Gold and the Kingdoms of Africa

Slavery as an institution in Africa is as ancient as African societies. Communal ownership of tribal land did not permit the rise of plantation agriculture, but the importance of farming in Africa made its peoples highly skilled in agricultural work. Exploitation and expansion of African slavery by the Portuguese and Spanish in the fifteenth and sixteenth centuries developed in conjunction with the rise of the modern nation-state in Europe and the emergence of colonial empires.

The great West African kingdom of Mali rose to power following the collapse of the kingdom of Ghana in the thirteenth century A.D. The wealth and prosperity of Mali depended largely upon control and expansion of the trans-Sahara slave trade and the mining and trading of gold, salt, diamonds, and copper. Those mining operations almost certainly made extensive use of slave labor. Mali reached the zenith of its development in the fourteenth century under the rule of Mansa Musa and was the richest empire in Africa. So great was the production of gold that Musa is famous for his pilgrimage to Mecca with a train of one hundred camels, each carrying three hundred pounds of gold. His generous donation disrupted the gold standard in Cairo for more than twelve years (Ilahi, 1997).

Gold mining on the Gold Coast of West Africa, the coastal region of present-day Ghana, depended on slave labor, especially with increased demand after the arrival of the Portuguese in 1471. Slavery and the slave trade were institutions in the area before the arrival of Europeans, but the Portuguese expanded that trade throughout the region and to the New World. The increased demand for gold and a shortage of labor to work the mines led to a new pattern of slave trade. The Portuguese imported slaves to the Gold Coast from other areas of Africa and sold them for gold. The slaves were sent to the mines to produce more gold, which was sold to obtain more slaves. The Portuguese monopoly in the Gold Coast ended in the seventeenth century with the arrival of other European companies, which built numerous castles and forts to protect their interests. The rapidly expanding sugar industry of the New World drew slave labor away from Gold Coast mining.

Slavery and Mining in the New World

Before the arrival of Europeans, indigenous cultures of the Americas practiced slavery, usually on a limited basis. Some North American native cultures enslaved captives taken in battle, but the captives were for the most part humanely treated, and captivity was generally of limited duration. Aztec society in Mexico included a large class of commoners known as *mayeque*, essentially serfs bound to the land. Slaves, or *tlacotin*, were often debtors or criminals, but the institution also was an economic refuge for poorer commoners during periods of drought or famine. Aztec slaves retained many social rights and privileges and were allowed to purchase their freedom. Captives in warfare were likewise slaves but were usually sacrificed shortly after enslavement. Although no serf class existed, the Maya owned many slaves, called *p'entacob*. Most were taken in battle and were owned by the soldier who captured them. Slave status was hereditary, and there was a large established regional slave trade at the time of European contact. Spanish conquerors in Central and South America often adopted and continued the existing chattel slavery and serfdom to exploit the resources of those regions.

The primary goal of global colonial expansion from the fourteenth through the nineteenth centuries was the extraction of wealth for European governments, companies, and entrepreneurs. That extraction was achieved through seizure of property, especially precious metals, from indigenous people and longer-term exploitation of natural resources. Seizure of property was usually accomplished through military conquest, intimidation, and genocide. Protracted exploitation of natural resources included development of large agricultural plantations and mining, especially for precious metals; both depended largely upon slave labor. That pattern of extractive exploitation was especially characteristic of colonial expansion in the Americas. The remarkable profitability of large agricultural plantations in the New World created much of the rapidly growing market for agricultural slaves and established the transatlantic slave trade. Although the Portuguese and Spanish are credited with initiating that trade, other European powers quickly became involved in the profitable commerce. The British gained control of the trade by 1800, transporting 90 percent of all African slaves through Liverpool, future birthplace of much of the Industrial Revolution in Britain, thanks largely to proceeds from the trade.

It is generally accepted that European slave traders exported some 11.5 million African slaves to the Americas between 1500 and 1870. About 10 million reached the New World alive. One in ten died under the extremely inhumane conditions, which included forced overland marches, crowded underground prisons, and sea transport in the cramped, filthy, sweltering holds of slave ships. Perversely, mortality was even higher among Europeans involved in the trade. It is estimated that 90 percent of seamen sent to work on English slave ships did not return.

About 40 percent of African slaves imported to the Americas went to Brazil; another 40 percent were dispatched to British, French, and Dutch colonies in the Caribbean; and less than 10 percent went to North America. Although fewer than five hundred thousand African slaves were brought to the United States, population growth raised that number to four million by 1860. Slavery in North America was concentrated in the agricultural plantations of the South, but the slave trade was largely controlled through ports in Massachusetts, Rhode Island, and Connecticut and built the fortunes of prominent northern families. The expansion of agricultural slavery in the South was originally based on sugar, rice, and tobacco. After the Industrial Revolution, that slavery was largely driven by demand for cotton for textile mills in the northern states and in England, resulting in the spread of plantation slavery into Mississippi, Louisiana, and Texas. Southern plantations produced 75 percent of the world's cotton by the 1860s.

Colonial mining in the New World usually employed either indigenous or imported slave labor. Spanish conquerors in the Americas promptly enslaved indigenous populations through a continuation and expansion of existing servitude and the introduction of even harsher exploitation, including the *encomienda* system of forced Indian labor, as well as outright chattel slavery. The Spanish came to the New World with extensive knowledge of lode-mining, which had been an important industry in Spain since the time of Carthage and Rome and was revived after Charles V took the throne in 1519. The placer deposits worked by the pre-conquest populations were exploited, then traced to their source lodes. Once the Spanish located the lode deposits, the native population was put to work in underground mines. Unfortunately, European diseases, excessive work, and cruel treatment, in addition to migration as a means of escaping slavery, quickly diminished local native populations. More than eight million people died in the Valley of Mexico within a few years of Cortez's arrival—over 30 percent of the population.

The discovery of the Espíritu Santo silver lode northwest of Mexico City in 1543 set off a silver rush that lasted for ten years. Other discoveries during that period included the lodes at Zacatecas, Guanajuato, Real del Monte, and San Luís Potosí. Spanish prospectors claimed, or "denounced," a discovery to the local authorities and agreed to pay one-fifth of the gross profit (the *quinto*) to the king. A work force was recruited by force or intimidation from the local inhabitants, often with military assistance. Production of approximately twenty-two million pesos of silver was reported from 1543 to 1553, more than all of the silver seized during the conquest.

The Spanish sometimes valued experienced Indian miners more than other laborers, and those miners actually enjoyed more liberty and some compensation. About five thousand native miners were working in the silver mines of the Zacatecas District of Mexico in the mid-1620s (Bakewell, 1971). They were paid wages, received a share of the mined ore, and were allowed to relocate to seek better compensation. In 1550 all mine slaves in Mexico were emancipated by order of

Viceroy Luís de Velasco, and their lot progressively improved to the point that in the early nineteenth century they were among the most esteemed and best-rewarded miners in the world.

The Spanish in Peru adopted the Incan *mita* system of compulsory rotational labor, under which all able-bodied Indian males were required to serve for a period of six months every seventh year in either the silver mines of Potosí (now part of Bolivia) or the mercury mines at Huancavelica. Conditions were so severe that 80 percent of the laborers died in the first year of service (Breuer, 1997). Each year, approximately 1,675 Indians worked under very harsh conditions at Huancavelica as rotational draft laborers, producing mercury that was used in the amalgamation-recovery process in silver-mining districts such as Zacatecas (Bakewell, 1971). Many miners died of mercury poisoning, but Spanish mine owners accumulated tremendous wealth and sent great portions of it to the king in Spain. Some villages near the mines avoided *mita* obligations by paying an agricultural tribute called *tasa* to feed the miners, and many Indians, hoping to secure less-dangerous jobs, left the villages to become full-time wage laborers at the mines. The loss of population through migration and death did not lessen a village's or region's *mita* obligations, and the people who remained were forced to work ever harder to supply the required service.

Like Portuguese and Spanish plantation owners in Central and South America, mine owners often found the use of native peoples as slaves to be a disaster. Native people were unused to a captive existence and the heavy work demanded by their captors. Abusive treatment and European diseases decimated native populations. Ironically, in 1513 Bartolomé de Las Casas, a Spanish priest sympathetic to the plight of the Indians, successfully pleaded with King Charles V of Spain to end the enslavement of native Americans and replace them with African slaves. The Africans gradually supplanted native Americans on agricultural plantations and in many of the gold, silver, and diamond mines of Central and South America.

Ten thousand Africans a year were being shipped to the New World by 1540, and more than nine hundred thousand had been sent there by the end of the century. Portuguese colonists had little knowledge of gold mining when they arrived in Brazil to establish agricultural plantations, and it has been suggested that the discovery and mining of gold in the sixteenth century was based on the knowledge and experience of African slaves. Alluvial diamonds were discovered during placer mining for gold in 1725, and Brazil was the world's major supplier of diamonds from 1730 to 1870, almost entirely through slave labor. Some slaves used in the mines smuggled out enough gold and diamonds to purchase their freedom. Alluvial diamond mining continued with free miners after Brazil became the last Western nation to abolish slavery in 1888.

The English Colonies and Colonial America

The national myth of British colonization of eastern North America to escape religious intolerance is only partially true. The goal of many early colonial efforts was the short-term extraction of wealth—from indigenous peoples and from the

Earth. Expectations of success were based on the results achieved by the Spanish in Central and South America and false reports of gold and silver in eastern North America. Precious metals were not in abundance along the Atlantic seaboard, and initial prospecting was disappointing. English colonists had little or no knowledge of prospecting and mining. Early leaders often complained that colonists neglected crops and providing for basic subsistence to pursue prospecting, and myths of undiscovered civilizations rich in precious metals "to the west" persisted.

Slavery was an accepted institution among the British of North America. Native Americans were frequently captured and sold as slaves, and the first African slaves arrived in 1619. Nevertheless, slave labor in the mining industry was slow to develop because the industry itself was slow to develop. The first such use of slaves did not occur until the mid-eighteenth century. The Principio Company in Maryland was the first American ironworks to use slaves, with 150 in its work force at the height of its operations (Johnson, 1959). The practice spread with the development of the colonial iron industry throughout the eighteenth century.

Before the American Revolution, development of domestic mining was discouraged by cheap imports from abroad. With the onset of hostilities, lead for bullets became a strategic necessity, and there was a major effort throughout the colonies to locate lead deposits and get them into production. Although there was prospecting and minor production from New York, Connecticut, Pennsylvania, and North and South Carolina, most of the lead used by the colonial army came from the Kanawha River area in present-day Montgomery County, Virginia (Mulholland, 1981). Lead mining at the site began in 1759 on a thousand-acre tract purchased by a company of prominent men, including the governor of Virginia. Operations languished between 1768 and 1775 with a small force of slaves.

The workings were expanded in 1775 to meet the growing demand of the Continental army, and the state of Virginia took over operations in 1776 (Henning, 1821). Thirty men produced up to sixty tons of lead per year with a stamping mill and smelter located across the river from the mines. The miners appear to have been mostly free laborers, exempt from military service. The strategic importance of the mine increased as the war shifted southward in 1780: the state militia provided security for it in the wake of British-incited attacks by hostile Cherokee Indians and a tory uprising that threatened production.

The main vein at the mine appeared to be giving out in the spring of 1781, creating a crisis in the lead supply. Gov. Thomas Jefferson urged the use of additional laborers, including slaves, to expand the workings and increase production (Ford, 1892-1899). The crisis was averted, and just under thirty tons of lead were produced in 1781 (Lynch, 1781). Lead production from the mines of Virginia, in part through the use of slave labor, helped maintain the revolutionary effort culminating in the surrender of Cornwallis at Yorktown on October 10, 1781.

Gold in the Carolinas

The elusive, all-but-forgotten dream of gold in the American colonies—Benjamin Franklin had declared in 1790 that "Gold and silver are not the produce of North America, which has no mines" (Sparks, 1840)—became a reality in 1799. In that year, twelve-year-old Conrad Reed discovered a seventeen-pound gold nugget in Cabarrus County, North Carolina. His father, farmer John Reed, was a former German conscript who had deserted from the British army during the Revolution. In 1803 the elder Reed and three partners began to work placer deposits on his farm with the help of slaves, and the following year a bondsman named Peter found a twenty-eight-pound nugget. Similar placer operations developed on farms throughout the Piedmont as word of Reed's success spread. Those finds precipitated the first gold rush in North America, and numerous discoveries were made throughout Piedmont North and South Carolina (as well as Virginia, Georgia, and Alabama) over the next fifty years. Mining was initially conducted by placering, with lode mining introduced in 1825.

Many mining operations employed slaves, some leased from their owners for as little as a half-dollar per shift (McCarthy, 1918). With sufficiently cheap labor, even low-grade placer and lode deposits could be worked for a profit. The discovery of gold in Burke County, North Carolina, in 1828 precipitated a rush to the Blue Ridge Mountains, and an estimated five thousand slaves were working in Burke County's gold mines by 1833 (Steiner, 1918). Robinson (1955) estimates that at times gold mining in North Carolina alone may have employed as many as thirty thousand men (though not all of them were slaves).

Slavery in the California Gold Fields

Slave labor did not characterize the California gold rush, but slaves were present in the diggings. Although the Native American population of the area had been largely exterminated through disease, starvation, and genocide, its surviving members remained subject to legal indentured servitude on cattle ranches and in mines, as well as to outright chattel slavery. The native population of California declined from about 150,000 to 30,000 between 1845 and 1880 (Franzius, 2000).

African American slaves were likewise present in the California mines. The entry of California into the Union as a free state in 1850 did not diminish that presence. There were an estimated two thousand African Americans in California in 1852. Many were free blacks or escaped slaves, but some came as slaves with their masters from the South. In the California gold fields, the slaves worked in defiance of the law. Wage laborers, along with free miners, strongly opposed slave labor for reasons of economic self-interest.

Robert Dickson of Burke County, North Carolina, joined the rush to California in 1852, accompanied by at least four slaves and his cousin, S. M. McDowell, who also brought slaves (Rohrbough, 1997). In the autumn of 1852 they put their slaves to work, despite the fact that California was a free state. When the profitability of Dickson's claims proved less than hoped, he hired out his slaves for $75 each per month, or all four for $300 per month. McDowell's slaves made a "raise" and recovered gold worth fifteen hundred to eighteen hundred dollars in a three-week period. McDowell returned to North Carolina, leaving his slaves behind to continue working in the gold fields. The slaves, apparently allowed to keep a portion of the profit, sent most of the money earned to their master in North Carolina.

Gold and the Gulag: Slavery in Siberian Gold Mines

The most extreme twentieth-century slavery is associated with the resurgence of particularly brutal forms of chattel slavery immediately preceding, during, and following World War II. Prior to and during the war, the Japanese sexually enslaved between seventy thousand and two hundred thousand women, known as "Comfort Women"—about 80 percent of them Korean—and used indigenous peoples and prisoners of war as slave laborers. Most heinous was the Holocaust, the enslavement and genocide of more than six million Jews and other minorities in Europe by Hitler and the Nazis between 1938 and 1945. In addition, the populations of conquered nations and Russian prisoners of war were extensively used as slave labor.

Equally horrific was the genocide and wholesale enslavement of Soviet citizens that marked the Stalin regime in the USSR. The former Soviet Union has long been a source of gold, from the land of Colchis in modern Georgian S.S.R. to the nomadic Scythian tribes north of the Black Sea. Gold mining in Russia became the domain of the czar in the eighteenth century. Placer and lode mining began in the Ural Mountains by 1774, and placer deposits in the Altai region of Siberia were in production by 1820. Mining began along the Lena River by 1829, with more than four hundred mines operating by 1900. Gold was discovered in western Siberia by 1840 and along the famous Kolyma River District, the Amur River, and in the Far Eastern maritime region about 1870. Gold mined in eastern Siberia totaled nearly one-fourth of annual world production by 1900 (Nagaev, 1959).

Prior to the Russian Revolution of 1917, mines were often operated by capitalists under government license. Working conditions of wage miners were often harsh in Siberia, and the czarist government operated gold mines with slaves. Although worker unrest led to protective regulations in the 1860s, the mining companies often ignored the laws (Melancon, 1994). In February 1912 three thousand gold miners on the Lena River went on strike to protest harsh conditions and high food prices. On April 4 czarist forces shot five hundred strikers, killing more than half of them (Koval'chuk, 1987). Following the massacre, worker unrest increased dramatically throughout Russia, and Lenin referred to the incident as the beginning of the Russian Revolution.

After the revolution, Stalin established the Central Gold Monopoly. Former mine producers and developers, viewed as capitalists, were liquidated, and many were bound over as slaves to the state. An enormous bureaucracy was established to restart and expand gold production. Political prisoners were utilized as slave labor in many state industries, including Siberian gold fields. Initially, miners received better treatment to optimize production, but that benign policy ended in 1937.

Slave labor accounted for up to 25 percent of the Soviet economy during the 1930s. It was managed by the Gulag ("Chief Administration of Camps"), a department of the Soviet secret police. The population of the Gulag peaked at eight to fifteen million prisoners in the late 1930s, about 5 to 10 percent of the entire Soviet population and 15 percent of all adult males (Kort, 1993). More than five hundred thousand people mined gold along the Kolyma River. The mortality rate in the Gulag was 10 to 20 percent each year. Three million people died in the Kolyma camps between 1937 and 1953. The Kolyma District was known as the "land of white death" (Kowalski, 1999). Miners who did not meet quotas were executed. Those who somehow met the quotas were starved until they no longer could do so and then executed. Life expectancy in the Eighth Apparatus Gold Mine was one month.

Most of the people condemned to the Gulag did not survive, and many never even reached the camps. Slaves for the mines often made a difficult rail journey to Vladivostok, where they were packed onto slave ships for a sea voyage lasting up to eleven days. Crowding, poor sanitation, scant food, and bitter cold resulted in numerous fatalities. An infamous slave ship named the *Dhzurma* became stuck in Arctic ice with twelve thousand slaves aboard bound for the Kolyma mines (Kort, 1993; Kowalski, 1999). The Soviet government refused outside offers of aid. When the *Dhzurma* arrived in port, all of the prisoners were dead of exposure and starvation.

Conclusion

The discovery by the bondsman Peter of a twenty-eight-pound gold nugget at the Reed mine in North Carolina in 1804 was part of a tradition of slave labor in mining that extends back perhaps five thousand years and ended less than fifty years ago. Mining, one of the earliest human industries, antedated large-scale agriculture. Underground mining for exploiting flint deposits developed during the Neolithic period, but mining remained a largely localized, occasional, small-scale enterprise by free laborers.

The rise of large, structured societies with centralized authority by 3000 B.C. increased the scale of agriculture and the consumption of natural resources, as well as the individual and societal accumulation of material wealth. That trend led to a division of labor and the development of specialized skills and industries. Technological innovation accelerated, further increasing demand and contributing to the development of large-scale trade and conquest to acquire and control mineral and other natural resources. The loss of subsistence labor to specialization and administration was offset by expansion of the institution of slavery, already present in limited capacity.

Rapidly growing demand for metallic resources encouraged large-scale placer and lode mining operations, which required a considerable labor force. Ancient mining was difficult, dangerous work that relied upon primitive techniques and often occurred in remote areas under harsh conditions. Through centralized authority and military force, the labor needs of the mining industry were often filled by slaves, who mined much of the gold and silver, copper and tin, and later iron that built the foundations of the great ancient civilizations, filled the burial chambers of Egypt and the coffers of Athens, and adorned the temples of Rome.

The European age of discovery, the growth of empires, the Renaissance, the Industrial Revolution, and the ultimate ascendancy of democracy were made possible to a significant degree by the destruction of indigenous societies and civilizations in Africa and the Americas and the enslavement and exploitation of native peoples through plantation agriculture and mining. Ultimately, the development of the new industrial technology eliminated the value of slave labor to the mining industry. Technically skilled wage laborers became more productive, and more profitable, than legions of slave laborers. The rise of modern forms of democracy, growing prosperity, and the accompanying evolution of social consciousness resulted in the rejection of the institution of slavery by most modern nations. It can indeed be argued that Western civilization, passed from Greece to Rome, to western Europe via the Middle East, and finally to America, is strongly indebted to the institution of slavery—a debt seldom acknowledged and impossible to repay.

Mineral resources and the great tradition of mining remain a basic, integral foundation of modern civilization. Slavery, tragically, also remains an aspect of modern human society. Although large-scale societal use of slave labor in mining is largely a thing of the past, it is the recent past. The potential for renewed utilization of slave labor in mining remains, and the practice may continue locally in isolated areas of the world. It will continue to exist so long as slavery exists—and so long as there exist societies that embrace elitism, nationalism, and racism and are willing to disenfranchise and degrade any race, creed, or color of fellow human being for the benefit of another.

References

Bakewell, Peter J. *Silver Mining and Society in Colonial Mexico: Zacatecas, 1546–1700*. Cambridge: Cambridge University Press, 1971.

Bloch, David. "Salt Made the World Go Round" (1998). *www.salt.org.il*.

Bruer, Kimberly H. "Mita." In *The Historical Encyclopedia of World Slavery*. Santa Barbara, Calif.: ABC-CLIO, 1997.

Burnham, Barry C. "History of Mining: Pre-19th-Century Mining Activity." In Alwyn E. Annels and Barry C. Burnham, eds., *The Dolaucothi Gold Mines*. Cardiff, Wales: University College, 1986.

Cunliffe, Barry, ed. *Oxford Illustrated Prehistory of Europe*. New York: Oxford University Press, 1994.

Finley, Moses I. *Economy and Society in Ancient Greece*. New York: Viking Press, 1981.

Ford, Paul L., ed. *The Writings of Thomas Jefferson*, 10 vols. New York: G. P. Putnam's Sons, 1892–1899, 5:191–192, 199, 263, 367.

Franklin, Benjamin. *The Works of Benjamin Franklin*, 10 vols. Ed. Jared Sparks. Philadelphia: Childs and Peterson, 1840, 2:347.

Franzius, Andrea. "California Gold—The Diary and Letters of James M. Burr, 1850–1855" (2000). *www.duke.edu/~agf2/history391/index.html.*

Glass, Brent D. "King Midas and Old Rip: The Gold Hill Mining District of North Carolina." Ph.D. diss., University of North Carolina at Chapel Hill, 1980.

Henning, Walter W., ed. *The Statutes at Large, Being a Collection of All the Laws of Virginia, 1619–1792*, 13 vols. Richmond, Va.: J. and G. Cochran, 1821, 9:237–238.

Iglesias, Fidel. "New Spain." In *The Historical Encyclopedia of World Slavery*. Santa Barbara, Calif.: ABC-CLIO, 1997.

Ilahi, Au'Ra Muhammad Abdullah. "Mali." In *The Historical Encyclopedia of World Slavery*. Santa Barbara, Calif.: ABC-CLIO, 1997.

Johnson, Keach. "The Baltimore Company Seeks English Markets: A Study of the Anglo-American Iron Trade, 1731–1755." *William and Mary Quarterly* [third series] 16 (January 1959): 37–47.

Kort, Michael. *The Soviet Colossus: Rise and Fall of the USSR*. 3d ed. New York: M. E. Sharpe, 1993.

Koval'chuk, O. O. "75-Richchia lens'koho rozstrilu [75th Anniversary of the Lena Shootings]." *Ukrains'kyi Istorychnyi Zhurnal* [USSR] 4 (1997): 135–138.

Kowalski, Stanislaw J. "Kolyma: The Land of Gold and Death" (1999). *www.personal.psu.edu/users/w/x/wxk116/sjk/kolyma.html.*

Lauffer, S. "Die Bergwerksslaven von Laureion." In Akad[emiscer Auslauschdienst] Mainz, *Der Wiss[enschaft]und der Literatur, geistes-undsozialwissenschaaftliche Klasse*, abhandlungen 11 and 12 (1955–1956).

Lynch, Charles. "Accounts of Col. Charles Lynch, Superintendent of the Mines, Annual Reports, 1779–1782" (folder). Auditor of Public Accounts [APA 661], Archival Collections, Library of Virginia, Richmond.

McCarthy, Edward T. *Incidents in the Life of a Mining Engineer*. New York: E. P. Dutton, 1918. P. 11.

Melancon, Michael. "The Ninth Circle: The Lena Goldfield Workers and the Massacre of 4 April 1912." *Slavic Review* 53 (fall 1994): 766–795.

Mulholland, James A. *A History of Metals in Colonial America*. University, Alabama: University of Alabama Press, 1981.

Nagaev, A. S., ed. "K istorii dvizheniia rabochikh zolotykh priiskov vostochnoi sibirii v pervoi polovine XIX V" [On the History of the Labor Movement in the Gold Mines of Eastern Siberia in the First Half of the Nineteenth Century]. *Istoricheskii Arkhiv*, 1959, 5. Pp. 216–223.

Pareti, Luigi. *The Ancient World: 1200 B.C. to A.D. 500*. New York: Harper and Row, 1965.

Poss, John R. *Stones of Destiny: Keystones of Civilization*. Houghton, Mich.: Michigan Technological University, 1975.

Rickard, Thomas A. *A History of American Mining*. New York: McGraw-Hill Book Company, 1932.

Robinson, Blackwell P., ed. *The North Carolina Guide*. Chapel Hill: University of North Carolina Press, 1955.

Rohrbough, Malcolm J. *Days of Gold: The California Gold Rush and the American Nation*. Berkeley, Calif.: University of California Press, 1997.

Shean, John F. "Philip II of Macedonia (382–336 B.C.). In *The Historical Encyclopedia of World Slavery*. Santa Barbara, Calif.: ABC-CLIO, 1997.

Steiner, Bernard C., ed. "The South Atlantic States in 1833, as Seen by a New Englander (Henry Barnard)." *Maryland Historical Magazine* 13 (September, December 1918): 346–348.

African Americans and Gold Mining in North Carolina

Jeffrey P. Forret

Jeffrey P. Forret holds degrees in history from St. Ambrose University in Iowa (B.A.) and the University of North Carolina at Charlotte (M.A.) and is studying for his doctorate at the University of Delaware. He has been a librarian, a teacher, and a researcher, as well as the assistant to the editor of the Journal of Urban History. *His article "Slave Labor in North Carolina's Antebellum Gold Mines" appeared in the April 1999 issue of the* North Carolina Historical Review.

From the meager beginnings of gold mining in North Carolina, African American slaves helped Piedmont farmers carry on haphazard placer mining operations. Along with their masters, they inspected creek beds, panned in streams, and dug at random into the red North Carolina clay. When John Reed, whose son made the first authenticated discovery of gold in the state, formed a small partnership in 1803, each of his three associates contributed two slaves to hunt for gold. One of those slaves, named Peter, discovered a twenty-eight-pound nugget that year.[1] Conducting a scientific survey of the Carolina gold region in 1805, the architect Dr. William Thornton of Washington, D.C., posited that "Blacks who have been accustomed to the hoe will consider the search more as amusement than work."[2] Even when underground gold-mining operations began in the 1820s, slaves remained a vital part of the mining work force.

But what did gold mining mean for slaves and their masters in North Carolina? Employing slaves in nonagricultural activities posed a distinct set of challenges to southern society. The unusual setting of the gold mines offers a lens through which to view some of the largest issues in southern historiography, including antebellum race relations, slave resistance, rebellion, and slave law. As the case of gold mining illustrates, the hiring of bondsmen for industry eroded the institution of slavery. The slave miners' own efforts to alleviate their condition and assert their independence, through stealing and running away, further weakened the master-slave relationship. North Carolina's gold mines were also the scene of the ultimate sign of slave disaffection—rebellion. The alleged slave miners' conspiracy of 1831 highlighted the dangers inherent in industrial slavery but, paradoxically, left the white population more assured of its place in society than before the incident. Despite the problems arising from hired slaves, thefts, fugitives, and purported rebellions, the employment of slaves in the mines, both before and after industrialization, never seriously threatened the social structure of antebellum North Carolina.

Historians have shown that slaves worked in industries throughout the antebellum South—in coal and iron mines, textile mills, tobacco factories, and even chemical works. Their studies demonstrate that slavery could be used profitably in industrial pursuits, both in skilled and unskilled positions.[3] In the latter decades of the antebellum era, 5 percent of all slaves in the South worked in industries such as coal, iron, textiles, and mining.[4] The limited evidence available suggests that gold-mining companies seem to have preferred renting slaves from local farmers to buying them outright. At Gold Hill, North Carolina's most significant gold-mining endeavor, slaves composed somewhere between 25 and 50 percent of the labor force, depending upon the particular gold mining company, and worked almost strictly as unskilled laborers, hauling or washing the ore.[5]

Early placer mining operations helped keep slave ownership profitable. Masters sought to avoid slave idleness and underutilization by keeping them constantly employed in various pursuits. Typically, North Carolina farmers conducted amateur prospecting on a seasonal basis so that a complementary rhythm of work developed. They used their slaves in the fields during the spring and summer, then, after the harvest, "When the crops were laid by, the slaves and farm hands were turned into the creek-bottoms, thus utilizing their time during the dull seasons."[6] The slaves, then, remained busy all year long, supplementing the farmers' income by hunting gold. Expert mining engineer J. Humphrey Bissell reported that gold mining provided "profitable employment" to "that large proportion of the slaves . . . who have been for years past a burdensome expense, instead of a source of profit, to their owners."[7] Some slave owners even pondered shifting their bondsmen entirely to mining. Describing how a recent hailstorm destroyed the remnants of his drought-stricken crops, one North Carolina farmer expressed his belief that "I shall . . . do better by diging [*sic*] of gold. . . . [I]f I can do better to keep my negroes and work gold minds [*sic*] I am willing to do so."[8] On a single Saturday in 1830, one fortunate miner made $40 through the efforts of twelve hands at one mine, $140 with forty at another. About that same time, over a two-week stretch, a miner named Charles Hill employed four hands and earned proceeds totaling $163.20.[9]

The advent of underground mining and the rise of mining corporations in the 1820s and 1830s caused no interruption of the harmonious relationship between mining and slave-based agriculture. Incorporation greatly expanded the scope of operations and increased the demand for slaves, so gold mining companies often tried purchasing or renting them from local farmers. English mining agent Capt. John E. Penman placed an advertisement for mining hands, stating that "I WISH to hire from 15 to 25 NEGROES, to be employed in the Gold Mines near Charlotte. The highest prices will be given for good hands; and those having some experience in the business will be preferred. Gentlemen having slaves whom they wish to hire advantageously, will please call on me."[10] The partnership of Bissell and Barker employed fifty-nine slaves in 1830,[11] while Charlotte's Capps mine used "38 negroes, 10 of whom are women."[12]

Hiring slaves for the incipient gold-mining industry proved difficult, however. Competition for industrial slaves was stiff in the antebellum years. The demand from the burgeoning railroads and the coal, iron, and textile industries exceeded the available supply. Some slave owners refused to hire out their slaves, fearing mistreatment during the rental period or the potential death of valuable property.

According to one historian, "Gold mine operators in particular, eager to make quick profits, sometimes made slave hirelings dig directly into hillsides without propping up the roof. The earth often gave way and crushed the workers, whose full value the employer was compelled to pay to the owner of the slaves."[13] A cave-in at the Christian mine in Montgomery County killed a slave belonging to Thomas Lilly and seriously wounded another.[14] Explosions and falls posed other threats to the slaves' safety. At one gold mine, an unexpected blast injured two bondsmen "engaged in blowing up rock, . . . putting out the eyes of the negroes."[15] Cognizant of the dangers in the mines, some North Carolina slave owners even stipulated in contracts with the railroads that their slave hirelings "shall not work . . . in the goldmines."[16]

While some farmers exhibited a reluctance to lease their slaves to mining companies, the fact that they maintained the option to shift slaves between mining and farming demonstrates that slaves represented a highly flexible form of capital. When cotton prices were high, farmers worked their slaves in the fields. But when overproduction of cotton drove prices down, the mines absorbed surplus labor from the farms. "I raised last year 27 bags of cotton & got for it only something over $400," complained Lincoln County farmer Robert Hall Morrison. He pondered "planting only about half as much cotton" in the coming season. Instead, he decided, "I will keep some of [the] hands working a pretty fair gold vein on my land." Ever distrustful of leasing to the mines, Morrison admitted that he "would not think of mining if it were not on my land."[17] But whether on a farmer's own land or not, the mines offered an economic safety valve for the slaveholders. Whenever they wished, or whenever economic necessity compelled them, they could redirect their slaves from agriculture to the mines.

Sound economic reasoning and a little reassurance surely convinced some masters to rent slaves to the mines. "I wish to make some arrangements to hire hands for another year," wrote miner James Scott to his brother Dr. John Scott of Salisbury. He enlisted his brother's aid in securing "8 or ten" slaves by "private contract," including "one or two half grown Boys & one or two stout young women at 30 or 40 dollars a piece according to quality." Scott suggested to his slaveholding brother that "perhaps it would be to your advantage to hire me some of yours. I will give you seventy dollars a piece for good hands." A persuasive pitchman, Scott also attempted to alleviate any fears of renting slaves for mining. "There is general prejudice against letting hands work in the mines," he conceded, "but it is without foundation, my hands are very healthy and well satisfied."[18] Although little evidence is available, the Scott brothers seem to illustrate the larger pattern that saw the vast majority of slaves hired to the mines come from within the mining region itself, where neighborhood and kinship networks fostered the bonds of trust necessary for a master to rent out valuable human property.[19]

Hiring slaves appeared to pose the perfect solution to the gold-mining industry's labor crisis, but the practice itself eroded the institution of slavery when used in industrial settings.[20] Some mine owners, such as Capt. John Penman, gave their slave employees a set wage. Penman promised in an advertisement for "100 COLORED or

WHITE MINERS" that he would pay "liberal and punctual wages" for their services.[21] Other owners granted slaves a percentage of the gold they mined. Hired slaves often gave a mutually agreed upon amount of gold to the master at regular intervals, retaining for themselves any gold beyond that amount. Explained one observer at the Reed mine, "hands hired by the mining parties were paid with pieces of gold pounded up in a common hand mortar."[22] In addition to receiving payments in gold or cash, slaves got Sundays and holidays off from work. At one mining enterprise, for example, Saturday, June 2, 1832, was decreed "Negros day."[23] Some slaves took advantage of their free time to hunt for gold, which they could keep for themselves. In the setting of the North Carolina gold mines, then, the institution of slavery assumed some of the trappings of free labor.

None of this implies, however, that mining offered a "better" form of slavery than that which plantation slaves experienced. Some mine operators expressed little concern for the slaves. In addition to the dangers of cave-ins, explosions, and falls, slave miners sometimes faced overwork, undernourishment, and harsh treatment from employers or masters enamored with visions of mountains of gold.[24] Traveling through the Burke County gold-mining region in 1833, one Francis M. Goddard of Monticello, Kentucky, noted that "thousands of slaves was [*sic*] engaged in these mines under cruel masters who had a great thirst for filthy Lucre."[25]

Industrial slaves, like those engaged in agriculture, took proactive measures—stealing and running away—to ameliorate their condition.[26] Those strategies of day-to-day resistance worked as well in the industrial setting as they did on the plantation or farm. Slave miners stole gold and hid it in their hair or in seams of clothing, or buried nuggets and retrieved them later when no one watched. At the Reed mine, according to one observer, gold "was concealed or stolen at night by the negroes."[27] During his tour of the North Carolina gold region, English geologist George W. Featherstonhaugh observed that slaves engaged in the washing process "appeared to be submissive in their manners and to work very hard," but overseers "closely watched [them] to prevent their secreting any pieces of gold they might find."[28]

What did the slaves do with the gold that they either earned, discovered, or stole? According to a report in the *United States Telegraph* of Washington, D.C., "several [slaves] have found so much as to pay for the freedom of themselves and families."[29] Self-purchase likely occurred in only rare instances, however. Slaves found it difficult to save or invest the gold that they accumulated. They had no secure depository for their gold, increasing the risk of theft.[30] They also probably sought to dispose of their contraband gold as quickly as possible before being detected. In those instances in which slaves did purchase their own freedom, it seems most likely that the master agreed with the slave to withhold an additional amount of the slave's wages, in effect creating an ad hoc savings account for the slave.

Slaves usually geared their spending more toward immediate gratification at a local grogshop, a practice that made sound economic sense under the circumstances.[31] They often sold their gold—contraband or otherwise—to a white buyer in exchange for cash or a desired product. "At the surrounding groceries and liquor

African Americans such as this man using a rocker often labored alongside white workers in North Carolina's antebellum gold mines. *Engraving courtesy North Carolina Collection, University of North Carolina Library, Chapel Hill.*

shops" near the Reed mine, "the negroes employed at the mines" would "exchange for a pint of brandy or whiskey a pint of gold, which they abstracted from their masters, or collected at night; for a single 'drink' they gave pieces varying from the size of a pigeon's to that of a hen's egg, and considered it quite a favor to have some sugar thrown in."[32] The slave miners' illicit activities, such as drinking, gave them a limited sort of freedom, helping them ease the severity of their condition.

The slaves' own clever, assertive efforts to improve their lives would have proven fruitless, however, without the complicity of local whites. Contraband gold was largely useless unless traded for something else, such as spirituous liquors. Thus, the crime of stealing automatically linked slaves to lower-class whites who operated the grogshops and groceries in the mining district, catering to a laboring clientele. Slaves perceived the differences between classes of whites and recognized which individuals might be willing to strike a deal.[33]

If such clandestine trafficking challenged slavery as an institution, so too did the slave act of running away. Slave miners often demonstrated their aversion to the peculiar institution by fleeing from their employers. Motivated by a desire to see their family members, some of these slaves surely sought only to return to the household of the master who had rented them to the mines. Jailer John Woods captured a fugitive named Nelson, who explained that "he belongs to Sarah Brooks, of Caswell county, North Carolina, and that he was hired to John Blackwell, to work at the Gold mines in Burke." In advertisements of runaway slaves in local newspapers, concerned owners often explained the various circumstances surrounding a slave's disappearance. John Mayhew and Nathaniel Hobbs placed an advertisement for a twenty-year-old "Negro man named Nicodemus," who fled the Capps gold mine in Charlotte, "passing himself by different names." Jacob, a thirty-year-old slave, ran away from the Capps mine in December 1830. Owner William J. Alexander offered a twenty-five-dollar reward for his return, while William D. Henderson promised twenty dollars to anyone who delivered his slave Harry to "the Mines in Burke county or to Dr. Samuel Henderson, at Charlotte."[34]

Other slaves used the gold mines as sanctuaries, perhaps there finding a brief respite from plantation work or poor treatment. One Cabarrus County master complained that "MY boy LEWIS left my plantation . . . without any cause, to my knowledge." Lewis, his master predicted, "will sculk [*sic*] about the gold mines in this county and Mecklenburg." Likewise, the owner of "a negro man named RANDALL" presumed his slave "to be lurking about some of the Gold Mines."[35]

That so many slaves sought sanctuary in the gold mines suggests that the mines may have offered more than simply a place to hide. Jeremiah Cureton's detailed advertisement for "my Negro Boy Jeff" vividly portrayed one slave as a conscious schemer, exploiting the local mines to his advantage. "From information, I am led to believe," Cureton explained, "that [Jeff] has been at work at different Gold Mines through the lower end of Mecklenburg County . . . in an underhanded way, such as carrying off the Ore after night and washing it when best suits . . . convenience from being detected." Jeff's exploits imply that the gold mines represented for the slave an unusual opportunity in a land of otherwise circumscribed activity. Slaves seemed to recognize that not only did the labyrinth of dark, underground tunnels offer a place to hide and that the bustle of mining camps provided a certain degree of anonymity, but also that the gold mines themselves produced a precious ore that the slaves could use to assert their independence.[36]

Once again, lower-class whites of corrupt moral character violated racial boundaries as they lent assistance to runaway slaves. Jeremiah Cureton implicated Jeff's friend John Underwood, a white man, for harboring the fugitive and aiding in the covert activity. A scandalous and mysterious figure, "John Underwood, at this time, is absent from his family," Cureton related. He "possesses a mean look and is a shoemaker by trade, is fond of spirits and lying in negro kitchens."[37] Slaves might also use the gold they acquired to elicit the assistance of unscrupulous whites in a flight to freedom. Sources suggest that this may have been the case with a "vile raskal [*sic*]" named Tom, a slave who fled his Burke County master William A. Burwell.[38] When captured, Tom disclosed that "a Mr. Griffin who keeps a grog shop or a confectioners house was the man that gave him the pass" to facilitate his escape.[39] North Carolina's network of gold mines, with the real potential for mitigating the oppression of slavery or for proving useful to slaves who attempted to escape it entirely, may have served as valuable depots of a literal underground railroad.

Rebellion, however, marked the ultimate expression of slave discontent and the greatest internal threat to the southern social structure. Rumors and imagined plots of slave insurrection haunted the South persistently. In that atmosphere of constant uneasiness, any news of assertive slave activity piqued whites' interest, and after Nat Turner's insurrection in August 1831, in which fifty-seven whites were killed, a wave of hysteria swept the South.[40]

In the immediate wake of the murders, sensitized whites perceived plots in every slave quarter. The contagion of fear following the Turner insurrection infected the mining region as well. In late September 1831, newspapers revealed a plot that originated among the gold-mining slaves in Burke and Rutherford Counties. Allegedly, slave miners at James'-Town, Brackett-Town, and Brindletown had conspired to rebel. Seeking support, a slave miner named Fed intimated the details of the impending rebellion to a black woman in Rutherfordton, who exposed the plot. One phalanx of slaves purportedly planned to attack Morganton, with another to descend upon Rutherfordton. Upon destroying the two principal towns in the area, the bondsmen would overrun the countryside and gain their freedom. According to the local press, the slaves set a date in early October to implement the scheme. News that a suspicious congregation of between sixty and eighty slaves had met to elect governmental leaders at the Rutherford County muster grounds the week before the plot was unearthed further fueled white hysteria. The Rutherfordton paper recommended vigilance as it recorded, "The excitement amongst the people is considerable." Authorities arrested Fed, who, though not necessarily the leader, was well aware of the plans.[41]

Fed nevertheless conformed to the pattern of other leaders of slave revolts. Gabriel Prosser, Denmark Vesey, and Nat Turner were all literate and privileged slaves. Prosser worked as a blacksmith, Vesey as a sailor, and Turner as a foreman. To varying degrees, religious overtones characterized each of their crusades.[42] Fed, a slave preacher, had been known to expound on Scripture, the Bill of Rights, the constitution of North Carolina, and a mysterious "book from the West." In addition to his literacy, Fed, like other slave leaders, occupied a privileged position within the black community. His owner, a Mrs. Ware of Rutherford County, permitted him to hire out his own labor, giving him both the mobility to make contacts and the sense of autonomy necessary to foment rebellion.[43]

Thus, Fed may have taken advantage of his master's circumstances. Exceptional in the patriarchal society of the Old South, female slaveholders confronted added difficulties in the management of slaves. According to census figures, Didema Ware headed a household of twelve. Whether her husband had died or abandoned her remains a mystery, but by 1830, at the age of forty at the latest, she found herself a single mother of seven boys, the oldest of whom could have been fifteen. She also commanded four black slaves—a male (presumably Fed) and female between the ages of twenty-four and thirty-six and two younger girls, one under ten and one between ten and twenty-four, both probably the slave couple's children. Faced with the economic hardship of supporting her extensive family, Mrs. Ware likely sent her only male slave away from home to help make ends meet, in effect granting him a degree of latitude far exceeding his social position.[44]

Despite Fed's suspicious credentials, a few level heads in the community confidently discounted the rumored revolt. Mining entrepreneur and master of ninety slaves in 1830, Burke County resident Isaac T. Avery observed that "a Negro Preacher . . . charged with meditating an insurrection, in the neighbourhood of the mines," incited "a state of uncommon excitement." Certainly, Avery conceded, "The great number of slaves employed about the mines, would render vigilance among managers and patrollers, proper at all times; but from the information I can get," he asserted, "there is no just ground for the Panic that has existed. The Negroes of the County, are as orderly and submissive, as I ever knew them. & I really believe that the alarm is alltogether [*sic*] imaginary . . . we feel no apprehension . . . of danger." Significantly, that assessment came from one of the largest slaveholders in the western portion of the state, a man with the most to lose from a successful slave conspiracy.[45]

An anonymous writer to the *Raleigh Register* also expressed the belief that the area's citizens had overemphasized the magnitude of the Fed affair. According to the author of the letter, "The much talked of conspiracy in Burke and Rutherford, was confined to very few, if indeed it extended beyond the . . . instigator himself." Fed "has made a confession and endeavoured, with a view as I believe to procure some favor to himself, to implicate a few others; and against two or three of these there are some slight corroborating proofs." The writer hypothesized that "Fed's plan . . . was . . . to obtain whatever money he could from the negroes, and more by plunder, then make his escape, and leave his poor deluded followers to shift for themselves." Furthermore, the editorial expressed doubts that Fed planned to act with "concert abroad" at an appointed time. "He will probably pay the deserved penalty for his wicked and rash temerity," the writer concluded, "and here likely the affair will end."[46]

Roswell Elmer Jr., editor of the *North Carolina Spectator and Western Advertiser* of Rutherfordton, assessed the situation differently. "If the writer designed to create an impression that there has been no danger," Elmer countered, it will take more than an editorial in a far-off newspaper "to satisfy the people of Burke and Rutherford that their danger has not been real and imminent." As proof, Elmer reported that Judge Joseph J. Daniel refused to discharge Fed and his alleged accomplices or to set bail for them, despite "a most strenuous effort of counsel." Daniel ordered them to jail to await the spring term of the Burke County Superior Court.[47]

Elmer articulated the opinion of most of the citizens of Rutherford County who convened at Rutherfordton for a meeting at noon on October 7 to issue a series of resolutions. The assembled citizens established a committee of five prominent local men to recommend rules and regulations designed to "suppress and put down the spirit of insurrection" and "restore among the white people, tranquility and security."

Assuming the loyalty of their own slaves—part of the paternalistic myth—participants in the meeting expressed the desire to limit the introduction "of negroes from abroad, [or] of bad or suspicious character" to the mines. Furthermore, they offered the hope that the state legislature would craft a law to restrict the number of "foreign negroes" used in the mines and that county authorities would "prohibit entirely" from the mines any slave who had participated in or been influenced by previous slave conspiracies. The citizens also authorized the appointed committee to ask the governor what military support the state might offer to defend the county in the event of future rebellion and commanded local militias to remain vigilant until regular patrols could be established.[48]

Seemingly embarrassed that "reports of the approach of large bodies of insurgent negroes" caused "great terror and alarm . . . when in fact no such danger existed," the assembled citizens passed two additional resolutions aimed at preventing future outbreaks of mass hysteria. First, they advocated criminal trials and punishments of future "evil disposed" rumormongers who disregarded "the peace and quietude of society."[49] According to one report, four white males occupied the jail in Lincolnton "for having wantonly and mischievously circulating [*sic*] false reports of approaching bodies of negroes."[50] Second, they commanded the Rutherfordton postmaster to correspond with other area postmasters, sharing early and accurate information on slave unrest. Through those measures, the citizens of Rutherford County hoped to prevent a repeat of the unwarranted panic and alarm that accompanied the unsubstantiated initial rumors of revolt in the county's mines.[51]

Area residents left the decision of what to do to the alleged conspirators in the hands of the court system, even though, in the Old South, the community usually dispensed justice itself. The management of slaves, southerners argued, was the proper domain of the masters, whose responsibility it was to dole out the appropriate punishments on their own plantations. When slave crimes—such as a purported conspiracy—extended beyond the household, however, to affect the security of the entire region, the legal system assumed jurisdiction over the slave.[52] Unlike the masters, courts did not recognize black bondsmen as fully subordinate to whites, leaving the door open for verdicts favorable to the slaves.[53]

The slave-conspiracy trial commenced at the March 1832 session of the Burke County Superior Court. Giles and Billy, two bondsmen belonging to Col. Thomas Turner, faced felony charges of attempted insurrection among the gold-mining slaves, a capital offense. They pleaded not guilty. The court subpoenaed two white men from Rutherford County, Garland and Mark Dickinson, nonslaveholders apparently familiar with the two slaves, "to appear & give evidence in behalf of the defendants." Turner countered whatever testimony the Dickinsons may have proffered, admitting that Giles's and Billy's dispositions made them likely candidates to rebel.[54] The slave preacher Fed himself turned state's witness and "testified that an insurrection was contemplated and that the defendants were instigators in the plot." The court discounted Fed's testimony, however, because, "having made many contradictory statements," Fed demonstrated "that he was a 'rogue' . . . unworthy of belief."[55] The twelve-man jury—which included Isaac T. Avery—found Giles and Billy not guilty and discharged them. The court freed Fed too.[56] The twin verdicts in favor of the alleged conspirators exasperated Roswell Elmer of the Rutherfordton *Spectator and Western Advertiser*. "Thus," the editor lamented, "by the chicanery of counsel, the whole of the negroes confined for conspiracy, have been set at liberty. Little doubt, however, exists in the minds of the community but that a design of this kind was contemplated."[57]

The proceedings of the trial of Giles and Billy, and their ultimate acquittal, exemplified the constraints that slave resistance placed upon a social system over which masters ostensibly maintained absolute control. The acts of running away or committing real or perceived crimes, although unorthodox methods of political participation, inherently produced political consequences as they forced the legal system to acknowledge slaves' humanity. If the courts did not deal with slaves as human beings, the bondsmen could not be held accountable for their illegal activities. But when the courts intervened in slave criminal cases, they recognized a slave identity independent of the master, thereby superseding the master's authority and redefining the master-slave relationship. What is important about the Giles and Billy trial, then, is not that the slave Fed's testimony was disregarded but that it was heard at all. The court dismissed Fed's testimony not because he was a slave but because his story was inconsistent. If slaves had no rights whatsoever, no trial would have been held. There would have been no *State v. Giles & Billy*, in which two white men testified on behalf of two black slaves. That the legal system recognized slave rights undermined to some limited extent the complete power of masters over their slaves, opening the slaves' status to debate.[58]

The insurrectionary panic served several functions, however, that validated the social order of the Old South. In the Carolina mining region, hired slaves, runaways, and the slaves' means of resistance had to a limited degree eroded racial barriers. The apparent conspiracy helped reestablish the social status quo by affirming blacks' and whites' proper racial roles.[59] Roswell Elmer of the Rutherfordton newspaper congratulated the white race for taking such "prompt, efficient, and uniform exertions . . . to put down this insurrectionary spirit."[60] Elmer also made explicit what the white response meant for blacks. While the ruling race's collective action "augurs favorably to the future security of the country, and gives confidence to the whites," he explained, "it strikes dismay, & argues a forlorn hope to the miserably deluded negro."[61] Black testified against black at the trial of Giles and Billy, exemplifying to whites division within the slave community.[62] Conversely, even though authorities in the allegedly besieged mining region sent four whites to jail for disseminating false information, whites did not view that act as a sign of weakened racial solidarity as much as an exercise in social control. Authorities dealt with the troublemaking whites by removing them temporarily from active participation in a society in turmoil. By contrast, the slave miners accused of conspiracy were released, following a court trial, back into bondage. The alleged conspiracy thus resulted in feelings of communal solidarity as it reasserted traditional values and restored the social equilibrium to a society fraught with tension over the process of industrialization and the appropriate role for slaves in the changing economy.[63]

The strange juxtaposition of panic and calm surrounding the Fed affair highlighted the viability of slave labor in North Carolina's gold mines, despite the many fears and concerns that hired slaves, day-to-day resistance, runaways, and potential rebellions generated. Amid the widespread public hysteria, extralegal vigilante justice succumbed to the power of the legitimate legal system. The court remained level-headed in its execution of the law and dismissed the charges against the accused slaves. Even in the gold mines, where slave control should have been at its weakest, whites preserved their dominance. Historians have noted that blacks and whites displayed a striking degree of teamwork in the integrated setting of the mines. They dug together, in many instances side by side, with little evidence of overt racial hostility.[64] That point underscores the fact that slaves, regardless of their resistance, did not threaten the social framework of the Old South; rather, they played an integral part in that framework.[65] In a slave society such as that of the Old South, masters possessed a familiarity with slaves, and they knew how to manage them. "Concessions" that masters granted slaves, particularly in industrial slavery, were precisely that: "part of a system of controlling slaves rather than of liberating them."[66] By those concessions, industrial slaves achieved a sense of autonomy and independence uncommon among plantation slaves. But while the slave miners' privileges and incentives eroded the master-slave relationship somewhat, the masters retained the ultimate control, solidifying their position atop the social hierarchy.

Notes

1. Richard F. Knapp, "Golden Promise in the Piedmont: The Story of John Reed's Mine," *North Carolina Historical Review* 52 (January 1975): 4–5.

2. William Thornton, *North Carolina Gold-Mine Company* (Washington, D.C.: n.p., 1806), 10.

3. See, for example, Ronald L. Lewis, *Coal, Iron, and Slaves: Industrial Slavery in Maryland and Virginia, 1715–1865* (Westport, Conn.: Greenwood Publishing Company, 1979); Charles B. Dew, *Bond of Iron: Master and Slave at Buffalo Forge* (New York: W. W. Norton Company, 1994); Ernest McPherson Lander Jr., *The Textile Industry in Antebellum South Carolina* (Baton Rouge: Louisiana State University Press, 1969); Joseph Clarke Robert, *The Tobacco Kingdom: Plantation, Market, and Factory in Virginia and North Carolina, 1800–1860* (Gloucester, Mass.: Peter Smith, 1965); and T. Stephen Whitman, "Industrial Slavery at the Margin: The Maryland Chemical Works," *Journal of Southern History* 59 (February 1993): 31–62. For a general overview, see Robert S. Starobin, *Industrial Slavery in the Old South* (New York: Oxford University Press, 1970).

4. Starobin, *Industrial Slavery in the Old South*, vii, 11–12; Walter Licht, *Industrializing America* (Baltimore: John Hopkins University Press, 1995), 36.

5. Brent D. Glass, "King Midas and Old Rip: The Gold Hill Mining District of North Carolina" (Ph.D. diss., University of North Carolina at Chapel Hill, 1980), 135, 137–138, 143–144; Glass, "The Miner's World: Life and Labor at Gold Hill," *North Carolina Historical Review* 62 (October 1985): 430.

6. Henry B. C. Nitze and H. A. J. Wilkens, *Gold Mining in North Carolina and Adjacent South Appalachian Regions* (North Carolina Geological Survey Bulletin 10) (Raleigh: Guy V. Barnes, 1897), 29.

7. U.S. Congress, House, *Assay Offices, Gold Districts, N. Carolina and Georgia, February 15, 1831 Report*, 21st Cong., 2d sess., 1831, H. Rept. 82, 22.

8. Spotswood Burwell to William A. Burwell, August 3, 1832, Burwell Family Papers, Southern Historical Collection, University of North Carolina Library, Chapel Hill.

9. *North Carolina Spectator and Western Advertiser* (Rutherfordton), March 26, May 14, 1830.

10. *Charlotte Journal*, September 25, 1835. Penman continued to use slave laborers in his mines in the 1850s. See his advertisements for white or colored miners in the *North Carolina Whig* (Charlotte), June 8, October 4, 1853. The first advertisement ran almost weekly for nearly three months; the second, weekly for a full year.

11. Fifth Census of the United States, 1830: Mecklenburg County, North Carolina, Population Schedule, National Archives, Washington, D.C. (microfilm, J. Murrey Atkins Library, University of North Carolina, Charlotte).

12. *North Carolina Spectator and Western Advertiser*, January 29, 1831.

13. Starobin, *Industrial Slavery in the Old South*, 47.

14. *Western Democrat* (Charlotte), August 31, 1858.

15. *Western Democrat*, August 30, 1859.

16. Starobin, *Industrial Slavery in the Old South*, 48.

17. R. H. Morrison to Rev. James Morrison, April 4, 1845, Robert Hall Morrison Papers, Southern Historical Collection.

18. James Scott to Dr. John Scott, November 12, 1833, Dr. John Scott Papers, Private Collections, State Archives, Division of Archives and History, Raleigh.

19. The assertion that most hired slaves came from within the mining region derives primarily from runaway-slave advertisements that appeared in eleven western North Carolina newspapers between 1824 and 1841. Census records from 1830 do not include slave schedules that could confirm or deny that impression. A few masters in the eastern portions of North Carolina surely did rent slaves to the mines, however. For an example of a runaway hired from outside the heart of the mining region, see the *Yadkin and Catawba Journal*

(Salisbury), March 19, 1832. For a discussion of eastern slaves sent to mountainous Burke County to search for gold, see Edward W. Phifer Jr., "Champagne at Brindletown: The Story of the Burke County Gold Rush, 1829–1833," *North Carolina Historical Review* 40 (October 1963): 489–500.

20. Clement Eaton, "Slave-Hiring in the Upper South: A Step toward Freedom," *Mississippi Valley Historical Review* 46 (March 1960): 663, 677–678; Eugene D. Genovese, *The Political Economy of Slavery: Studies in the Economy and Society of the Slave South* (New York: Pantheon Books, 1965), 23, 53, 181, 225; Elizabeth Fox-Genovese and Eugene D. Genovese, *Fruits of Merchant Capital: Slavery and Bourgeois Property in the Rise and Expansion of Capitalism* (New York: Oxford University Press, 1983), 374; John Ashworth, *Slavery, Capitalism, and Politics in the Antebellum Republic*, 1 vol. to date (New York: Cambridge University Press, 1995–), 1: 109–110; Fabian Linden, "Repercussions of Manufacturing in the Ante-Bellum South," *North Carolina Historical Review* 17 (October 1940): 323.

21. *North Carolina Whig*, October 4, 1853.

22. August Partz, "Examinations and Explorations in the Gold-Bearing Belts of the Atlantic States," *Mining Magazine* 3 (August 1854): 165.

23. William A. Burwell Account Book, folder 206, Burwell Family Papers, Southern Historical Collection.

24. David Williams, *The Georgia Gold Rush: Twenty-Niners, Cherokees, and Gold Fever* (Columbia: University of South Carolina Press, 1993), 86; Charles B. Dew, "Disciplining Slave Ironworkers in the Antebellum South: Coercion, Conciliation, and Accommodation," *American Historical Review* 79 (April 1974): 402.

25. Diary of Francis M. Goddard, November 14, 1833, Special Collections, University of Kentucky Library, Lexington.

26. Eugene D. Genovese, *Roll, Jordan, Roll: The World the Slaves Made* (New York: Vintage Books, 1976), 599–609, 648–657.

27. Partz, "Gold-Bearing Belts of the Atlantic States," 164.

28. George W. Featherstonhaugh, *A Canoe Voyage Up the Minnay Sotor with an Account of the Lead and Copper Deposits in Wisconsin; of the Gold Region in the Cherokee Country; and Sketches of Popular Manners*, 2 vols. (1847; reprint, St. Paul: Minnesota Historical Society, 1970), 2:333.

29. *United States Telegraph* (Washington, D.C.), November 11, 1826; Fletcher Melvin Green, "Gold Mining: A Forgotten Industry of Ante-Bellum North Carolina," *North Carolina Historical Review* 14 (January 1937): 15.

30. Peter H. Wood, *Black Majority: Negroes in Colonial South Carolina from 1670 through the Stono Rebellion* (New York: Alfred A. Knopf, 1974), 209.

31. Wood, *Black Majority*, 209.

32. Partz, "Gold-Bearing Belts of the Atlantic States," 165.

33. Winthrop D. Jordan, *Tumult and Silence at Second Creek: An Inquiry into a Civil War Slave Conspiracy*, rev. ed. (Baton Rouge: Louisiana State University Press, 1995), 161.

34. *Yadkin and Catawba Journal*, March 19, 1832; *Raleigh Register*, January 23, 1829; *Western Carolinian* (Salisbury), February 14, 1831; *Carolina Watchman* (Salisbury), February 23, 1833.

35. *Yadkin and Catawba Journal*, June 9, 1829; *Miners' and Farmers' Journal* (Charlotte), October 11, 1834. See also *Charlotte Journal*, September 9, 1836, January 20, March 3, 1837, and *North Carolina Whig*, May 16, 1854.

36. *Miners' and Farmers' Journal*, September 14, 1833.

37. *Miners' and Farmers' Journal*, September 14, 1833. The November 10, 1857, issue of the *Western Democrat* reports that a Joseph Underwood, from the same area as John Underwood and a possible relative, was arrested for offering to write a pass for a slave.

38. Spotswood Burwell to William A. Burwell, August 3, 1832, Burwell Family Papers, Southern Historical Collection.

39. Spotswood Burwell to William A. Burwell, July 31, 1832, Burwell Family Papers, Southern Historical Collection.

40. John Scott Strickland, "The Great Revival and Insurrectionary Fears in North Carolina: An Examination of Antebellum Southern Society and Slave Revolt Panics," in *Class, Conflict, and Consensus: Antebellum Southern Community Studies*, ed. Orville Vernon Burton and Robert C. McMath Jr. (Westport, Conn.: Greenwood Press, 1982), 57–58 (hereafter cited as Strickland, "Slave Revolt Panics").

41. *North Carolina Spectator and Western Advertiser*, September 24, 1831.

42. Genovese, *Roll, Jordan, Roll*, 593; Douglas R. Egerton, *Gabriel's Rebellion: The Virginia Slave Conspiracies of 1800 and 1802* (Chapel Hill: University of North Carolina Press, 1993), x, 20, 179–181; Jordan, *Tumult and Silence*, 187–188. Despite Egerton's assertion that Gabriel himself was not a messianic figure, Jordan makes a compelling case that a certain religious impulse guided some members of the uprising.

43. *North Carolina Spectator and Western Advertiser*, September 24, 1831. Gabriel Prosser also hired out his own labor. Egerton, *Gabriel's Rebellion*, 24.

44. Fifth Census of the United States, 1830: Rutherford County, North Carolina, Population Schedule, National Archives (microfilm, J. Murrey Atkins Library); Egerton, *Gabriel's Rebellion*, 24.

45. Fifth Census of the United States, 1830: Burke County, North Carolina, Population Schedule, National Archives (microfilm, J. Murrey Atkins Library); Isaac T. Avery to Selina Lenoir, September 26, 1831, Lenoir Family Papers, Southern Historical Collection.

46. *Raleigh Register*, November 3, 1831, reprinted in *North Carolina Spectator and Western Advertiser*, November 12, 1831.

47. *North Carolina Spectator and Western Advertiser*, November 12, 1831.

48. *North Carolina Spectator and Western Advertiser*, October 8, 1831.

49. *North Carolina Spectator and Western Advertiser*, October 8, 1831.

50. *North Carolina Spectator and Western Advertiser*, October 15, 1831.

51. *North Carolina Spectator and Western Advertiser*, October 8, 1831.

52. Bill Cecil-Fronsman, *Common Whites: Class and Culture in Antebellum North Carolina* (Lexington: University Press of Kentucky, 1992), 158–162; Bertram Wyatt-Brown, "Community, Class, and Snopesian Crime: Local Justice in the Old South," in Burton and McMath, *Class, Conflict, and Consensus*, 195.

53. Bertram Wyatt-Brown, *Southern Honor: Ethics and Behavior in the Old South* (New York: Oxford University Press, 1982), 363–365, 401; Bertram Wyatt-Brown, "Community, Class, and Snopesian Crime," 192–197.

54. *State v. Giles and Billy*, Burke County Superior Court Minutes, March 1832, pp. 70–71, State Archives; Fifth Census, 1830: Rutherford County. The census lists the surname of the white witnesses as Dickerson, not Dickinson. For information on North Carolina slave codes, the state supreme court's interpretations thereof, and the subsequent ramifications of those interpretations on the master-slave relationship, see Thomas D. Morris, *Southern Slavery and the Law, 1619–1860* (Chapel Hill: University of North Carolina Press, 1996), especially 190–193, 271, and 278–281, which discusses *State v. Mann* (1829), *State v. Tom* (1830), and *State v. Will* (1834).

55. *Miners' and Farmers' Journal*, April 17, 1832.

56. *State v. Giles and Billy* and *State v. Fed slave*, Burke County Superior Court Minutes, March 1832, pp. 70–71, State Archives.

57. *Miners' and Farmers' Journal*, April 17, 1832.

58. James Oakes, *Slavery and Freedom* (New York: Vintage Books, 1991), 138–139, 155, 160; Genovese, *Roll, Jordan, Roll*, 26–30, 46–47.

59. Strickland, "Slave Revolt Panics," 72–73.

60. *North Carolina Spectator and Western Advertiser*, October 1, 1831.

61. *North Carolina Spectator and Western Advertiser*, October 15, 1831.

62. Wyatt-Brown, *Southern Honor*, 433.

63. Strickland, "Slave Revolt Panics," 73; Wyatt-Brown, *Southern Honor*, 403–406, 432. See also Laurence Shore, "Making Mississippi Safe for Slavery: The Insurrectionary Panic of 1835," in Burton and McMath, *Class, Conflict, and Consensus*, 99–100.

64. Green, "Gold Mining," 15; Starobin, *Industrial Slavery in the Old South*, 138; Dew, "Disciplining Slave Ironworkers in the Antebellum South," 394.

65. Genovese, *The Political Economy of Slavery*, 222.

66. Robert William Fogel, *Without Consent or Contract: The Rise and Fall of American Slavery* (New York: W. W. Norton Company, 1989), 194.

Recent and Future Gold Rushes in the Carolinas

Dennis J. LaPoint

Dennis J. LaPoint, registered geologist and owner of Appalachian Resources, a consulting firm that explores for and manages precious and base-metals properties and programs in the eastern United States, holds a Ph.D. from the University of Colorado. He is active in a number of professional associations and has led geological projects in Alaska, as well as the South, the West, and the Midwest. He has published nearly two dozen professional geological contributions, including material on gold found at the Haile, Kings Mountain, and Ridgeway deposits in the Carolinas.

Introduction

When historians discuss gold in North Carolina, they often use the phrase "mined out," leaving the impression that gold mining is a thing of the past in North and South Carolina. Journalists write with a touch of nostalgia about the old prospectors and the glamour of the few modern prospectors dredging in the creeks or digging in their small mines. The newest and largest gold rush in terms of gold produced began in the late 1970s after the price of gold had been decoupled from the monetary standard, citizens were once again allowed to purchase gold, and gold prices had begun to rise. By the late 1980s South Carolina was the sixth largest gold-producing state and was turning out approximately 200,000 to 300,000 ounces of gold per year from four active mines. The overall production from those mines in the last twelve years eclipsed the total prior production of gold from North and South Carolina combined (table 1).

Table 1: Historic Gold Production Compared to Recent Gold Production (in total ounces)

LOCATION	PRE-WORLD WAR II	1987–1999
North Carolina	720,000	-0-
South Carolina	255,000	1,775,000
Total, both states	975,000	1,775,000
Ridgeway mine (S.C.)	-0-	1,400,000

Sources: James R. Craig and J. Donald Rimstidt, "Gold Production History of the United States," *Ore Geology Reviews* 13 (November 1998): 407-464; Arthur H. Maybin III, personal communication with author, 1999.

The recent gold rush was unlike the first one in that geologists associated with large and small mining companies, rather than independent prospectors, were the prime movers in finding new mines. The key to success (besides good fortune) was exploration drilling, persistence, geologic personnel who believed that additional mines could be developed in the Southeast, and support of management to allow time and funds for exploration and development.

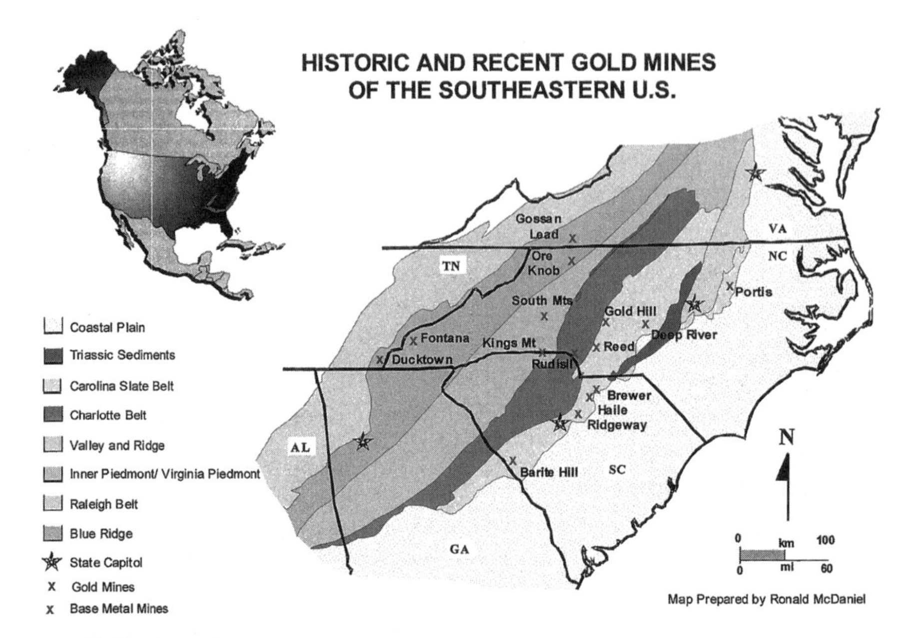

Map supplied by the author.

The latest gold rush is now at an end, but not because there is no further gold to be found. Rather, it is a victim of low gold prices and the desire of mining companies to avoid the "NIMBYism" (not in my back yard) of some Americans. For many mining companies, the opposition to mine development engendered by active, well-funded, national environmental groups is a greater liability than the economic risk of developing mines in politically unstable third-world countries lacking an infrastructure for development of mineral resources. For those who do not grasp the continuing need for minerals, out of sight, out of mind is the rule.

The Carolinas (and Virginia, Georgia, and Alabama too) are not mined out—indeed, they contain excellent potential for large, new discoveries similar to major finds elsewhere in the world. The key to future discoveries lies in the ability to conduct exploration on private fee land in spite of the changing land use of the southeastern United States. Explosive population growth is rapidly fragmenting the landscape of the Carolinas and "sterilizing" the two states' mineral resources. Farm and timber lands are being replaced by mini-estates, subdivisions, shopping centers, golf courses, and roads. Whether or not exploration for new discoveries will be forthcoming in the region will depend on the ability of mining concerns to work within environmental constraints, the price of gold, overcoming the reluctance of western-based mining companies to work in the eastern states, maintaining and increasing support for state and federal geologic surveys to provide new geologic maps and other information needed for exploration, and educating the general public and elected officials on the role of mining in our society and the ongoing need to produce natural resources.

The 1980s Gold Rush

World War II closed the last few mining operations in North and South Carolina—the Portis and Haile mines (Pardee and Park, 1948; Bryson, 1936; Carpenter, 1992). After the war, there was some sporadic mining in the 1950s and 1960s, including underground mining at the Star and Howie mines in North Carolina. Typically, those operations were hampered by a lack of operating capital and ore reserves. In the 1960s, Bear Creek Exploration, part of Kennecott Mining Company and at that time one of the largest and most aggressive mineral-exploration groups, was often at the forefront of exploration in new areas. In 1965 it initiated mineral exploration in the Carolina slate belt (Worthington, 1993). That exploration provided ideas and training for geologists associated with the later gold boom. In the 1960s, exploration for gold was a by-product of exploration for base metals associated with massive sulfide deposits and igneous intrusive rocks (porphyries). Until the price of gold began to rise during the 1970s, there was little incentive for gold-only exploration or significant research studies on the precious metal.

Evolution of Geologic Ideas

Joseph Worthington and Irving Kiff, in an article published in 1970 (and in subsequent papers written in collaboration with others), attributed the gold mineralization at the Sawyer mine in Randolph County, North Carolina, and the Haile mine in South Carolina to fluids accompanying hot-spring deposits that were forming in late Precambrian-age rocks associated with active volcanism. The work was one of the first papers on mineral deposits in the Southeast that connected the region's gold mineralization to modern hydrothermal processes. Since then, debate has waxed and waned between hot-spring-related gold deposits (referred to as syngenetic and/or epithermal deposits) and mineralization linked to later mountain-building processes (referred to as orogenic deposits). Some geologists have concluded that the gold deposits were formed by deep-seated fluids derived from deformation during metamorphism, probably many millions of years after the volcanic activity (Tompkinson, 1988; Hayward, 1992). Recent age dates by the U.S. Geological Survey and other workers and mapping by Ken Gillon at Ridgeway (Gillon, 1998) demonstrate that the gold mineralization is nearly contemporaneous with the volcanic activity. Because volcanic rocks and hot hydrothermal fluids occur in an active tectonic setting, structural activity must play an important role, and later faulting and metamorphism during mountain-building processes have certainly modified the deposits. This debate is important in the search for new gold deposits because if the mineralization is related to volcanic rocks and processes, one explores those regions first, whereas if zones of ancient faulting and metamorphism are the source of gold deposits, exploration should focus initially on those regions.

Geographic Focus of Exploration

In the early 1980s, much of the exploration for gold focused on mineralization along the trend of the historic Haile-Brewer mines and locations elsewhere that contained similarly altered rocks. Such sites included the Sawyer mine in North Carolina and the new discovery at Ridgeway, South Carolina (figure 1). This type of deposit is attractive to large corporations whose business is the production of gold. Deposits like these can be mined by low-cost bulk-mining methods, and low grades of ore can be profitably extracted. At other locations, such as the Barite Hills mine in South Carolina, exploration focused on base and precious mineralization associated with sea-floor, hot-spring deposits ("black smokers"). One of the major historic gold producers of that type was Gold Hill in Rowan County, North Carolina. Nearby, at Silver Hill (likewise a historic mine, in Davidson County), geologists defined a small zinc-lead-silver-gold reserve in the 1960s. Since then, many companies have prospected that trend, but surprisingly little drilling has actually been completed when compared to similar geologic environments in Canada and the western United States, where major new discoveries have been made. These "black smokers" often contain high grades of gold associated with copper and zinc mineralization. Open-pit or underground mining methods or both are used, depending on the distribution of the mineralization.

Some areas of the Carolinas have not received much exploration for various reasons. The Charlotte area (including the Reed mine) now features too many people and expensive buildings. The Howie mine near Monroe is a classic geologic setting for developing a long-term underground mining operation, but it is being surrounded by costly homes. Much of the potential at the Kings Mountain mine on the south side of Kings Mountain lies at depth under a large factory complex (LaPoint, 1992). The South Mountains gold region was not considered promising for bulk mining of gold because past production there was from thin quartz veins and placer deposits that would not be economical by modern mining methods. The region remains interesting because of the quantity of gold previously produced and the presence of diamonds. The eastern slate belt of North Carolina (figure 1), lying northeast of Raleigh, has been a focus of interest from large corporate mining concerns such as Texasgulf, BHP, Asarco, and Kennecott because of the grade of gold found there, the presence of numerous historic mines such as the Portis that existed within it, and the large quantities of gold encountered by panning in creeks that flow through it. Yet, because of a thin cover of younger Coastal Plain sediment, deep weathering, and poor rock exposures for mapping, the gold mineralization is well hidden. That region is the most challenging that I have explored and will require a commitment to exploration and significant amounts of drilling if a large discovery is to be made there.

Exploration for gold in the Carolinas reached its peak in the early 1980s with dozens of mining companies investigating sites at any particular time. But even then, the amount of funds expended for the systematic drilling necessary to make new gold discoveries was a small percentage of that spent elsewhere in North America by mining companies. Besides the discoveries that led to mines, other companies discovered or rediscovered areas of mineralization that have not quite achieved economic viability as mines but do contain published reserves (table 2).

Table 2: Reported Original Reserves at Various Mines and Prospects

Mine/Prospect	Owner (million tons)	Initial Reserves	Grade (optimum)	Total Ounces	Date Mining Commenced
Ridgeway (S.C.)	Kennecott	56.2	0.032	1,800,000	1988
Haile (S.C.)	Piedmont	1.8	0.048	86,000	1985
Haile	Amax/	13.16	0.091	1,200,000	—
	Piedmont	6.849	0.101	690,000	—
Brewer (S.C.)	Brewer Gold	10.9	0.037	400,000	1987
Barite Hill (S.C.)	Nevada Goldfields	1.66	0.038	60,000	1991
Russell-Coggins (N.C.)	Piedmont Mining	4.13	0.051	210,000	—
Sawyer (N.C.)	Minefinders	1.65-3.3	0.05	80,000	—
Kings Mountain (N.C.)	Texasgulf/Altai	0.275	0.25	650,000	—

Some of the deposits listed in table 2 are too small for profitable large-scale mining or have received inadequate drilling to expand the known deposit size. Others suffered from inadequate exploration, often as a result of mining-company mergers and closures, changing priorities within mining companies, or junior companies having insufficient financial resources. For certain prospective areas, landowners were reluctant to sign mining leases.

Newly Discovered and Rediscovered Mines of the 1980s Gold Rush

As I conduct exploration programs, I am always impressed with the efficiency and thoroughness demonstrated by the old prospectors and miners of the nineteenth century in finding gold. The known belts of gold mineralization have changed very little from one hundred years ago (figure 2; table 3). There were still a few recent prospectors in the 1980s, such as John Chapman, who originally found Ridgeway and other new prospects, and Bill Bell, who explored much of eastern Georgia.

Table 3: Major Historic Gold Producers in North and South Carolina

State	Mine (district)	Dates Active	Historic Production	Deposit Type
South Carolina	Haile	1828–1942	360,000	Epithermal
South Carolina	Brewer	1828–1940	>22,500	Epithermal
South Carolina	Dorn (McCormick)	1852–1880	45,000	Massive Sulfide
North Carolina	Gold Hill	1842–1915	150,000	Massive Sulfide
North Carolina	Kings Mountain	1834–1892	65,000	Carbonate
North Carolina	Mills property	1828–1916	>50,000	Placer/Quartz vein
North Carolina	Iola	1901–1916	>50,000	Quartz-Carbonate veins
North Carolina	Howie	1840–1942	51,300	Shear hosted
North Carolina	Portis	1835–1936	>50,000	Rhyolite dikes
North Carolina	Reed	1799–1899	>50,000	Quartz veins
North Carolina	Russell-Coggins	1882–1926	>20,000	Epithermal/Shear
North Carolina	Rudisill	1829–1938	>50,000	Quartz veins
Virginia	Melville-Vaucluse	1832–1935	>50,000	Shear hosted
Virginia	Franklin	1825–1935	>50,000	Sheared Diorite

Companies now focus more on disseminated gold mineralization in the rock, which is of a much lower grade than the considerably smaller but higher-grade lode quartz veins mined in the 1800s. In an era of unstable metal-commodity prices, firms need large-scale mining operations to be low-cost and competitive producers. Instead of incurring the high unit costs per ton of ore associated with discovering and mining gold in the higher-grade underground veins, as was the practice in the nineteenth century, modern mining companies can economically mine and extract gold from many millions of tons of rock containing relatively low-grade ore. Geologists now prospect for rocks that may contain invisible, very fine gold that can be trapped with silica or sulfides. Early miners could not extract such gold economically if at all. At Ridgeway the average grade of gold mined is 0.03 ounces per ton (or one gram per ton). An underground mine in the 1800s would have yielded ore at least ten times as concentrated (0.3 ounces per ton).

In the 1960s prospector John Chapman discovered mineralization by panning and dredging the creeks that would later become part of the Ridgeway mine. The South Carolina Geological Survey even drilled some auger holes to test for gold along a roadway there. The geologists encountered altered rock and gold values very similar to what is mined at Ridgeway, but it was of such low grade that little attention was paid to the results. Following the then current belief, the gold values were considered a product of weathering, which would decrease with depth. The area was of little interest in the 1960s, and the report was filed away. In the late 1970s Irv Kiff, then a consultant, remembered the report and Chapman's prospecting and showed the area to William Spence of the Amselco company. During a field trip in 1980, at about the same time that interest in gold in the slate belt was increasing, Henry Bell, an experienced geologist with the U.S. Geological Survey, stopped near the future Ridgeway mine. Not knowing about the recent exploration, he wrote that the alteration appeared identical to that at the Haile mine and speculated that perhaps there was another gold mine close by. Unknown to Bell and many participants in the field trip, Robert Carpenter, a consultant to the Amax Gold Company, was finding gold in the adjacent creek and leasing land for Amax, and Amselco was following up on John Chapman's discovery. A modern-day land play was on; Amselco eventually won out with persistence from its geologists and management—as well as Amax's decision to focus on a new deposit in Nevada.

Ridgeway differed from the Haile and Brewer mines in that it exhibited little evidence of earlier prospecting. I attribute this lack of nineteenth-century activity to the relatively few exposed quartz veins containing visible gold, the low-grade and fine-grained size of the gold, a thin cap of Coastal Plain sands, and a near-surface depletion of gold resulting from extreme weathering and leaching of the rock below the Coastal Plain sands. Both the Haile and Brewer mines were among the earliest mines in South Carolina, and they likewise were among the major producing gold mines in the Southeast in the years leading up to World War II. In the 1970s Cyprus Mining conducted an exploratory drilling program at the Haile mine. Besides defining ore near the historic pits, Cyprus conducted a shallow auger-drilling program to sample below the Coastal Plain sand cover. Results were not encouraging enough for Cyprus when gold prices were low, but Earl Jones and Irv Kiff, who worked on the deposit for Cyprus, later acquired the rights to the property. In 1987, on a shoestring budget, their company (Piedmont Mining) started the first gold mine of the recent gold rush. By the 1990s, like many small mining companies, Piedmont could not raise sufficient capital to further explore and develop the property and instead formed a joint venture with a larger company, Amax Gold, to undertake a major drilling program. That venture was successful in discovering new, deeply buried deposits, but once again, when Cyprus Mining acquired Amax the project did not meet Cyprus's corporate objectives and was ordered to be sold. Recently, Cyprus sold all of the assets of Amax Gold to a Canadian company, Kinross, which has no current plans to proceed with mining and is instead reclaiming the Piedmont mine and related heap leach sites.

During the 1970s, John Carroll of New York leased the rights to the old Brewer mine and subsequently assigned them to Gold Resources, a small company based in Colorado, then to Nicor Mineral Ventures, which was acquired by Westmont Mining. Following drilling, Westmont put the Brewer mine into production as a small heap leach operation. Still later, a British company, Costain, purchased Westmont and began mining deeper into sulfide ore—in spite of the fact that heap leaching was neither an efficient nor an environmentally sound method of extracting gold from copper-rich sulfide ores. The mining of sulfides has created a long-term environmental liability from acid mine water. The Brewer mine pit has since been filled with waste rock and heap leach material and capped with a clay layer, but it remains an environmental problem. New discoveries have been and will be found in close proximity to the Brewer mine; examples include the Buzzard discovery for Cepeda Minerals by the author (*Northern Miner*, August 26, 1996), as well as a similar find for Inter-Rock Gold on the east side of the Brewer property by Chris Cherrywell.

Exploration Methods

The old saying that the best exploration method is to see the trend of head frames (or pits, nowadays) remains true. One of the best areas for new discoveries is along the trend of existing large deposits, and the trends along the Haile-Brewer and Ridgeway mines remain geologically favorable for large new discoveries. Trends of historic mines such as the Gold Hill-Silver Hill trend of gold-rich massive sulfides or the Russell-Coggins mines also remain excellent areas for new discoveries. New prospects may be based as well on a review of old publications or unpublished material in archives or systematic evaluation of a belt of rocks such as the eastern slate belt, the Kings Mountain belt, or the volcanic rocks of the Uwharrie Mountains. Since the boom in gold exploration, there has been much new literature on gold deposits and discoveries around the world, as well as summary papers on new model types. For geologists familiar with geology and mineral deposits in the Southeast, those recent descriptions of deposit types can be applied to similar deposits in that region, such as porphyry gold at Deep River in the Carolinas (Capps and others, 1997; LaPoint and others, 1999). The best tools for making new discoveries are an observing mind and a drill rig.

Reports on Gold Deposits of the Carolinas

During portions of the 1800s, both North and South Carolina conducted active state geological surveys with an interest in mining and agriculture. In North Carolina, Denison Olmsted produced the nation's first state geologic map—which also outlined the North Carolina gold fields—in 1825 (LaPoint, 1999b). Subsequent reports on North Carolina by Ebenezer Emmons (1856), Washington C. Kerr and George B. Hanna (1888), Henry B. C. Nitze and Hanna (1896), Nitze and H. A. J. Wilkens (1897), and Herman J. Bryson (1936), as well as reports on South Carolina by Michael Tuomey (1848) and Oscar Lieber (1857, 1858, 1859, 1860), are essential descriptions of the mines either when they were active or better exposed for description than currently. Work by J. T. Pardee and C. F. Park Jr. (1948)—actually concluded prior to World War II for the U.S. Geological Survey—is the most complete summary description of gold deposits of the southeastern states since the 1800s. Unfortunately, by the 1930s, when Pardee and Park completed their fieldwork, most mines had long since been closed, and there was little geologic information to be gleaned from many deposits. In North Carolina, P. Albert Carpenter of the state survey published a recent update (1993) with accurate mine locations and descriptions, and Camilla McCauley and Robert Butler (1966) did the same for South Carolina. Neither publication provides new information because of the poor exposures found in long-abandoned mines and a lack of mine records.

The U.S. Geological Survey (USGS) has had a long history of involvement in twentieth-century southeastern geology. Recent studies and exploration for gold have built on the active geologic mapping of the slate belt of North Carolina and the Charlotte 1:250,000 quadrangle in North and South Carolina by the USGS in the

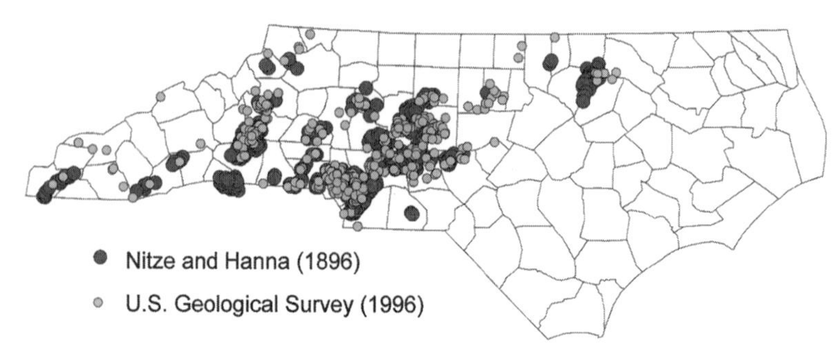

Map supplied by the author.

1960s and 1970s. The survey's Henry Bell was completing the publication of his geologic and geochemical studies in South Carolina at the time of his death in 1989, and much of his work is still unpublished. Similarly, Terry Offield was revising and preparing to publish his mapping of the structure and alteration of the slate belt at the time of his death in early 1999. Current work is just now being completed by Robert Ayuso, Nora Foley, and Terry Kline. Robert Schmidt (1985) was one of the first to recognize an association between porphyry-intrusive rocks and gold deposits in the central slate belt.

The 1980 field trips by Bell, Carpenter, and Geoffrey Feiss and a similar undertaking by Feiss in 1985 are indications of the state of knowledge at the start of this most recent gold rush, and the Society of Economic Geologists field trip lead by Doug Crowe (1995) and the Carolina Geological Society field-trip stop by Ken Gillon in 1998 best summarize the state of our knowledge at the end of this latest gold rush.

Recent Geologic Exploration and Mapping Methods

Irv Kiff was one of the first geologists in the Southeast to recognize the association of gold mineralization near the transition from felsic volcanic rocks to volcanically derived sedimentary rocks in the southeastern gold deposits (Worthington, 1993). This association is not unique to the Southeast but is nearly universal with many world-class gold deposits. Other geologic features that are associated with gold mineralization are the proximity to centers of volcanic activity; bimodal volcanism; and proximity to major, deep-seated, and long-lived structural zones. All such features are the result of active volcanic and tectonic processes within island arc and collisional settings related to plate tectonics. Thus, as geologists learn more about the geologic setting of the Carolinas, new areas for potential gold deposits can be related to recent plate tectonics on the Pacific rim and other modern active tectonic areas (Feiss and others, 1993).

There are few detailed published geologic maps for the Carolinas and fewer still that map the alteration associated with gold deposits. Geologic mapping is an essential exploration tool at all stages of exploration. For instance, regional geologic mapping by Amselco discovered the Stevens prospect near Lancaster, South Carolina, in 1983. Detailed geologic mapping at a scale of 1:12000 or 1:4800, usually on a grid, is important in selecting trench and drill sites. One focus of mapping is alteration. Alteration in slate belt rocks is characterized by the presence of silicification, sericite, gossans, aluminosilicate minerals, hydrothermal brecciation, and stockwork quartz veining. These alteration types result from the interaction of hot waters with the rocks through which they move. Exploration geologists must be good observers, open-minded, and good field mappers. Many mine workings are not reported in the literature but are found during mapping or sampling.

Geochemistry

Carrying on the method practiced by the old prospectors, pan-concentrate sampling of small drainage basins (panning) remains one the best methods for generating gold targets. Geologists try to be more systematic and standardized in their technique and focus more on the fine flour gold that is typically closer to the primary source rather than the coarse gold sought by the old prospectors. Some drainages near Ridgeway and Haile may contain more than one hundred colors in pan samples. Ridgeway was recognized by panning—first by John Chapman in the 1960s, then by Henry Bell for the USGS in panning for tin as the mineral casserite, then by Bob Carpenter for Amax in the late 1970s. In 1992, through the panning of one of my employees, Jerry Prosser, I found a new region of gold mineralization known as Copperhead. That find led to the discovery of the bedrock source by drilling beneath a thin surface veneer of Coastal Plain sands in 1996. Later, it was learned that nearby streams had been placered in the 1840s and by John Chapman in the 1970s.

In the western United States, sampling rock exposures has found many new deposits; but in the Carolinas, as a result of deep weathering and overlaying soil or Coastal Plain sands, there may be few rocks, and many altered rocks are not resistant enough to form outcrop. Soil sampling on a systematic grid is one of the most useful methods for defining trench and drill targets. Clusters of gold values greater than one hundred parts per billion often merit further work such as trenching and drilling. Analyses for pathfinder elements in rock and soil samples often help define targets. For the Ridgeway-Haile deposits, molybdenum and arsenic are the two best pathfinder elements. For other deposits, elements such as copper, lead, zinc, fluorine, and tellurium may be of use.

Geophysics

The alteration of rocks by hot fluids associated with mineralization often destroys the primary magnetic iron minerals in the rock. Thus altered, mineralized rocks may be represented by areas that are magnetically low when compared to unaltered rocks. The magnetic intensity can be determined by an air or ground magnetic survey. Silica is almost always associated with mineralization, and any geophysical survey that measures the conductivity of a rock unit, or its inverse (resistivity), is useful in defining the extent and trend of the silica-rich zone. Potassic alteration represented by potassium-rich minerals such as sericite or feldspar can be detected by radiometric surveys. Pyrite is conductive, and massive sulfide lenses or disseminated sulfides may be recognized through electromagnetic methods. However, this type of survey is subject to the cultural interference from power lines, fences, and buried cables.

Land Acquisition

Planning for land acquisition is as critical as planning for exploration. In the eastern United States, most land is private and owned by individuals or corporations. Mining companies must negotiate a mineral agreement—a lease—with the landowner. Soil grids, geologic mapping, and geophysical investigations are often completed without leasing the land but with the verbal approval of the landowner.

This is appropriate if the landowner is a farmer or a locally owned timber company. I find these local residents to be honest and trustworthy people who will honor their verbal agreements even if other companies are working in the area. Some large forest-products companies are somewhat bureaucratic and slow to negotiate an agreement. Most people in the country are friendly and willing to allow a geologist on their property. It is best to be honest in what you are doing; inaccurate and evasive answers will not promote later cooperation. The time and frustration it takes to acquire sufficient land for evaluation is often more discouraging to a company or a geologist than poor drill results!

Mineral Evaluation Methods

Prior to more expensive core drilling, trenching is one of the best low-cost methods of examining mineralization, alteration, and geology in areas with little or no Coastal Plain sand to be excavated. In areas with up to one hundred feet of such sand, rotary drilling has proven to be the most effective tool. This type of drilling is less expensive than core drilling but more subject to contamination and loss of sample in areas of high water flow. Core drilling collects solid cylinders of rock and provides an excellent sample for obtaining accurate chemical analysis and other information on the structural and stratigraphic controls for mineralization.

The key to discovery is a program of systematic and intensive drilling. For too many good gold targets in the Southeast, exploration has consisted of one to five holes scattered across the area of soil or geophysical anomalies. When ore-grade mineralization over significant thicknesses is not encountered, management quickly loses interest. Because gold deposits are complex in their distribution of gold, the first holes may provide only the geologic information needed to make a significant new discovery.

Mine development is a slow and costly process. Ridgeway required approximately seven years of effort and ten million dollars in exploration costs before mining started. Permitting is becoming more rigorous, and mine planning now includes provision not only for development and extraction of minerals but also for the cost of environmental cleanup. A new mine such as Ridgeway may require more than one hundred million dollars of capital expenditure and an additional twenty-five million in costs associated with mine closure!

Geologic Setting of Southeastern Gold Deposits

Our knowledge of the origin of gold deposits in the Carolinas parallels our increase in understanding of the processes that form the geology of the region and our ability to decipher past events using new techniques in scientific studies. Like the younger mineral deposits of the Pacific rim, those of the Carolinas are related to tectonic processes derived from the movement of plates. The tectonic events that appear to be related to gold deposits primarily occurred in early Paleozoic and late Precambrian time, 400 to 700 million years ago. Beginning in late Precambrian time, the Carolina Piedmont was not part of North America but rather a chain of volcanic islands, similar to the Aleutian or Carribean islands. Large volcanoes and active eruptions were frequent, and eroded material from the volcanoes formed an apron of volcanically derived sedimentary rocks. Some eruptions also occurred on the sea floor. There were several cycles of volcanic activity and sedimentation in both North and South Carolina, as well as a variety of tectonic and volcanic environments related to that tectonic setting. The copper-gold-lead-zinc-silver deposits at Gold Hill and Silver Hill in North Carolina and at the Barite Hill and Dorn mines in South Carolina represent "black smokers"—hydrothermal vents that deposited metals under anaerobic conditions in which iron sulfides were precipitated.The Red Hill pyrite pit at the Haile mine appears to be a similar feature but low in base-metal values. Similar deposits are found in the Kings Mountain belt and in older deposits associated with the initial rifting of the North American Craton in the Blue Ridge Mountains of North Carolina.

Slightly younger extensional events, possibly in a back-arc tectonic environment (Feiss and others, 1993), involved large volumes of silica-rich magma that formed large calderas and rhyolite domal fields that make up much of the Uwharrie Mountains. These caldera complexes were in part subaerial and in part submarine or lacustrine. This geologic environment forms many of the world's large epithermal gold deposits. (A similar active environment is the northern island of New Zealand, where active hot springs are precipitating gold, and slightly older hot-spring deposits are gold mines.) The Ridgeway-Haile-Brewer mines resemble the New Zealand deposits. The Russell-Coggins mines and many other historic deposits in central North Carolina are associated with rhyolite domes and hot fluids associated with those domes. Similar silica-rich domes are thought to be responsible for the gold mineralization at the Kings Mountain mine, where such fluids reacted with the limestone to silicify the rocks intensely and precipitate gold, base metal, and lead telluride mineralization.

Mineralization associated with shallow-level intrusive rocks—porphyries—has been suggested but not confirmed until recent mapping at the newly discovered Deep River gold-copper-molybdenum prospect in central North Carolina. There, mineralization occurs in the upper portions of shallow-level igneous rocks that intrude and dome up the surrounding volcanoclastic and sedimentary rocks. Mineralization is associated with silification, sulfides, silica flooding, stockwork quartz veining, and gypsum veinlets. Fluorite-molybdenite-pyrite veinlets are also present. The size of that mineralized system is at least three miles by five miles. Robert Schmidt of the USGS (1985) thought that gold deposits of the gold pyrophyllite belt, such as Pilot Mountain, were also intrusive related. The Brewer mine in South Carolina represents a similar system, with well-documented multiple phases of hydrothermal breccia associated with silification, precipitation of sulfides, and gold mineralization.

The gold deposits of the Haile-Brewer-Ridgeway trend are the best exposed, because of mining, and the best studied. Worthington and Kiff first proposed that they were hot-spring-related in 1970, but as exploration proceeded at Ridgeway and with new studies at Haile, the pendulum swung toward shear-hosted, in that the rocks have undergone intense deformation (Tompkinson, 1988; Heyward, 1992). Now, new work by the USGS and others supports the idea of hot-spring and epithermal mineralization. Robert Ayuso (1998, 1999) has dated the rocks hosting mineralization at 567 million years old. Holly Stein (1996) has also dated the molybdenum associated with the mineralization and found an age similar to the age of the enclosing volcanic rocks. Nora Foley (1998) demonstrates that at the Haile mine there are many generations of early pyrite that indicate nearly coeval deposition of the gold and host rocks.

Where Are the New Discoveries?

The new discoveries will be made in several ways. First, by further exploration along the trend of mineralization defined by the Ridgeway, Haile, and Brewer mines. Second, by prospecting for precious-metal-rich massive sulfide deposits such as in the Gold Hill and Silver Hill areas of North Carolina and elsewhere in the slate belt and Blue Ridge Mountains of North Carolina. Third, by further drilling and evaluating the historic gold deposits—most of which have received very little or no drilling. As to specific cases in which drilling has taken place, the core and information about them have been lost or are stored in the depths of corporate files. Finally, as new gold deposits are found throughout the world and information about such deposits is published or spread, a geologist will make the connection to some rocks, an old mine, or a geologic description he read about or recalls in an area in the Carolinas. New discoveries will depend on the current land use of the region, public opinion concerning the relative importance of mineral resources, and regulators' willingness to permit new mining in light of the economic pressures on the mining companies to produce a favorable return on their investment.

We all are aware that minerals are a nonrenewable resource, unlike timber and crops. Minerals, including gold deposits, are fixed to the earth by the geologic processes that formed a particular site. Unlike shopping centers, roads, subdivisions, or landfills, whose locations are based on decisions by people and involve only the surface of the earth, the location of a mine is based on the presence of a mineral deposit below the surface of the land. Mining companies must work hard to develop a trust in maintaining good stewardship of the land before, during, and after mining is carried out. All of us should also realize that mines are unique and exciting educational tools that provide windows through which to examine the earth through time. The geologic legacy of mines should be preserved even after they are closed and reclaimed.

References

Ayuso, Robert A., et al. "Genesis of Gold Deposits in the Carolina Slate Belt, USA: Constraints from Comparative Mineralogy, Trace Elements, and Isotopic Variations." [Proceedings], Mining Engineering [section], Society for Mining, Metallurgy, and Exploration Annual Meeting, 1998, p. 62.

Ayuso, Robert A., et al. "Gold Deposits of the Carolina Slate Belt: Zircon Geochronology and Isotopic Characteristics." Geological Society of America, *Abstracts with Programs* 31 (1999): A-3.

Becker, George F. "Reconnaissance of the Gold Fields of the Southern Appalachians." U.S. Geological Survey Annual Report 16 (1895). Washington: Government Printing Office. Pp. 251–331.

Bell, Henry. *Geochemical Reconnaissance Using Heavy Minerals from Small Streams in Central South Carolina*. U.S. Geological Survey Bulletin 1404. Washington: Government Printing Office, 1976.

______. *Preliminary Geologic Map and Sections of the Kershaw Quadrangle, Kershaw and Lancaster Counties, South Carolina*. U.S. Geological Survey Open File Report 80–226. Washington: Government Printing Office, 1980. Map.

Bell, Henry, and P. Popenoe. "Gravity Studies in the Carolina Slate Belt Near the Haile and Brewer Mines, North Central South Carolina." *U.S. Geological Survey Journal of Research* 4 (1976): 667–682.

Bell, Henry, Robert H. Carpenter, and P. Geoffrey Feiss. "Volcanogenic Ore Deposits of the Carolina Slate Belt," in Robert W. Frey and Thornton L. Neathery, eds., *Excursions in Southeastern Geology: Field Trip Guidebooks to the Annual Meeting of the Geological Society of America, 1980*, 2 vols. Atlanta: American Geological Institute, 1980, 1:149–178.

Bryson, Herman J. *Gold Deposits in North Carolina*. N.C. Geological Survey Bulletin 38. Raleigh: North Carolina Department of Conservation and Development, 1936.

Capps, Richard C., Ronald D. McDaniel, and Jonathan A. Powers. "Porphyry Gold Mineralization at the Deep River Prospect, Carolina Slate Belt Lithotectonic Province, Moore and Randolph Counties, North Carolina." Late Breaking Abstracts (No. 60007) from Geological Society of America annual meeting, 1997.

Carpenter, P. Albert. *Gold in North Carolina*. N.C. Geological Survey Information Circular 29. Raleigh: North Carolina Department of Environment, Health, and Natural Resources, 1993.

Crowe, Douglas E., ed. *Selected Mineral Deposits of the Gulf Coast and Southeastern United States: Guidebook Prepared for the Society of Economic Geologists Field Conference I, 4–5 November 1995, II, 9–11 November 1995, pt. 2: Gold Deposits of the Carolina Slate Belt*. Vol. 24 of Society of Economic Geologists Guidebook Series. Littleton, Colo.: Society of Economic Geologists, 1995.

Duckett, Roy P., Peter H. Evans, and Kenneth A. Gillon. "Economic Geology of the Ridgeway Gold Deposits, Fairfield County, South Carolina." In D. T. Secor, ed., *Southeastern Geological Excursions*. Columbia, S.C.: Geological Society of America, 1988. Pp. 40–44.

Emmons, Ebenezer. *Geological Report on the Midland Counties of North Carolina*. New York: George P. Putnam and Co., 1856.

Feiss, P. Geoffrey, ed. "Volcanic-hosted Gold and High-alumina Rocks of the Carolina Slate Belt." Society of Economic Geologists, field trip, 1985.

Feiss, P. Geoffrey, R. K. Vance, and D. J. Wesolowski. "Volcanic Rock-hosted Gold and Base-metal Mineralization Associated with Neoproterozoic-Early Paleozoic Back-Arc Extension in the Carolina Terrane, Southern Appalachian Piedmont." *Geology* 21 (1993): 439–442.

Foley, Nora K., et al. "Remnant Colloform Pyrite: A Textural Key to the Genesis of Gold Deposits in the Carolina Slate Belt." Geological Society of America, *Abstracts with Programs* 30 (1998): 368.

Gillon, Kenneth A., and others. "The Ridgeway Gold Deposits: A Window to the Evolution of a Neoproterozoic Intra-Arc Basin in the Carolina Terrane, South Carolina." *South Carolina Geology* 40 (1998): 29–70.

Hayward, Nicholas. "Controls on Syntectonic Replacement Mineralization in Parasitic Antiforms, Haile Gold Mine, Carolina Slate Belt." *Economic Geology and Bulletin of the Society of Economic Geologists* 87 (January–February 1992): 91–112.

Kerr, Washington C., and George B. Hanna. "Ores of North Carolina." Chapter 2 of *Geology of North Carolina*. Raleigh: Edwards and Broughton, 1888.

LaPoint, Dennis J. "Geologic Setting of the Kings Mountain Gold Mine, Cleveland County, North Carolina." In Dennison, John M., and Kevin G. Stewart, eds., *Geologic Field Guides to North Carolina and Vicinity*. Chapel Hill: University of North Carolina,1992. Pp. 35–48.

_______. "The Gold Rush in North Carolina." *Geotimes* 44 (December 1999): 14–18.

LaPoint, Dennis J., Jonathan A. Powers, and Ronald D. McDaniel. "Deep River Prospect, a Gold-Copper Porphyry System in the Carolina Slate Belt of North Carolina." Presentation and abstract, Prospectors and Developers Annual Meeting, Toronto, 1999.

Lieber, Oscar M. *First Annual Report on the Survey of South Carolina . . .* [for 1856]. Columbia: R. W. Gibbes, 1856–1860.

_______. *Second Annual Report on the Survey of South Carolina . . .* [for 1857]. Columbia: R. W. Gibbes, 1856–1860.

_______. *Third Annual Report on the Survey of South* Carolina . . . [for 1858]. Columbia: R. W. Gibbes, 1856–1860.

______. *Fourth Annual Report on the Survey of South Carolina* . . . [for 1859]. Columbia: R. W. Gibbes, 1856–1860.

Nitze, Henry B. C., and George B. Hanna, *Gold Deposits of North Carolina*. N.C. Geological Survey Bulletin 3. Winston, N.C.: M. I. and J. C. Stewart, 1896.

Nitze, Henry B. C. and H. A. J. Wilkens. *Gold Mining in North Carolina and Adjacent South Appalachian Regions*. N.C. Geological Survey Bulletin 10. Raleigh: Guy V. Barnes, 1897.

Offield, Terry W. "Carolina Slate Belt Framework, pt. 1: Regional Setting and Internal Deformation Pattern." Geological Society of America, *Abstracts with Programs* 27 (1995): 78.

______. "Carolina Slate Belt Framework, pt. 2: Structure and Gold Deposits." Geological Society of America, *Abstracts with Programs* 27 (1995): 78.

______. "Lithotectonic Map and Field Data—Reconnaissance Geology of the Carolina Slate Belt and Adjacent Rocks from Southern Virginia to Central South Carolina." U.S. Geological Survey Open File Report 94–420 (two diskettes), 1994.

Overstreet, William C., and Henry Bell. *Geologic Map of the Crystalline Rocks of South Carolina*. U.S. Geological Survey Miscellaneous Geological Investigations, Map I–413. Washington: Government Printing Office, 1965.

Pardee, Joseph T., and Charles F. Park Jr. *Gold Deposits of the Southern Piedmont*. U.S. Geological Survey Professional Paper 213. Washington: Government Printing Office, 1948.

Scheetz, John W. "The Geology and Alteration of the Brewer Gold Mine, Jefferson, South Carolina." Master's thesis, University of North Carolina at Chapel Hill, 1991.

Schmidt, Robert G. *High-alumina Hydrothermal Systems in Volcanic Rocks and Their Significance to Mineral Prospecting in the Carolina Slate Belt*. U.S. Geological Survey Bulletin 1562. Washington: Government Printing Office, 1985.

Speer, W. Edward, and John W. Maddry. "Geology and Recent Discoveries at the Haile Gold Mine." *South Carolina Geology* 35 (1993): 9–26.

Spence, W. H., and others. "Origin of the Gold Mineralization at the Haile Mine, Lancaster County, South Carolina." *Mining Engineering* 32 (January 1980): 70–73.

Tomkinson, Marcus J. "Gold Mineralization in Phyllonites at the Haile Mine, South Carolina." *Economic Geology* 83 (November 1988): 1392–1400.

______. "Petrology of a Number of Gold Deposits in the Southern Appalachians." Ph.D. diss., University of Southampton, England, 1985.

Tuomey, Michael. *Report on the Geology of South Carolina*. Columbia: A. S. Johnston, 1848.

Worthington, Joseph E. "The Carolina Slate Belt and Its Gold Deposits: Reflections after a Quarter Century." *South Carolina Geology* 35 (1993): 1–8.

Worthington, Joseph E., and Irving T. Kiff. "A Suggested Volcanogenic Origin for Certain Gold Deposits in the Slate Belt of the North Carolina Piedmont." *Economic Geology* 65 (August 1970): 529–537.

Worthington, Joseph E., and others. "Applications of the Hot Springs or Fumarolic Model in Prospecting for Lode Gold Deposits." *Mining Engineering* 32 (1980): 73, 79.

The California Gold Rush: The Land of Wealth and Surprises

Malcolm J. Rohrbough

Dr. Rohrbough is professor of history at the University of Iowa. He studied at Harvard University (B.A.) and the University of Wisconsin (M.A., Ph.D.). His books include The Land Office Business: The Settlement and Administration of American Public Lands, 1789–1837 *(1968);* Aspen: The History of a Silver Mining Town, 1879–1893 *(1986);* The Trans-Appalachian Frontier: People, Societies, and Institutions, 1775–1850 *(1989); and the award-winning* The California Gold Rush and the American Nation *(1997).*

On the morning of January 24, 1848, James W. Marshall picked some flakes of mineral out of the tailrace of a sawmill on the American River. Marshall carried the particles to the house and laid them on a table with the comment, "Boys, by God I believe I have found a gold mine." Later, primitive tests proved that his initial reaction was correct. Marshall's discovery set in motion a series of events that would become known as the California gold rush, a phenomenon that would affect the history of California and the American nation. It would have an equally powerful impact on the experiences and expectations of individual Americans, their families, and their communities.[1]

The discovery of gold in California was to trigger the greatest mass migration in the history of the young Republic up to that time: some eighty thousand people in 1849 alone and probably three hundred thousand by 1854, an immigration—largely male and generally young but not exclusively either—by land across half a continent and by sea over thousands of miles of ocean to new and previously unimagined adventure and wealth. And it would become an international event, with immigration from throughout the world: farmers from China and lawyers from Paris, miners from Wales and merchants from Chile. California would become in a few short years the most cosmopolitan place in the world, and San Francisco the most cosmopolitan city.[2]

Although California was half a continent distant, gold there was readily accessible to the most inexperienced mining novice with the simplest and most inexpensive kinds of equipment. In short, this was a game that anyone might play, and the stakes necessary to participate were small ones, at least as they existed through 1848 and 1849. The early mining exercise may be easily described. Gold was found in the nooks and crannies of old stream beds and in the bottoms of existing watercourses, where it had been left by thousands of years of movement of water, which carried the mineral downstream to a point at which the movement of water was insufficient to support the weight. Thus, water was a crucial agent in early gold mining, both as an instrument in transporting and dropping the mineral and as a force for separating it from the dirt around it. The first miners quickly mastered the primitive techniques by which moving water flowing through a tin pan separated the gravel, then carried it off from the heavier gold particles, which would sink to the bottom of the pan, where they could be easily retrieved and stored in a small sack. The widespread presence of gold in the streams draining the western slope of the Sierras and in beds of former streams—where it might be uncovered simply by removing a layer of topsoil—offered a degree of accessibility that made the early gold rush an open exercise for everyone. All that was necessary to participate was a shovel and a pan.[3]

Furthermore, the search for gold was uninhibited by institutional influences. Access to the rivers, streams, and valleys was uncontrolled, and no licenses were issued by any kind of central authority (although the new state of California made an attempt to levy a tax on alien miners in 1850). The issue of who actually owned the land remained in abeyance, for the land was largely unclaimed in a formal western European sense (except for the tracts of businessman John Sutter, Marshall's employer). The California Indians were so weakened by two generations of dependence upon missions and the trauma associated with their destruction that they could offer little or no resistance. Indeed, the access to the gold in California may be said to be a pure representation of American economic democracy in action in the middle of the nineteenth century: large numbers participated; some did substantially better than others; where rules did not exist, miners met in time-honored American fashion to hammer out rules by popular consensus; and, finally, on several occasions groups of miners united by ethnic identity or jealousy or greed (or variations of all three) banded together to expel Native Americans, Mexicans, French miners, and Chileans from especially attractive claims, laying the groundwork for a future in which the Chinese (only barely visible by 1851) would later become the universal victims of such vigilante activities.

Modifications or advances in technology employed in California came quickly in the form of the "rocker" or "cradle," in use earlier in the gold mines of Georgia and North Carolina. California territorial governor Richard B. Mason described the machines in June 1848: "in the full glare of the sun, washing for gold, some with tin pans, some with close-woven Indian baskets, but the greater part had a rude machine known as the cradle. This is on rockers six or eight feet long, open at the foot, and at its head has a coarse grate and sieve; the bottom is rounded with small cleets nailed across." The principle was the same—namely, to let the water do the work of separating the gold from the gravel. The difference was the economy and efficiency provided by a machine that would allow men to pool their labor into larger units. The first cradles required the labor of three or four men. One dug the gravel from the river bank or dry stream bed; the second carried the gravel to the cradle and emptied it into the grate; the third poured water or directed water from the stream itself through the machine; and the fourth agitated the cradle back and forth by a handle to produce the rocking motion that propelled the gravel through the machinery and out the lower end while trapping the heavier gold nuggets or flakes in the cleats on the bottom.[4] The pan and the cradle—both relying upon the washing action of the water and the weight of the gold, both simple to purchase or easy to make from a few boards of lumber—provided the technology for the first two years of the California gold rush. The widespread availability of both instruments—in spite of rising prices in accordance with supply and demand—ensured that every prospective Argonaut could have access to the latest technology available in the gold fields. But I anticipate the story here.

News of the discovery of gold reached an American nation sharply divided by sectional issues. The focus of sectional tensions in the fall of 1848—once over the issue of the tariff and antislavery agitation, then continued with debates over the annexation of Texas—was now on a quarrel over the territories taken in the recent war against Mexico. At the center of this division was the so-called Wilmot Proviso, an amendment attached to a war appropriations bill in 1846 by David Wilmot of Pennsylvania that specified that no territory acquired from Mexico as a result of the war would be open to slavery. Although defeated in Congress, the Wilmot Proviso reopened the issue of the extension of slavery in an infinitely expanding nation (or so it seemed at mid-century).

In this context of rising sectional tensions, the response to James W. Marshall's discoveries was remarkably similar in all parts of the nation. Gold in California transcended the issue of slavery in the territories. It did not unite Americans in any more than a superficial way, but it did provide a kind of cease-fire in the sectional tensions. Everywhere in the nation, men (and some women) prepared to go to California in pursuit of well-advertised riches, and, for a period of several months at least, politicians, public figures, ministers, journalists, and others who contributed to public discourse had something new and exciting to talk about. And talk they did, with raised voices, stretching the credibility and use of adverbs and adjectives to describe the new wealth in the most western territory of the new, continental nation.

Editors vied with one another in their large-type fantasies. In addition to stories of men picking up nuggets of pure gold, reports connected with children were especially popular, for the image of a child harvesting gold in California dramatized the ease with which the opportunity was open to everyone. Another dimension was the financial success enjoyed by servants. That men and women of the servant class (although not slaves, of course) could succeed in the search for this new, instant wealth served to confirm the universal availability of the bounty. The *Raleigh Star and North Carolina Gazette*, for example, in 1849 reported two instances of finds by servants in language that characterized such happenings as remarkable for their frequency. In the first case, two servants of an official went to the mines and returned with seventy-five thousand dollars' worth of gold; in the second, the gold was "so abundant that there is not necessity for washing the earth; $700 per day is the amount [found] by each man."[5]

"Gold Fever," as it became known, spread up and down the Atlantic coast and inland to the remote reaches of the cotton kingdom. Charles Harlan used the phrase "the California Mania" to describe conditions in Mississippi. Prominent citizens and yeoman farmers alike in Georgia, Tennessee, and South Carolina talked of nothing else for six months. One correspondent in Missouri, writing in the summer of 1849, noted that "all Creation appeared to be in an uproar" and added, in reference to the exodus to California, that "the like was never known since the Isrealites left old Egipt [*sic*]." Within the South, there were comments that suggested diversity of opinion. In a different take on the emigration to California, one correspondent commented that the exodus from Virginia would be especially great because that state "possesses a greater number than any other state of men too lazy to labor and too proud to remain contented in poverty."[6]

Citizens and editors of Texas, the great state of the Old Southwest and potential gateway to California, joined the chorus. "CALIFORNIA THE GOLD FEVER," announced the headlines of the *Corpus Christi Star*. The editor of that journal compared the "Gold Fever" to other rapidly spreading diseases: "never did Cholera, yellow fever, or any other fell disease rage with half the fury with which the gold fever is now sweeping over our land." The epidemic devastated "communities in one fell swoop, sparing neither age sex nor condition. It is the rage."[7] The editor of a New Orleans newspaper advanced a military analogy. (For a nation still celebrating its triumph over Mexico in the recent war—a triumph that would result in the cession of California to the Republic—a military analogy that compared the enthusiasm of war to the ardor displayed for gold came easily to mind.) The editor, writing of the California emigration, declared: "It has had much of the effect of a war which elicits the popular sympathy. The young and the ardent fly at once to arms, and the old and cautious deliberate and ponder, whether it is best to yield to the natural impulse; and in the end they do yield, and go out to meet the foe. There has been a marvelous resemblance between the formation of gold seeking companies and the recruiting of a volunteer force for an aggressive war."[8]

Southern editors, politicians, and public figures did not ignore the issue of slavery in their discussions of the gold rush. To the contrary, the questions came tumbling out one after another. What was the applicability of the institution of slavery to gold in California? What was the future status of slavery in California? Or, as the editor of the Helena, Arkansas, *Southern Shield* put the issue, "will it be safe to take negroes to California[?]"[9] Then there was the question of the suitability of slave labor to the nature of the task—that is to say, mining as it was understood in the first months of the gold rush. A Georgian had a solution to the first query. He suggested that three hundred to five hundred Georgians emigrating to California, each accompanied by from one to five slaves, would force the admission of California as a slave state. A consensus gradually emerged that the status of slavery in California was uncertain, but editors and public officials urged southerners to take their bondsmen to California, where they would labor as both servants and as miners.[10]

The specter of fortunes to be made, buttressed by news of the organization of companies, blared forth from the pages of local newspapers, dominated talk at the store and after church, and hung like a miasma over the conversations in numerous parlors. The lure of California gold was self-evident. At a time when agricultural laborers made one dollar a day for work in the fields from sunrise to sunset and skilled labor in shops perhaps half again as much, farmers-, hired hands-, and mechanics-turned-miners harvested twenty dollars a day in the gold diggings in California, washing dirt and picking nuggets from the bottom of a pan. In the years from 1849 to 1855, the forty-niners extracted some three hundred million dollars' worth of gold from the California Sierra.[11]

To go west, even to a sure fortune in 1849, was not easy and was certainly not cheap. By sea or overland, it was a formidable undertaking. The costs ranged from twelve hundred dollars up for travel by sea, depending on the accommodations and whether the forty-niner went around Cape Horn or by way of Panama. Overland, the voyage to the West cost half that sum, but six hundred dollars was still a large sum for most families—in most cases, the savings of a lifetime. How were such sums to be found for such an expedition? They might be generated internally within the family from savings or relatives, perhaps in part as a payment of eventual divisions of estates. Or they might be sought externally from members of the community in the form of loans or, on a more impersonal basis, through banks or strangers. Virginia politician and planter David Campbell urged his nephew to invest five hundred dollars usefully "by fitting out some young man, on whose integrity you could rely and who had means of going himself—and agree to divide the profits with him." The sum would support "some poor but hardy man" for two years in California with the prospect of a good return for both parties.[12]

Besides sheer cost, there were other difficulties and constraints to undertaking such a journey. Fortune seekers traveling overland were obliged to depart from the "outfitting" towns of Independence, Missouri, and Council Bluffs, Iowa, early in the spring, when grass was available on the prairie for the draft stock; and they had to cross the passes of the California Sierra before the first snows of the fall closed them to travel. Yet, these were only *physical* challenges to the overland trip. There were other equally daunting and complex dimensions, namely those connected with families. Internal family dynamics had been set in place over years. They pertained to questions of authority, responsibility, earning, and spending. In many families, those dynamics spanned three generations and involved important reciprocal duties. The numbers were infinite, but we might refer to a few specifics: marriage, which involved the union of two families; children, and the consequent responsibilities and obligations involving them; aging parents, and responsibility for them; and, surrounding all of those human connections, the question of overall family resources. Who was entitled to what when? What represented a reasonable share for the individual? For the family? The sudden appearance of gold in California fractured those delicate relations that had evolved over generations.[13]

The resolution (or the lack thereof, as the case may be) of those familial matters was not limited to the South, but it is useful to examine some of the known cases that involved southerners and a southern context (and in this connection, I take "southern" to mean those states in which slavery was legal). Consider the range of options that exists. One of the common features of the gold rush was that it cut across sectional and class lines. At least one suitor threatened to go to California in order to force the reluctant object of his affections to a decision. She married him, and the expedition to California was canceled.[14] In another case, a father and son talked constantly of going to California for reasons apparently related to ongoing family quarrels.[15]

Conclusions reached in weighing the question of going to California were almost always set in the context of the family. Individuals would go to dig gold, but all members of the family would benefit. Each generation would be advantaged, if in a different way. Men and women who had labored in field, shop, or store in endless drudgery for years would find relief; children who had been unable to go to school because their labor was so necessary for the survival of the family would have an education; aging parents who lived their final years in penury would find leisure and relief. Yet, these rosy or golden scenarios were not without cost. For an entire family to benefit, some of its members would have to depart. And by far the largest proportion of those family members were men in the prime of life, exactly that time when they contributed most to the family in terms of physical, economic, and emotional support. Hence many lengthy discussions focused on how the gold seekers' familial responsibilities were to be met or, at the very least, redistributed among remaining family members. For every forty-niner who made the decision to go to California in search of a fortune, two or three prospective Argonauts decided to stay at home. Those who elected not to venture to California do not immediately command our attention, but, at the same time, the cohort that remained at home represents something of a "control group."[16]

Most of the people who went to California in 1849 or in the next half dozen years did so in companies. Traveling in large groups had many advantages, whether the journey was by sea or overland. To begin with, there were economies of scale. Groups traveling together could do so at less expense. For those going by sea, individuals banded in companies could charter ships; for overlanders, traveling companions provided security against the hazards of the plains and the mountains, namely, the threat of hostile Native Americans. And, for both groups, to make the voyage with friends and neighbors enhanced its appeal, in that it offered a strong safety net in case of accident or illness and assured contact with home in case of disaster to individuals. Thus, men across the breadth of the nation at mid-century banded together in companies, elected officers, pooled their resources, purchased supplies, made arrangements for departure, and then received tearful good-bys from their families.[17]

California was seen as a place where traditional values of loyalty eroded or vanished in the headlong rush for gold; it was also a place full of disease, injury, and misinformation, and establishing a support system to protect individuals was a crucial part of preparing for the journey—as important as wagons, oxen, bacon, or flour. Travelers received blessings and advice from their families and communities. Family members asked them to go forth safely, to make their fortunes quickly, and to return promptly—and unchanged. Members of their communities admonished them to behave as upstanding representatives. Ministers, politicians, and editors contributed their wisdom to these departures. This collective counsel emphasized mutually respectful conduct and the observance of local standards of honor. On the occasion of the departure of a company that called itself the "Mississippi Rangers," for example, the Honorable S. Adams, apparently a local leader, addressed the departing parties in these words: "You are leaving us with the reputation of the most moral, elevated company of gentlemen that ever left this State on an expedition of this kind; this high character you must, you will sustain, wherever you go."[18] When a Mr. Foster addressed another of the departing companies, he told its members that "men in California were different from what they was [*sic*] at home." He asked the departing travelers "if we did not Calculate to stand by each other in sickness and in health when we should be in foreign climes far away from home and friends." When he asked for a show of support, men raised their hands in a gesture of solidarity with one another.[19]

The California gold rush was an endless series of surprises. It began with fabulous stories of first discoveries, subsequently discounted as exaggerated misinformation but in turn followed within a few short months by legitimate accounts of a new El Dorado, the literal reincarnation of the first Spanish gold discoveries. Next came the confusion and missteps involved in actual plans to go to California, then the surprises of the journeys themselves. For those going by sea, the great shock was probably the boredom of six months of ponderous travel. What began as a great adventure turned out to be endless days of watching infinite expanses of water. For those going overland, the elements of surprise were embodied in unexpected hazards—the lack of forage, the breakdown of draft animals, and the harsh heat of the desert encountered during portions of the journey. For many individuals, the dangers involved river crossings, accidental gunshot wounds within companies heavily armed for protection against depredations by Native Americans, and the vestiges of the cholera epidemic of 1849, which followed the forty-niners up the Missouri River valley to the outfitting towns and then out onto the plains. The journey across the plains provided most forty-niners with their first view of "the elephant," that mythical figure that represented everything new, dangerous, and different about the gold rush. And the challenge of this journey led those who completed it to think of themselves as forever enshrined in the halls of America's pioneers who had crossed the plains as part of the great national adventure.[20]

The forty-niners headed across the plains with guidebooks in hand. They knew about distances, about time, about equipment and foodstuffs, and about traveling light in order to preserve their draft animals. That they sometimes failed to do so and paid the price was part of the series of surprises of the California trail. The forty-niners found the route jammed with like-minded travelers. John Milner of Alabama wrote to his family: "There are thousands of men going along the road; in fact, it looks like the wagons hauling cotton to Macon just after a rise in the staple. I believe there are wagons stretched in sight of one another for 500 miles."[21]

Once they had arrived in California, the forty-niners found themselves confronted with a wide range of questions: what to buy, how to travel to the mines, and where to mine. But the most important one was probably with whom to mine. One of the surprising aspects of the California gold rush was that mining—associated in the popular mind with a lonely prospector accompanied by a solitary burro—turned out to be a very collective enterprise. Thus, the first search in California (if not conducted beforehand) was not for gold but for trustworthy companions with whom to work and live. The choice was a very personal one. With new partners—six to eight men was the most common unit for living and mining—the forty-niner would establish a new allegiance. The "company," or "mess"—both terms were in use—also offered support in case of illness, and if death followed, these companions would send word to the family in the East and close the accounts of the deceased. It was a personal connection in what for all forty-niners was a very crowded, competitive, impersonal world.[22]

The forty-niners soon had to come to terms with two other unanticipated factors associated with the California gold rush. The first was simply that mining was hard, repetitive, exhausting work. Contrary to the popular view of nuggets lying around to be harvested at leisure, gold in California was claimed in small increments by backbreaking labor. E. Gould Buffum, who toured the gold regions in 1849, offered this comment on the intensity of the work: "The labour of gold-digging is unparallelled by any other in the world in severity. It combines, within itself, the various arts of canal-digging, ditching, laying stone-walls, ploughing, and hoeing potatoes." One Argonaut described the labors of mining in these terms: "Digging for gold is the hardest work a man can get at—I have dug holes 20 feet deep and about 6 feet square—and I have seen holes dug 100 feet deep."[23]

Still another unforeseeable matter surrounded the definition of who "succeeded" and who "failed" in the California gold rush. Americans at mid-century—coming from generations of experience in which estates were built slowly by the hard work of all members of a family—always assumed that honesty, morality, and diligent labor brought success. The California gold rush was a glaring exception. There, fortune rewarded miners at random. Two mining companies would work equally hard in adjacent claims. One would average twenty-five dollars a day per man; the second, twenty-five cents. Much mining turned out to be a large-scale lottery. And morality had little to do with it. Honest miners might make ten dollars a day (if they were lucky); men who sold whiskey and ran gambling establishments almost surely made twenty times as much. Prentice Mulford, one of the most insightful of gold-rush observers, commented that the gold dust that made its way from the mines to farms, villages, and cities of the East "was not always dug by the moral Argonauts, from whom most was expected. It was often the gathering of some of the obscurer members of our community."[24]

The activities of some southerners and their enslaved African Americans in the gold fields also had some unanticipated results. Although California entered the Union as a "free" state in 1850, enslaved African Americans worked in the mines long thereafter. Consider, for example, the Dickson family of North Carolina. Robert Dickson left his close-knit family in Burke County to join the California gold rush in 1852. Accompanying him to California were at least four slaves, as well as his cousin, S. M. McDowell, who likewise brought slaves. Apparently neither man felt the slightest uneasiness about taking slaves to California.[25] In late fall 1852 Dickson and his slaves began work in Woods Diggings. Dickson reported that his claim grossed from four to seven dollars a day per hand. He had an interest in five other claims about four miles away on Montezuma Flat. When the claim in Woods proved less productive than he had hoped, Dickson—despite California's status as a free state—hired out his slaves for seventy-five dollars per month each, or three hundred dollars a month for the four.[26]

In a series of letters to Dickson's family, McDowell pronounced himself disappointed with mining prospects in California: "I cannot clear more than one fifth as much as I expected to when I left home," he wrote. "Still there is much more money here than there is in North Carolina, and notwithstanding all the high expenses in this country much more money can be made here, I think, than at home." And, he added, "much more will be spent." Although mining was the "principal employment in this section of the country," McDowell remarked, many forty-niners worked elsewhere. "Merchandising" continued to be a good business. McDowell noted that some excellent young men of "judgement, health, and enterprise" had done well and might continued to do well, but most of the miners "work long and hard for moderate wages." Opportunities for wealth in California, he concluded, were "fluctuating and uncertain." It was a guarded assessment, emphasizing equally the generally favorable moral character of the forty-niners and the mixed economic opportunities that awaited them—and him.[27]

In spite of his mixed feelings about opportunity in California, McDowell and his enslaved African Americans had a run of very good luck. His slaves ("his boys," as McDowell wrote of them) "made him some fifteen hundred or eighteen [hundred] Dollars" over three weeks. Lagan and Lawson were the most productive of the slaves, and in one letter McDowell singled them out for praise, although there is no evidence that they were further rewarded. The assumption among most southerners was simply that the master owned the slaves and the product of the slaves' labor, although, as will be shown, there were alternative arrangements.[28]

S. M. McDowell took his money and returned to North Carolina; Robert Dickson stayed on with his slave John. From the opening of Dickson's correspondence with his family, he mentioned John frequently. He consistently noted John's health, varied occupations, and earnings and conveyed John's good wishes to the family. John often sent his love to his mother and asked the plantation slaves to write letters to him, for "all the Negroes gets letter[s] from home but him." Robert Dickson sometimes closed his letters by sending "howdys" to the "Black ones."

Beginning in the spring of 1853, times became hard for Dickson and John. Sometimes Dickson thought of himself as unlucky; at other times, he continued to have hopes for the future. By the spring of 1854 he was prospecting more than mining, testing from eight to ten claims a week, making wages or less on all of them. Dickson finally concluded he must try something other than mining, a transition that would offer more security but less chance of making a substantial sum. His luck did not improve and instead took a dramatic turn for the worse in April 1855 when he was found murdered in his cabin. In the style of so many forty-niners, close friends—in this case a cousin—took care of the estate, which was encumbered by debts and scarcely large enough to pay the expenses of a burial.[29]

When S. M. McDowell had returned to North Carolina, he left slaves in California to continue mining his claims. Several letters concerning the disposition of money passed between masters and mistresses on the plantations and slaves on the watercourses of the sierra. Should the slaves send the gold dust by the usual bank draft, or should they exchange the dust for paper money and bring the money east when they returned?

Albert McDowell, apparently S. M. McDowell's servant, remained in California for two years after his master returned to North Carolina. In May 1855 he wrote to ask permission to remain in the gold fields for another year. He thought that he could make four or five hundred dollars in that time "if God Gives me Health" and pledged to remit that money to his master. Albert McDowell had sent $400 to his master the previous year, making a total of $950 "since Master Samuel [S. M. McDowell] left for home." Albert also forwarded $200 to his wife, apparently part of an arrangement under which he would remit the largest portion of his earnings to his master but retain a portion of them for himself. With his earnings for the year (presumably after expenses—a considerable item in the gold fields) calculated at $600, Albert McDowell kept one-third for his wife. That was a substantial sum, but so was the larger amount that he forwarded to S. M. McDowell. Albert McDowell surely enjoyed the degree of independence he found in the mines. The presence of various members of the Dickson/McDowell families in the neighborhood probably offered him some degree of protection, and the continued stay of his wife on the plantation in North Carolina ensured his return.[30]

The story of Robert Dickson, S. M. McDowell, John, and Albert McDowell and their activities in California has a distinctly southern flavor. But in engaging in mining itself and in their approaches to entrepreneurship, Robert Dickson and S. M. McDowell were simply two more forty-niners intent on seizing the prospects for wealth offered by the Golden State. Where their activities differed from those of most other forty-niners was in their use of African American slave labor. On the other hand, their slaves benefited from an unusual degree of independence. Mining involved a repetitive work routine in search for incremental bits of wealth, and only the most rigid supervision could ensure absolute honesty, suggesting that what seems to have been the arrangements these slaveholders made—trust and sharing a portion of the returns—were probably the best for effective use of slave labor. Or perhaps the white southerners had great confidence in the fidelity of their black property. The financial arrangements further suggest that this particular extended family inculcated in their slaves a strong degree of personal incentive.

Among the vicissitudes of the California gold rush, perhaps none was more unanticipated than the indecision concerning the proper time to return home. Most forty-niners intended to return home at the first available opportunity. They had promised to do so when initially seeking their family's permission to join the gold rush, and they reiterated that commitment in letters on a regular basis. The question of returning home was the one most often asked by their families; it was in every letter running from east to west. It was more frequently asked than questions about gold. Of course, the response was clear and unequivocal. Yet, as the Argonauts settled into the camps in California and went to work, the issue seemed less and less clear. In the words of one forty-niner writing to his family, "You wanted me to say when I was Coming home. That I cannot say for when a person gits to California It is hard to say and tell when he gets away."[31]

How had such a straightforward decision become so problematic? In the final analysis, like the sacrifice that was ostensibly the reason for going in the first place, to remain in California and at work in the placers under the hardest conditions for minimal returns validated the determination and sacrifice of the forty-niner. The sense of mission that began in the parlor intensified in the gold fields, particularly if the returns were disappointing. Wrote another forty-niner: "If I had gone home then [last fall] without looking in other places, I should not have been satisfied or thought I had done my duty to you or to my children."[32]

But there were other, equally pressing reasons, many of which involved the fear of disappointing loved ones by surrendering to outright failure. To the consternation of almost every Argonaut, gold turned out to be scarce, and its scarcity increased over time in direct proportion to the numbers of miners at work. At the close of 1849 there were 40,000 miners in the gold fields; in the fall of 1850, 50,000 were at work; and by the fall of 1852, 100,000 miners were crisscrossing the same gold-bearing streams. In many places, the sheer number of prospectors had reduced individual mining claims to ten feet square. But the increasing scarcity of gold in the placers was not obvious in the countryside, towns, and cities of the eastern half of the continent. There, stories of wealth from California continued to fill newspapers and the dreams of potential gold seekers. In light of such expectations, it was hard to return with little or nothing. Indeed, to come home, it was soon argued, was the easy way out. These Argonauts of '49 were made of sterner stuff. The challenge of California was to stay in the diggings and to persevere in the name of their families.

Reinforcing such determined views was an ill-disguised contempt for those who gave up easily the chance to do something important for their families and returned too soon. One forty-niner expressed that disdain with these words: "We know you would cheerfully assist us at home, if we were there, but we now have an opportunity to which if we improve aright will be of great advantage to us. . . . But with anything like luck we cant fail, & Lord only knows what sends so many Georgians home, except the reflexion of how 'comfortably I may live at home at my father's hard earnings.' I know many who refuse large wages here & go home. If necessary I would give names."[33] The only conclusion possible was that families themselves should willingly accept the sacrifices of the forty-niners and their determination to labor on in the face of considerable hardships as the more difficult choice. It was better to be a casualty in the struggle for a golden future than to be branded a coward for desertion.

The difficulties of going home or even contemplating a return touched on a range of questions at the heart of the gold rush. Everyone had gone to California to find gold, or to enrich themselves through the economic oppportunities associated with gold. But what constituted enough gold with which to come home? What represented success, an acceptable return for so many years in the placers? And who was prepared to return home with little or nothing? In short, who was ready to admit that he had failed in a California awash with gold? For such a failure would involve a great loss of face not only for an individual but for an entire family as well.

For the Argonauts of '49, returning home was the final step in coming to terms with failure. There might have been temporary setbacks in California, but these were always seen in the context of brighter prospects for the next season. To return home with little or nothing was to admit there would be no further seasons, notwithstanding the many stories of dazzling successes told about gold-rush California. One forty-niner wrote that he could not consider coming home because of "mortification" over his failure. For several months, he had dug and washed in the mines, worked as a carpenter, and clerked in a post office, and yet he had nothing to show for his endless labor and had been unable to send his wife sufficient funds to ensure her support. Another prospector wrote to his mother: "I dislike the idea very much of returning without making any thing . . . besides I know that a good deal would be said about me as every person in the states think that no person that is industrious can come out here without making a fortune."[34] These Argonauts did not wish to measure their thin money pouches against the grandiose expectations in their communities at the time of their departure. One of them wrote: "I cannot think of returning home from the Land of Gold with Nothing, notwithstanding the chords [*sic*] of Fraternal, Conjugal and Paternal Love are drawing me toward home." He continued: "Would you want me to return without making every exertion to better our condition even though I might be obliged to remain here longer, than I intended to stay[?]" And, he observed, "you know what I said to my friends when I came away—that when I came back they might know that I had more money than when I went away."[35]

Most of the forty-niners eventually did return home, where they generally found a loving and warm welcome from their families. When they talked about the experiences they encountered in California, they were met with quizzical looks. When they tried to explain the difficulties of mining to their creditors, they met incredulous stares. For the Argonauts of 1849, a lasting legacy of their gold-rush experience was the unanticipated and sometimes harsh reality of California leavened, perhaps, with a lifetime of memories and stories about their participation in one of the great American adventures of the nineteenth century.

Notes

1. On the original gold discoveries, see Rodman W. Paul, *The California Gold Discoveries: Sources, Documents, Accounts, and Memoirs Relating to the Discovery of Gold at Sutter's Mill* (Georgetown, Calif.: Talisman Press, 1966), especially the introduction.

2. Malcolm J. Rohrbough, *Days of Gold: The California Gold Rush and the American Nation* (Berkeley: University of California Press, 1997), introduction and chapter 1 set the stage for discoveries and the spread of the news around the world.

3. Rodman W. Paul, *California Gold: The Beginning of Mining in the Far West* (Cambridge: Harvard University Press, 1947), 140–141.

4. Mason quoted in Paul, *The California Gold Discoveries*, 91–92. See also Paul, *California Gold*, 52–53.

5. *Raleigh Star and North Carolina Gazette*, March 28, 1849.

6. Charles Harlan to Julia LeGrand, December 8, 1848, Charles T. Harlan Papers, Bancroft Library, University of California, Berkeley; Joel and Christina Thomas to Nathaniel and Katharine Comer, July 15, 1849, Nathaniel Comer Papers, Special Collections Library, Duke University, Durham; William Cook to John Rutherford, March 28, 1849, John Rutherford Papers, Special Collections Library, Duke University.

7. *Corpus Christi Star*, January 13, 1849.

8. *New Orleans Daily Picayune*, March 4, 1849, Bieber Collection, Henry E. Huntington Library, San Marino, California.

9. *Southern Shield* (Helena, Arkansas), February 24, 1849.

10. The *Louisville Daily Journal* of May 9, 1849, lays out the Georgia plan to enforce slavery on California. Bieber Collection, Huntington Library.

11. Paul, *California Gold*, 345–438.

12. David Campbell to his nephew, April 22, 1850, Campbell Family Papers, Special Collections Library, Duke University.

13. Rohrbough, *Days of Gold*, chapter 3, discusses the family dynamics.

14. Rebecca Camfield to Octavia Milligan, February 18, May 13, 1849, Milligan Family Papers, Southern Historical Collection, University of North Carolina Library, Chapel Hill.

15. Charles Anthony Hundley Papers, passim, Special Collections Library, Duke University.

16. Rohrbough, *Days of Gold*, chapter 3.

17. David M. Potter's introduction to *Trail to California: The Overland Journal of Vincent Geiger and Wakeman Bryarly* (New Haven: Yale University Press, 1945) analyzes the importance of the company as an organizing unit of the gold rush.

18. *Monroe Democrat* (Aberdeen, Mississippi), January 9, 1849, Bieber Collection, Huntington Library.

19. Benjamin Baxter, Journal, March 25, 1850, Huntington Library.

20. The introduction to Potter, *Trail to California,* is excellent on the unforeseen hazards of the California trail. On the "elephant," see John Phillip Reid, *Law for the Elephant: Property and Social Behavior on the Overland Trail* (San Marino, Calif.: Huntington Library, 1980), ix–x.

21. John Milner to his sister, May 18, 1849, John Turner Milner Correspondence, Alabama Department of Archives and History, Montgomery.

22. Rohrbough, *Days of Gold,* 75–80, discusses the search for reliable companions.

23. E. Gould Buffum, *Six Months in the Gold Mines: From a Journal of Three Years' Residence in Upper and Lower California, 1847–8–9* (Philadelphia, 1850), 180; Christian Miller to his family, March 23, 1851, Christian Miller Letters, Bancroft Library.

24. Prentice Mulford, *Prentice Mulford's Story: Life by Land and Sea* (New York, 1889), 4.

25. The William Dickson Family Correspondence at the Southern Historical Collection contains a wealth of material about this family, with an emphasis on affairs in North Carolina. The exchanges suggest the degree to which life continued on in very customary ways, even as distant relatives reported the exotic experiences they encountered in California. The correspondence of two young people, filled with references to the California gold rush and its influences on their lives and the larger society around them, is in Elizabeth Roberts Cannon, ed., *My Beloved Zebulon: The Correspondence of Zebulon Baird Vance and Harriet Nowell Espy* (Chapel Hill: University of North Carolina Press, 1971).

26. Robert Dickson to "My Very Dear Sister," December 10, 1852, Dickson Family Correspondence, Southern Historical Collection.

27. McDowell to "My Very Dear Uncle, Aunt & Cousins," undated but presumably written in early 1853, Dickson Family Correspondence, Southern Historical Collection.

28. Robert Dickson to "My Very Dear Father and Mother," January 24, 1853, Dickson Family Correspondence, Southern Historical Collection.

29. Junius Gates to William Dickson, April 26, 1855, R. M. Dickson to "My Dear Brother," April 24, 1853, R. M. Dickson to William and Margaret Dickson, May 27, 1854, Dickson Family Correspondence, Southern Historical Collection.

30. Albert McDowell to "My affectionate Mistress," July 13, 1854, and to "My Dear Master," May 15, 1855, Woodfin Letters, Southern Historical Collection. A case in which a slave owner from Mississippi who took three slaves (later freed) to California for the gold rush sued for their return is analyzed in Ray R. Albin, "The Perkins Case: The Ordeal of Three Slaves in Gold Rush California," *California History* 67 (1988): 215–227.

31. Ephriam Delano to Jane Delano, April 24, 1853, Ephriam Delano Letters, Huntington Library.

32. John Fitch to his wife, February 16, 1852, John Fitch Letters, Huntington Library.

33. John Milner to his father, January 7, 1850, John Milner Letters, Alabama Department of Archives and History.

34. James Burr to Caroline Burr, December 22, 1850, Wright and Green Family Papers, Southern Historical Collection; Daniel Horn to his mother, November 1, December 14, 1850, Daniel Horn Letters, Southern Historical Collection.

35. Levi Hillman to his family, December 9, 1852, Levi Hillman Letters, Minnesota Historical Society, St. Paul.

The Cripple Creek Mining District: The First Ninety Years

Ed Hunter

Since graduating from the Colorado School of Mines in 1953, Ed Hunter, E.M., has worked as a miner, engineer, and manager for mining companies from Vanadium, New Mexico, to Nome, Alaska. Hunter, a past chairman of the board of trustees of the Western Museum of Mining and Industry, Colorado Springs, Colorado, continues to serve as a lifetime trustee of that body. He is also a member and a past council member of the Mining History Association, Denver, Colorado.

For hundreds of years, Colorado's Cripple Creek/Victor gold deposit, one of North America's great goldfields, lay hidden under grasslands high on the west flank of Pikes Peak. The gold seekers rushing to California in 1849 had bypassed the area, for far better transcontinental routes lay to the north and south of Colorado. As early as 1874 H. T. Woods, a surveyor with the government topographical survey of the western Pikes Peak area under the direction of F. V. Hayden, noted the presence of gold in the area. The Hayden survey party mapped the Pikes Peak region for future homesteading, but its final report generated little interest in Woods's initial observation.[1] The reputation of the region as a gold prospect suffered as those who came west in 1859 under the banner of "Pikes Peak or Bust" were forced to return "busted." The only winners were the miners from Georgia who found gold in Gregory Gulch, a few miles west of Denver.[2] To add insult to injury, in 1884 merchants in the vicinity of Canon City, Colorado, allegedly hired "Chicken Bill" Lovell (also known as S. J. Bradley) of Leadville to "salt" some claims about thirteen miles west of what is now known as the Cripple Creek Mining District. The merchants prospered by selling supplies to the reported hundreds of prospectors who rushed in. "Chicken Bill" escaped just ahead of the noose when the hoax was discovered. A bottle of gold chloride, a favorite substance for instantly enriching or salting barren gold samples, was found on one of his partners. For many years afterward, memories of the hoax, which was incorrectly identified as having taken place at Mount Pisgah, adjacent to Cripple Creek, coupled with the specter of the "busted 59er's," discouraged almost all prospective gold seekers from wasting their time in the Pikes Peak region.[3]

The district then lay dormant for several years save for ranching and part-time prospecting by Bob Womack, a ranch hand for the Broken Box Ranch. In addition, both B. F. Requa, a Manitou Springs, Colorado, merchant, and Henry Cocking, a Cornish miner from Central City, Colorado, drove short adits in the area but failed to find gold.[4] Finally, in 1890, rock from Womack's diggings on the El Paso Lode, assayed in Colorado Springs, yielded twelve troy ounces of gold per ton of ore.[5] Those results led to the formation of the Cripple Creek Mining District on April 5, 1891.[6] An additional reason for the late discovery of gold in the area was the lack of "free" gold (gold released from parent rock and concentrated elsewhere). Some free gold was found just northwest of the present city of Cripple Creek on Mineral Hill and subsequently at two or three other small deposits in the district. However, a great deal of the gold in the Cripple Creek district was chemically tied up as a gold telluride, a compound of gold and tellurium, far beyond the ability of the gold seekers of that time to extract.

When small placer deposits that were present were traced to an outcrop, machinery was required to break up the rock to free the gold. The experienced California and Gilpin County, Colorado, miners brought in stamp mills to aid recovery.[7] Two hundred and seventy stamps were operating in the district by 1892. Very poor gold recovery resulted, however, for most of the free gold bore a coating or film of iron telluride, which resisted amalgamation.[8] High-grade ores yielding several ounces of gold to the ton could be sent to smelters in Pueblo and Denver for recovery. At first, that process entailed hauling the rock by horse and wagon to the Colorado Midland Railroad, twenty miles to the north. The evident need for better transportation quickly resolved itself. In 1894 the Florence and Cripple Creek Railroad reached the district from the Arkansas River to the south.[9] The Midland Terminal Railroad arrived during the same year, coming south from the Colorado Midland Railroad at Divide.[10] By the end of the century, the two railroads merged, creating a virtual monopoly.[11] The inevitable rate increase that followed antagonized the mine operators, and in 1901 those men completed the Colorado Springs and Cripple Creek Railroad or "Short Line" around the south flank of Pikes Peak into the district.[12] Competition returned, and price wars ensued.[13]

But what could be done with the bulk of the Cripple Creek ores, which did not contain sufficient values to pay for transportation to the smelters, let alone the smelting charges? Miners from the Black Hills of South Dakota brought into the Cripple Creek district the chlorination process, which became a favored method for recovery. The Economic Mill, just outside Victor, was an early user of chlorination.[14] By 1893 the Brodie Mill in Squaw Gulch, near Victor, held a license for the MacArthur cyanide process, but it took the work of metallurgist Philip Argall, superintendent of the Brodie Mill, to develop a practical application for the Cripple Creek ores. By 1895 Argall had improved the process of recovery by cyanidization sufficiently that David Moffat, mining and railroad mogul, hired Argall to build the four-hundred-ton Metallic Extraction Mill along the Arkansas River at Cyanide.[15]

Although several small chlorination and cyanide mills were in operation in the district by the late 1890s, problems had arisen. Fuel for mine and mill boilers was nearly nonexistent in the area. The few trees that had been present were soon used for timber and firewood. Coal and fuel oil were available from the coal mines and an oil field in Florence, thirty-two miles to the south and five thousand feet lower in elevation.[16] Nevertheless, it was cheaper to haul ore downhill than to haul uphill the large quantities of fuel needed. Water for mills and homes was rather scarce inasmuch as the region has an annual precipitation rate of just over nineteen inches.[17] Such factors led to construction of larger, more efficient mills closer to sources of fuel and large quantities of water. At first, those newer mills, such as the Metallic

Extraction Mill, were situated along the Arkansas River between Florence and Canon City. Later Colorado City, just west of Colorado Springs, became the milling center for the ores of Cripple Creek.[18] The chlorination and cyanidation processes competed for the Cripple Creek market into the early 1900s. Chlorination, highly favored at first, soon lost out to the more efficient cyanidation. By 1911 all of the chlorination mills had been closed, many by fire (blamed on that newfangled invention, electricity).[19] The Cripple Creek eruptive breccias (with the original rocks broken up and re-cemented) hosted much of the gold mineralization, which was emplaced along structurally prepared zones. The harder, usually barren, clasts (larger material within the breccias) were surrounded by gold mineralization in the fine-grained material. Screening and ore-sorting became a critical part of the recovery system.[20] Crude screening evolved into sophisticated plants using trommels, followed by hand-sorting. Reduced ore volume after sorting cut transportation costs (railroads were never known for charity to miners).

The Cripple Creek Mining District flourished between 1891 and 1900. Oddly enough, the financial panic of 1893 helped to accelerate a rapid rise in production. In the wake of the panic, one-third of the railroads in the United States were in receivership, 600 banks failed, 15,000 commercial establishments went bankrupt, and the nation could no longer issue gold certificates inasmuch as its gold reserves had fallen below the million-ounce limit. The final blow to the Colorado mining economy came when Congress repealed the Sherman Silver Purchase Act. That action, coupled with the cessation of silver buying by the mint in India, resulted in an immediate plunge in the price of the metal.[21] Many of the great silver mines at Creede and Leadville virtually shut down, and unemployed silver miners flocked to Cripple Creek. Gold production increased, and in 1899 nearly nine hundred thousand ounces of gold were recovered. In 1898 Winfield Scott Stratton, an itinerant carpenter and part-time prospector, sold his great Independence mine to the English Ventures Corporation for eleven million dollars.[22] The original lone prospectors were giving way to companies with the financial resources to sink deeper shafts and expand operations. Flourishing stock markets in Colorado Springs, Cripple Creek, and Victor raised capital for the mines and mills—and an occasional swindler.

Like most mining camps, Cripple Creek and Victor were fodder for extensive fires just before the turn of the century. Cripple Creek burned twice in three days, Victor only once. Both towns rose overnight from the ashes into modern cities of brick.[23] The wood-framed surface buildings of the Gold Coin mine were destroyed in Victor's fire but were rebuilt in brick; the new structures featured stained glass windows in the hoist house! Three railroads and two trolley lines tied the surrounding small settlements to Cripple Creek and Victor, creating a metropolitan area. All sorts of foods and the latest fashions came in by rail. A would-be miner could board a train on the East Coast and step off in Cripple Creek. Contrast that relative degree of convenience with the challenges faced by the gold seeker traveling to the Klondike over Chilcoot Pass. What a difference! Good times rolled. The district boasted an amusement park served by both trolley and train at Cameron.

The park's zoo had rock dens (which still exist) built to house the bears. Entertainment reached a new high (or low) when the only recorded bullfight in North America was held in Gillett.[24] The local humane society stopped the action, but the entrepreneur, one Soapy Smith, went on to entertain miners traveling to the Klondike.

A tapering-off of production began after the turn of the century and continued until World War I, primarily because of factors associated with increased mining depth. Such factors included decreasing ore grades, narrowing of ore veins, and increasing quantities of water underground (despite its relative absence on the surface). The Cripple Creek district has been likened to a granite bottle with a large sponge inside. Thousands of years of rain and snow melt had saturated the "sponge"—the gold-bearing interior.[25] In 1900 underground mine pumps were inefficient at best. The great Cornish beam pumps had reached their epitome at the Comstock. Centrifugal pumps had yet to come into their own, and the steam pumps of that era were expensive to operate. Meeting the challenge of the excess underground water required the use of a technique that had been employed at least as early as A.D. 1100 at the Rammelsberg mine in Germany.[26] There drainage tunnels, or adits, were driven to dewater the mines. The topography of the Cripple Creek area was conducive to that procedure inasmuch as the district lies between 9,500 and 10,500 feet in elevation, above much of the surrounding terrain. Tunnels could be driven from drainages at lower elevations into the heart of the district. The process has been compared to drilling a hole through a granite bottle and letting the water drain by gravity. By 1919 four major drain tunnels had been driven at successively lower elevations to aid in the drainage as mining progressed. Processing of previously mined waste rock became more prevalent as operators, with improved technology, used the dumps to improve gold production. In 1905 metallurgist Philip Argall proposed that a client place fifty thousand tons of waste rock on a concrete pad and that a low concentration of cyanide be allowed to percolate through the heap, discharging a gold solution at the bottom. The client passed up Argall's farsighted proposal.[27]

The evolving recovery techniques that helped shape the district were paralleled by emerging mining practices. Initial prospectors' pits gave way to small shafts, serviced by hand-powered windlasses or horse-driven whims. As the mines became deeper and more extensive, mechanical hoists were installed, and skips in which ore was hoisted to the surface supplanted the older method of hoisting loaded one-ton cars to the surface for dumping. Several of the larger mines such as the Portland, the Independence, and the Elkton used reel-type, flat rope hoists that, in later years, gave way to the conventional double-drum, round rope hoists. Electric power eventually superseded steam power.

As a result of two periods of labor unrest (1893-1894 and 1903-1904), the Cripple Creek Mining District became union-free after 1904. Government troops were called out in both instances, but the most violence took place in the 1903-1904 period.[28] Albert Horsley, better known as Harry Orchard, a dynamiter for the Western

Federation of Miners (WFM), detonated a bomb in the Vindicator mine, resulting in the death of the superintendent and a foreman. Shortly thereafter, Orchard blew up the Florence and Cripple Creek Railroad station in the town of Independence as the night shift from the nearby Findley mine waited to go home. Thirteen miners died, and many were severely injured. He then fled to the Coeur d'Alene mining district of Idaho, where he set another bomb, fatally wounding the former governor of that state.[29] Orchard was apprehended for that tragic act and put on trial along with the officers of the WFM, who, incidentally, were taken by force from Denver to Idaho to stand trial. Orchard was convicted and spent the rest of his life in prison. The work of the defense team, including prominent attorney Clarence Darrow, successfully defended the WFM hierarchy of William "Big Bill" Haywood, Charles Moyer, and George Pettibone against complicity in any of Orchard's crimes.[30]

One bright spot in the history of the district occurred in 1914 on the 1,200-foot level of the Cresson mine. Miners broke into a 15-foot by 20-foot by 28-foot-high vug (a small unfilled cavity within a lode), much like a giant geode, that contained sixty thousand ounces of gold. Iron doors were installed and armed guards were posted while trusted employees bagged the gold for shipment to the smelter. Even at the low (by historical standards) present market price of gold, the vug's content would be valued at more than fifteen million dollars.[31]

In spite of the relief in heavy operating costs afforded by installation of the drain tunnels, production fell dramatically after World War I through the 1920s and early 1930s. A brief upswing occurred in 1934 when the gold price was officially raised from $20.67 to $35.00 per ounce.[32] By the late 1930s, however, it had become evident that regardless of the price increase, something would have to be done to drain potential production levels below the eight-thousand-foot elevation level if mining were to continue. The Southern Colorado Power Company entered the discussions, promoting the use of newer pumps (with electricity supplied by it, of course) against the unknown ground conditions likely to be encountered with a deeper drain tunnel.[33] The prime mover in the district, the Golden Cycle Company, nevertheless indicated a preference for a drain tunnel at about seven thousand feet in elevation as the best option. When Golden Cycle failed to obtain federal funding for the project, it financed the drive internally at a cost of approximately 1.2 million dollars. The actual tunnel construction began in July 1939. The cream of the district's miners, under the direction of veteran tunnel driver Long John Austin, drove each mile of the tunnel faster than the previous one. The total advance of 6.11 miles through granite was completed in two years and five days at an average rate of 47.24 feet per day. That pace of progress set records for drill-and-blast tunnel advances from a single heading and accomplished the drainage of the mine workings down to the seven-thousand-foot elevation by 1941. The Carlton Tunnel, although a masterpiece of underground mining construction, turned out to be a prime example of poor timing.[34]

During World War II the government restricted mining of nonessential metals, and the district's gold mines were shut down for the duration. Miners were sent to mine copper, lead, and zinc for the war effort or to serve in the armed forces. Golden

Cycle was able to mine a small tonnage for a short time in return for agreeing to convert part of its Colorado Springs gold mill to a facility to recover lead and zinc being mined in various areas of Colorado.[35] In 1953, during testimony at congressional committee hearings on compensation for the shut-down mines, the president of Golden Cycle testified that he knew of only four or five miners from the entire Cripple Creek district who ever became copper miners. Although the wartime restrictions were declared unconstitutional in 1954, no compensation for damages was ever awarded.[36]

After the war, the district did not immediately recover. The idle mines had deteriorated, and former miners did not return for one reason or another. Postwar inflation drove up the cost of materials and supplies. The uranium boom on the western slope of Colorado drew many of the district's miners to better-paying jobs. In an effort to revive the district, the Golden Cycle Company shut down the only surviving mill in Colorado Springs and erected the Carlton mill in the center of the mining district. Because the relocation eliminated the freight cost to Colorado Springs, Golden Cycle hoped that that incentive, as well as improvements in the efficiency of the mill, would boost district production. The mill officially commenced operations in 1951 when Lowell Thomas, author, commentator, and world traveler, threw a switch to energize the mill via international radio.[37] At the time, Carlton, then processing one thousand tons of ore per day, was the largest custom gold mill in North America and only the third facility in the nation to utilize the chemical process of carbon adsorption for collecting gold.[38] During its ten years of operation, the mill produced 437,077 ounces of gold before shutting down in 1961.[39] In that year, with gold still pegged at thirty-five dollars per ounce and operational costs per ounce running at that or higher rates, most gold-mining operations (with the exception of Homestake at Lead, South Dakota) closed. With the closure of the Carlton mill in 1961, the Cripple Creek Mining District lay dormant once again, dampening the hopes and dreams of many people associated with it. Although sporadic exploration took place there throughout the 1960s, it was not until the federally imposed ceiling price of thirty-five dollars per ounce of gold was removed in the early 1970s that interest in the district began to revive.[40]

Notes

1. Waldemar Lindgren and Frederick L. Ransome, *Geology and Ore Deposits of the Cripple Creek District, Colorado*. USGS Professional Paper 54. (Washington D.C.: Government Printing Office, 1906), 130.

2. E. Merton Coulter, *Auraria: The Story of a Georgia Gold Mining Town* (Athens: University of Georgia Press, 1956), 112. The gulch may have been named for John Gregory, former Georgia gold miner.

3. Marshall Sprague, *Money Mountain* (Boston: Little, Brown and Company, 1953), 22–27, 304; Thomas A. Rickard, "The Cripple Creek Goldfield," *Transactions of the Institute of Mining and Metallurgy* [London] 8 (November 15, 1899): 50; N. E. Guyot, "Cripple Creek, An Inside Story—1," *Engineering and Mining Journal-Press* 118 (December 13, 1924): 933.

4. Lingren and Ransome, *Geology and Ore Deposits of Cripple Creek*, 130–134.

5. Robert Guilford Taylor, *Cripple Creek Mining District* (Palmer Lake, Colo.: Filter Press, 1973), 23.

6. Sprague, *Money Mountain*, 65.

7. Rickard, "The Cripple Creek Goldfield," 79.

8. Lingren and Ransome, *Geology and Ore Deposits of Cripple Creek*, 138.

9. Tivis E. Wilkins, *A History of the Florence and Cripple Creek and Golden Circle Railroads*. Colorado Rail Annual, No. 13. (Golden: Colorado Railroad Museum, 1976), 35.

10. Edward M. McFarland, *The Cripple Creek Road: A Midland Terminal Guide and Data Book* (Boulder, Colo.: Pruitt Publishing Company, 1984), 15.

11. Morris Cafkey, *Rails Around Gold Hill* (Denver: Rocky Mountain Railroad Club, 1955), 94.

12. Tivis E. Wilkins, *The Short Line to Cripple Creek: The Story of the Colorado Springs and Cripple Creek District Railway*. Colorado Railroad Annual, No. 16. (Golden: Colorado Railroad Museum, 1983), 41.

13. Cafkey, *Rails Around Gold Hill*, 121–124.

14. Lindgren and Ransome, *Geology and Ore Deposits of Cripple Creek*, 142.

15. Robert Spude, "Cyanide and the Flood of Gold: Some Colorado Beginnings of the Cyanide Process of Gold Extraction." Essays and Monographs in Colorado History, No. 12. (Denver: Colorado Historical Society, 1991), 18–21.

16. H. Lee Scamehorn, "In the Shadows of Cripple Creek: Florence from 1895–1910," *Colorado Magazine* 55 (1978): 219.

17. Cripple Creek & Victor Gold Mining Company, Weather Records, CC&V Administration Building, Victor, Colorado.

18. Cafkey, *Rails Around Gold Hill*, 122.

19. Charles W. Henderson, *Mining in Colorado—A History of Discovery, Development and Production*. USGS Professional Paper 138. (Washington: Government Printing Office, 1926), 60.

20. Lindgren and Ransome, *Geology and Ore Deposits of Cripple Creek*, 136–137.

21. Joseph J. Hagwood, *The California Debris Commission: A History* (Sacramento: U.S. Army Corps of Engineers, Sacramento District, 1981), 30.

22. Frank Waters, *Midas of the Rockies, Stratton Centennial Edition* (Denver: Sage Books 1948), 214–227.

23. Sprague, *Money Mountain*, 188–198, 315.

24. Sprague, *Money Mountain*, 177–187.

25. Albert H. Beebe, "Carlton Drainage Tunnel," *Mining Congress Journal* 27 (January 1941): 32.

26. Wilhelm Bonhardt, *The History of the Rammelsberg Mine*, trans. T. A. Morrison (Berlin: Prussian Geological Society, 1931; reprint, London: The Mining Journal, Ltd., 1989), 221. (Information in text from author's personal copy of manuscript, 1988.)

27. Report, Philip Argall to Stratton Estate, July 1905, Cripple Creek & Victor Gold Mining Company, company records, Victor, Colo.

28. Benjamin McKie Rastall, "The Labor History of the Cripple Creek District: A Study in Industrial Revolution," *Bulletin of the University of Wisconsin*, No. 198 (February 1908): 119.

29. Stewart H. Holbrook, *The Rocky Mountain Revolution* (New York: Henry Holt and Company, 1956), 101–105, 117–123.

30. J. Anthony Lucas, *Big Trouble* (New York: Simon and Schuster, 1997), 722, 748.

31. Horace B. Patton, "The Cresson Bonanzas at Cripple Creek," *Mining and Scientific Press* 115 (September 15, 1917): 381-385.

32. Taylor, *Cripple Creek Mining District*, 119.

33. Golden Cycle Company Records, Cripple Creek & Victor Gold Mining Company, company records, Victor, Colo.

34. Ed Hunter, "The Carlton Tunnel: 'it never was a bore!'" *The Mining History Journal: The Fifth Annual Journal of the Mining History Association* (1998): 40.

35. Annual Report, Golden Cycle Corporation, 1960, 18, Western Museum of Mining & Industry Library, Colorado Springs.

36. "H.R. 4393: For the relief of the owners of gold mines as a result of operation of L-208," Official Stenographer's Minutes, Committee on War Claims, U.S. House of Representatives, Washington, D.C., 1945, 266; Stephen M. Voynick, *Colorado Gold—From the Pikes Peak Rush to the Present* (Missoula, Mont.: Mountain Press Publishing Company, 1992), 111.

37. Annual Report, Golden Cycle Corporation, 1960, 18.

38. C. E. McFarland and Noel W. Kirschenbaum, "The Cortez Story: 125 Years of Evolution and Innovation," *Minerals and Metallurgical Processing* 8 (May 1931): 57–59.

39. Report, Golden Cycle Corporation, December 1974, Cripple Creek & Victor Gold Mining Company, company records, Victor, Colo.

40. Taylor, *Cripple Creek Mining District*, 147.

Pikes Peak or Bust: A Continuum of Mining History, Cripple Creek Mining District, Colorado, USA

David M. Vardiman

David M. Vardiman holds the B.Sc. degree in geological engineering from the Colorado School of Mines and has been a miner, exploration geologist, production geologist, and exploration manager (both surface and underground) since 1976. During his career he has worked at the Homestake Gold Mine in Lead, South Dakota; the Bulldog Mountain Silver Mine in Creede, Colorado; the Eskay Creek Gold and Silver Mine in Iskuit River Valley, British Columbia; and the Cripple Creek and Victor Gold Mine in Victor, Colorado.

Introduction

The Cripple Creek Mining District flourished during its first ninety years of discoveries and normal production challenges to produce more than 21 million troy ounces of gold from underground ores. That production was made possible by continually adapting improved technologies to meet mining and processing demands. The Cripple Creek and Victor Gold Mining Company (CC&V) is presently extending the district's continuum of mining history from surface minable ores and, it is hoped, underground ores into a second century. Many of the challenges from the first century continue to exist alongside the significant additional problem of highly increased social expectations on the part of environmentalists, historic preservationists, and the public generally that the effects of mining activity on adjacent communities can be minimized.

The Regional Geologic Setting

The Cripple Creek Mining District is located on the southwestern flank of the 4,314-meter-high Pikes Peak massif (figure 1), within the front range of Colorado, approximately 120 kilometers south of Denver. The district is hosted by a Tertiary-age, alkaline volcanic complex dated at approximately 32 million years old,[1] which in turn is surrounded by Precambrian-age granites, granodiorites, and metavolcanic and metasedimentary units. The Cripple Creek volcanic diatreme complex is a unique geological volcanic system formed by explosive volcanic eruptions and subsequent subduction of surface materials to depth by large convection cells. The emplacement of this alkaline diatreme complex, located eighty kilometers on the eastern side of the Rio Grande Rift system, is genetically related to the extensional phase of tectonic development along this major mid-continent, north-south-trending rift system. The Pikes Peak region experienced significant uplift and local doming prior to onset of volcanism,[2] which developed the regional structural fabric and structural controls for emplacement of the diatreme complex.

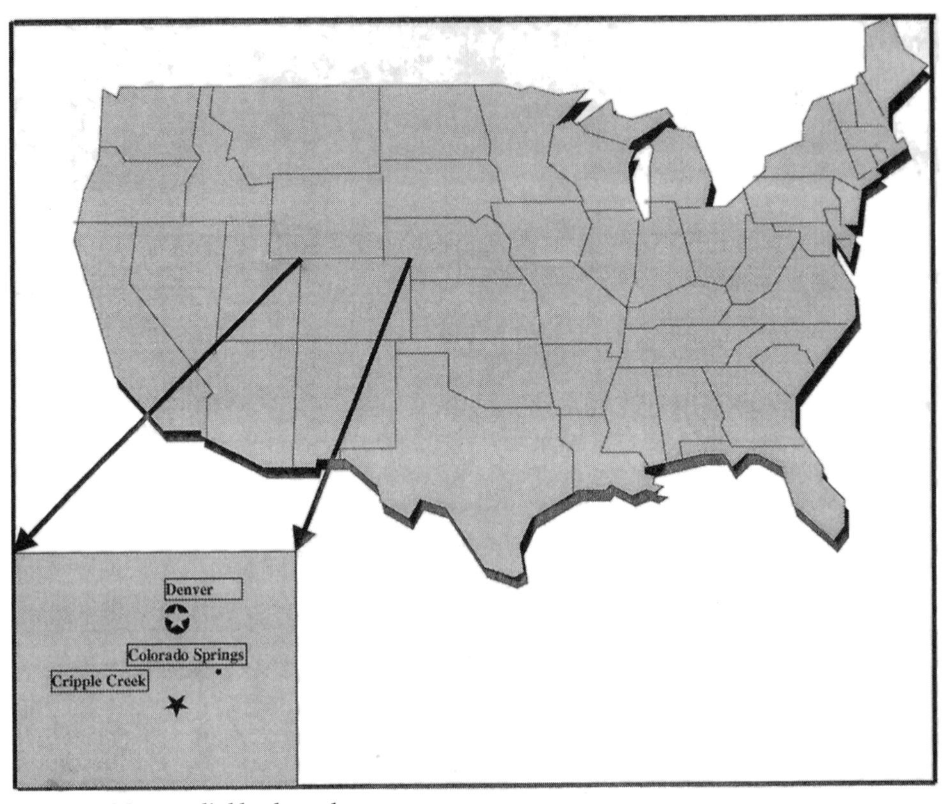

Figure 1. *Map supplied by the author.*

District Geology

The Precambrian lithologies surrounding the Cripple Creek district as described by Hutchinson and Hedge (1968)[3] are: Biotite gneiss, a 1.7+ Ga postulated volcanogenic rock; Granodiorite, at 1.7 Ga; Cripple Creek quartz monzonite, at 1.5 Ga; and Pikes Peak Granite, at 1.0 Ga. The Cripple Creek Mining District volcanic/diatreme complex is comprised of highly variable heterolithic diatremal breccias, volcanoclastic sediments, and bedded tuffs. Detailed mapping of the district has shown a complex sequence of intrusive, extrusive, and diatremal eruption events[4] derived from several postulated overlapping eruptive centers. These diatremal breccia units were subsequently intruded by a continuous petrographic and chemical progression of mafic phonolitic to highly differentiated peralkaline phonolitic dikes, sills, and/or cryptodomes[5] emplaced along major district structural zones and ancestral lithostatic boundaries.[6] Late-stage alkaline dikes occupy major district structural zones. Those structural zones exhibit minor postvolcanic dip-slip movement, with offsets commonly less than 30 meters.

Structure

Major district structures developed prepared sites for mineralization by forming cymoid structures or rhombochasms with localized sheeted tensional fracture development at intersections.[7] Emplacement of dikes, sills and/or cryptodomes within the diatremal breccia also generated excellent stockwork systems for subsequent mineralizing fluids peripheral to their emplacement boundaries. Late-stage alkaline dikes occupy major district structural zones, bearing 340°-350° and 20°-70° with subvertical dips.

The ore bodies located within the Cresson mine are the one exception to the typical district structural controls. The Cresson Pipe ore bodies, which are located primarily peripheral to the vertical Cresson Pipe lamprophyre intrusive body, appear to be controlled by the generation of void space at or near the contact of the lamprophyre body and the surrounding Cripple Creek breccia unit. The intrusion of the lamprophyre body is itself controlled by the larger district structural controls. The void spaces have been interpreted to have been generated by the devolatilization and subsequent shrinkage of the lamprophyre intrusive body during cooling.

Alteration

Alteration in the Cresson deposit ranges from strong, localized zones of potassic, argillic, and pyritic alteration to broad, weak zones of propylitic, sericitic, and variable carbonate alteration.[8] Wall-rock alteration adjacent to veins in volcanic units consists of strong groundmass replacement of volcanic breccia and phonolite by potassium feldspar, sericite, and pyrite. Sericite-pyrite alteration, representing the early alteration stages of host rocks, forms broad halos around stronger potassium feldspar-altered zones. Generation of secondary potassium feldspar is caused by more advanced, acidic alteration of phonolitic material, producing feldspar and quartz. Zones of strong potassium feldspar alteration are spatially associated with low-grade mineralization in disseminated and sheeted zones, but the alteration may not have occurred at the same time as deposition of the gold.[9]

Mineralization

Field relationships indicate that onset of gold mineralization began following the emplacement of the last intrusive event of lamprophyre dikes. Gold mineralization is comprised of gold-silver tellurides associated with pyrite. This mineralization is emplaced within subvertical narrow structural zones within the Cripple Creek breccia or more commonly along contacts of reactivated later-stage dikes, sills, and/or cryptodomes and the Cripple Creek breccia.

Mineralization consists of fluorite, opaline silica, kaolin with iron pyrites and other iron minerals, manganese oxides, and, more rarely, small quantities of galena, cerussite, anglesite, malachite, acanthite, tetrahedrite, stibnite, sphalerite, calaverite, native gold, oxidized tellurium minerals, gypsum, calcite, and numerous other minerals in smaller quantities.[10] As a result of oxidation deep along structurally prepared zones, oxidation of tellurides has generated native gold associated with iron oxides within 100 to 200 meters from the surface.

Mining History

The Cripple Creek Mining District lay dormant from 1962 until the 1970s, when United States citizens were once more allowed to possess gold. In addition, in 1973 the price of gold was deregulated from a fixed United States government standard of thirty-five dollars per ounce and allowed to float based upon marketplace supply and demand. This act, along with existing worldwide economic conditions, set the stage for an unprecedented demand for gold and, therefore, increases in its worldwide price.

The greatly increased price of gold in the late 1970s, coupled with new technological improvements to historic gold-leaching processes, which circumvented costs associated with milling of ores, prompted a rejuvenation of mining activities across the globe. New human and financial efforts were implemented around the world as modern-day phenomena to rival the original gold rushes. Those efforts focused on attempts to renovate old gold deposits or discover new ones. The Cripple Creek Mining District was soon recognized as a potential source of high-tonnage, low-grade surface production.

One final element of the successful revitalization of the district remained to be activated. Although a great deal of consolidation of district property had taken place during the first century, exploration was still limited by a variety of "patch-quilt" ownerships. The Cripple Creek and Victor Gold Mining Company through a series of joint ventures starting in 1976 consolidated more than 90 percent of the district property covering the diatreme by 1993.

In 1976 Golden Cycle Gold Corporation formed a joint venture with Texasgulf by creating the Cripple Creek & Victor Mining Company. That initial partnership was the impetus for district-consolidated exploration and therefore property consolidation. The company's main focus was exploration for classical underground ore bodies through renovation of the Ajax shaft. By 1985 CC&V was processing by cyanide vat leach technology the district's historical waste dumps of material originally derived from underground operations. The "new" technology of cyanide vat leaching, or heap leaching, was actually utilized by the Spaniards in 1752 to recover copper.[11] In the Cripple Creek Mining District, mining consultant Phillip Argall recommended in 1905 that fifty thousand tons of low-grade ore be placed on a concrete or asphalt pad and that a weak solution of cyanide be percolated through the heap. Gold would then be recovered by zinc precipitation from the gold-cyanide solution.

During 1988 CC&V began active mining at the Portland (Captain Stopes) and Goldstar surface mines. District-wide exploration for near-surface, large-tonnage, low-grade deposits was under way. Nerco Minerals purchased Texasgulf Metals and Minerals's CC&V interests in 1989 and continued exploration for low-grade deposits. By 1991, surface-pad leaching of ores derived from the Globe Hill and Ironclad surface mine had begun. Independence Mining Company purchased Nerco's interest in CC&V in 1993, and mining of the Globe Hill and Ironclad deposits began to decline, while exploration for the Main Cresson surface deposit commenced. The Main Cresson mine went into operation in 1994 and produced 920,000 ounces of gold from initiation to year-end 1999. In 1999 the Anglogold (Colorado) Company acquired Independence Mining Company's interests in CC&V.

Current Operations

Owner and Operator of the Cresson Project

At present CC&V is a joint venture of Anglogold (Colorado) Company (67 percent) and Golden Cycle Mining Company (33 percent). The involvement of Anglogold's parent company, Anglo Gold, South Africa (the world's largest single producer of gold at more than seven million ounces annually) allows for continued exploration, development, and operation of the most efficient and socially acceptable gold-mining operation in the United States (figure 2). The CC&V operation continues surface mining while enabling exploration for surface and underground ore reserves and improvements in technology to extend the life of the operation well past 2000.

Figure 2. *Photograph supplied by the author.*

Future Mine Activities

The Cripple Creek Mining District contains a published (1998) ore reserve of 127.9 million ore tons, which in turn contain 4.3 million ounces of gold. Two of four deposits—the Main and East Cresson mines—are currently being worked. The two mines produced 10.5 million ore tons and 231,000 ounces of gold during 1999. The potential for additional ore reserves exists throughout the district, with published resources of 165.1 million ore tons that in turn contain 3.15 million ounces of gold in addition to the 127.9 million tons of ore reserves mentioned above.

Employment and Economic Impact

CC&V directly employs some 288 "team members" and is the largest private employer in Teller County. The annual operating budget of the Cresson Project is about $41.2 million. Each year, CC&V's mining adds $20.1 million to earnings in the county alone and $46.5 million to the entire state's economy. Capital costs for the project are approximately $154 million.

Exploration

Significant potential exists for new near-surface mineral reserves, which can be mined with surface-mining techniques. CC&V has extended reconnaissance drilling to greater depths in order to find underground minable resources. While underground mining requires higher-grade ore reserves, the district's history and the current reconnaissance suggest that such reserves may still exist. If those reserves are confirmed, mining in the district could be extended with reintroduction of underground mining. Aggressive exploration and the discovery of new reserves are the lifeblood of CC&V.

Mining

Mining of the Main Cresson and East Cresson deposits employs sixteen 85-ton Caterpillar trucks, three 15-yard Caterpillar loaders, an 18-yard Hitachi front shovel, and three Ingersoll Rand DM45E blast hole drills. The Cresson and East Cresson mines produce some 80,000 tons per day with up to 40,000 tons daily going through the crusher. The crushing plant operates a 54-inch-by-75-inch primary gyratory, followed by two 7-foot standard secondary cone crushers. Crushing produces a nominal 1.5-inch product, which is stacked onto the leach pad.

Processing

The recovery plant operates two carbon trains, which have a combined capacity of about six thousand gallons per minute of solution pumped from a closed-circuit, nondischarging, multiliner valley leach facility. The facility has an extensive system of monitor wells and surface sampling points, plus operating requirements to enable it to comply with its zero (water)-discharge permit. The project also complies with stringent water-quality, air, seismic, and noise standards and with numerous other local, state, and federal governmental requirements.

Gold and silver are removed from crushed ore using the same basic process employed in the district since its early beginnings, and almost everywhere else. Sodium cyanide is added to the leach solution, causing naturally occurring metals (including gold and silver) exposed on the crushed ore surfaces to dissolve, forming a cyanide compound. At Cressson, the leaching occurs in a facility constructed in a valley, essentially as a very large double- and triple-lined bathtub.

The crushed ore is placed in 35±-foot layers in the facility, and a dilute and moderately basic solution of sodium cyanide (a "barren solution") is applied using drip irrigation tubes on top of the ore. The solution pH is elevated by applying lime to the ore to retain the cyanide in solution. The solution seeps toward the bottom of the "bathtub" and exchanges the sodium with gold and other metals to form metal compounds in solution. The solution, now termed "pregnant," is collected and circulated to the adsorption desorption recovery (ADR) facility. Cresson has no external ponds for leach solution. Rather, the solution is contained within the pore space of the ore until pumped out, much as groundwater is pumped from porous bedrock. Excess capacity to accommodate precipitation and varied pumping scenarios is built into the leach tub, which is triple-lined in areas in which solution is held within the pore spaces of the ore and also operated with a comprehensive water-monitoring system.

The "pregnant" solution enters the ADR facility and passes through activated carbon granules to adsorb the gold metal-cyanide compound. The leach solution is stripped of dissolved gold and sent back to the drip irrigators to continue the leaching cycle. Periodically the gold compound is processed to remove the gold using a hot alkaline strip solution. The gold-rich solution is piped to an electro-winning cell, where direct current attracts metals from the solution to a reusable stainless-steel-wool cathode to form a solid, or "mud." The "mud" is dried and sent to a furnace for melting into a gold-silver amalgam called doré, which is sold to refineries for final purification and fabrication.

Permitting

CC&V places high priority on environmental compliance and reclamation. Experienced professionals, along with the operation's entire work force, strive to ensure compliance with the many federal, state, and local environmental protection regulations. Mining is one of the most regulated industries in the United States. Extensive investigations, studies, and detailed planning must be carried out before permit applications are sought, and substantial base-line data must be collected to describe the pre-mining environment. All plans and data must be submitted before any construction or mining occurs, and there is no guarantee that a permit will be issued. All information submitted is available for public review.

All of CC&V's operations take place on private lands. County zoning and land-use restrictions govern mining. The state regulates wildlife protection, air quality, water quality and quantity, blasting, and noise. Management of wastes and recycled materials must likewise meet state and federal requirements. Storage and use of operating products and chemicals, as well as the tonnages of rock excavated and crushed, are regulated. Reclamation is required and guaranteed by posting an adequate financial warranty payable to the state should the operator not comply with permits. As of January 1999, the financial warranty.at Cresson was almost twenty-six million dollars.

As a result of federal involvement in permitting, extensive archaeological and cultural-resource surveys have been conducted on 2,545 acres of private lands and small areas of public lands. Deconsecration and collection of hunting-camp artifacts have been coordinated with Native American tribes. Inventories have been conducted regardless of whether or not there was planned disturbance. CC&V archaeologists have inventoried 3,250 acres of private land. The company has saved the headframes of the Cresson and Gold Sovereign mines and relocated them to the Victor Gold Bowl and the district museum respectively, successfully applied to have the Independence mine and mill site placed on the National Register of Historic Places, and donated significant examples of mining equipment to the Lowell Thomas Museum in Victor. CC&V has provided access to the refurbished American Eagles Mine historical display site and sponsored within the Victor-Goldfield area a series of walking trails telling the history of the district. The company has invested more than five million dollars in such activities. The archaeological inventories yielded an electronic database with text and photographic documentation for the surveys. The database enables the searcher to investigate features by location and to access both the results of mitigation research and photographic and other graphic materials.

The Cresson Mine Project was designed to comply with new and more stringent state statutes and regulations that were developed in 1993-1994. Among the new requirements were an environmental protection plan and the designation of certain mines, principally metal-producing facilities dealing with sulfide rocks and chemicals for mineral benefication, as "designated" mining operations. Furthermore, the technical specifications for the proposed construction and operation had to be detailed and submitted as part of an application. The leaching facilities at the Cresson Project are of advanced-design nondischarging triple- and double-lined containment and yet are accompanied by extensive environmental monitoring systems to forewarn of unanticipated problems and to ensure compliance with water-quality criteria.

The Cresson Project operates under a state air-quality permit pursuant to federal Clean Air Act and state mandates. Particulates ("dust") generated from exploration, mining, crushing, hauling the disturbed lands themselves, and reclamation must be quantified and controlled to meet ambient-air-quality standards outside the project site and at its boundaries. A relatively high in-situ moisture content and the coarse nature of the excavated material restrict the potential for the escape of dust into the air following blasting, digging, hauling, and dumping. A chemical compound is applied to roads to assist in dust control, and as much as 250,000 gallons of water are used in road and transfer-point dust control each day. Transfer points on the conveyor system that transports crushed ore to the leaching facility are enclosed. Stationary sources of potential pollutants at the lime silo, ADR, and laboratory are equipped with control devices such as baghouses, wet scrubbers, and carbon filters to control gaseous emissions. Monitoring stations along the perimeter of the site confirm that air-quality criteria are not exceeded. Speed limits on the property promote safety and help mitigate fugitive dust. Prolonged public access is also controlled.

CC&V meets federal and state storm-water-management criteria and stringent water-quality standards that protect existing and potential future uses. CC&V aims to retain storm-water runoff on-site long enough to remove sediment; all drainageways and slopes are designed with that goal in mind. Numerous detention ponds slow water down and allow settling of solids. The project does not discharge water from mining, crushing, or benefication operations. Water applied at the crusher is retained in the ore placed for leaching, joins the leach solution, and is continuously recycled. CC&V operates eighteen surface-water stations and fifty-three groundwater-monitoring wells. Some wells are sampled as often as daily.

External laboratories participating in the U.S. Environmental Protection Agency's quality control program perform all chemical analyses. Typical analyses include pH, nitrogen compounds, cyanide, numerous metals (Al, As, Cd, Cr, Cu, Fe, Hg, Mn, Ni, Pb, Se, Ag, and Zn), SO4, and common rock-related parameters such as Ca, Mg, K, Na, CO3, HCO3, TDS, Cl, F, and TSS. Stringent procedures prevent contamination and ensure the validity of samples and data.

Other permits with which CC&V must comply involve septic tank systems for domestic sewage; a temporary water supply (augmentation) plan to replace water in streams that would receive runoff from precipitation that falls on the nondischarging leach facility; identification numbers—from the EPA and the state for shipping potentially hazardous wastes, and from state and federal agencies for transporting diesel fuel, water solutions of sodium cyanide, explosives, and recycled materials (metals); and federal wetlands and waters.

As of 1998, CC&V is required to report annual releases of substances. At the Cresson, several chemicals and compounds require reporting. In fact, many of the substances (an example is silver) are not "released" but are moved from one place to another because they occur in the ore and surrounding overburden. Because an interim decomposition product of a portion of the sodium cyanide used to remove the precious metals from the ore is sometimes released from the ore, that interim product may be reported as a "release," even though it breaks down into water and oxide gases shortly after leaving the ore surfaces. CC&V moves all substances in accordance with explicit permit conditions and regulations. The new reporting reveals specifically what is moved, but unfortunately the amounts are not explained by agencies or advocacy groups in terms of whether there are risks. Therefore, the company encourages visitors to see how it manages those elements and compounds to ensure full compliance with health and environmental standards.

Reclamation

CC&V has an active concurrent reclamation program and conducts quantitative monitoring surveys and engages in other activities as part of a comprehensive environmental plan to ensure realization of that program. Some one hundred acres have been reclaimed since 1992, and those areas, by design and revegetation, now supplement the habitat occupied by many wildlife species. As a result of planting and bans on firearm hunting, the entire site is a wildlife refuge. CC&V intends to reclaim land to a beneficial use for grazing and wildlife habitat by planting grasses, legumes, forbs, trees, and shrubs (figure 3). Some transplants have been twenty-foot-tall

Figure 3. *Photograph supplied by the author.*

Bristlecone pines. Since 1995 a total of 6,176 transplants have been made on the site, many of them with the help of Scouts, 4-H'ers, and other youth groups.

As noted earlier, CC&V has posted with the state a financial warranty of twenty-six million dollars to ensure that the lands are reclaimed according to an approved plan. Only when stages of reclamation are completed and approved is the warranty reduced. CC&V also has an approved emergency response plan based on risk assessments that cover the potential for accidental releases of hydrocarbons, cyanide, propane, or other substances within the area.

Community Involvement

CC&V considers itself part of the community and views the community's well-being as important. The company encourages its employees, most of whom live throughout the Pikes Peak region, to participate in community and local government affairs and rewards those who do with additional vacation time; thus its team members are active in church, school, and community groups. It participates in emergency services and youth activities. In 1998 CC&V volunteers donated three thousand hours of work to more than fifty organizations. Employees donated more than one hundred thousand dollars to schools, local community organizations, and local charities. At Thanksgiving, volunteers deliver food baskets to local residents. Various parks and recreation programs have received substantial amounts of donated time and financial contributions from CC&V and individual employees.

Local students, supported by CC&V, staff the Lowell Thomas Museum in Victor during the summer. Under a grant matched by CC&V, exhibits there have been cleaned and reorganized. CC&V participated in several community Christmas projects. The company comprehensively supported the annual Gold Rush Days festival. That event was supported entirely by CC&V funds and included tents, prize money, equipment, and volunteer time for pre-event planning; it enabled the public to enjoy an afternoon of mining and drilling contests.

For more than one hundred years, gold in the Cripple Creek Mining District has supported communities. For more than twenty years, the Cripple Creek & Victor Gold Mining Company has helped to support local communities by being an active part of them.

Scholarships and Education

In addition to substantial donations to Cripple Creek and Victor schools, CC&V's and Anglogold (Colorado) Corporation's donations include numerous contributions to education in Colorado. A recently enhanced college scholarship program provides $180,000 in annual advanced educational scholarships to CC&V dependents. The program supports four-year and two-year accredited institutions, as well as certified technical and trade schools.

Taxes

CC&V pays three major types of taxes in addition to federal corporate income taxes: state and local sales and use taxes, property (surface) taxes, and a net proceed or "mineral severance" tax. The latter is a state tax based on the mineral estate (subsurface) extracted and benefits both the state general fund and the local communities affected by mineral extraction. CC&V's operations contribute annual property taxes, sales taxes, and severance taxes in excess of $1.115 million per year, or in excess of $15 million over the fourteen years of production currently projected for the Cresson project. The company's team members provide more than $500,000 each year to Colorado in income taxes, and more to the local and state treasuries from individual sales and property taxes.

Social Expectations

Mining is a necessity in modern world economies such as the United States. It is not an artifact or historical relic of the past but a viable industry that provides basic elements upon which other industries are developed. Nor is mining a fixed entity without growth or development. It is vibrant and dynamic and proactive in seeking answers to technical and social questions. Mining requires temporary disturbance of the environment, and such activity draws the attention of the environmentally conscious public. Mining and farming, unlike other industries, are linked inextricably to their locations. Ore deposits that make mining possible are not transient and cannot be relocated to environmentally or socially acceptable locations. As the old saying goes, "Gold is where you find it." The unbreakable bond of mining and its location requires that activities that surround extraction of minerals change in response to the social-political-biological environment in which mines operate. Mining, like many other industries in the United States, exists, even when located on private property, with the consent of the public. Such consent is evidenced through permitting requirements by city, county, state, and federal regulatory bodies. At the present time, the long-term success of a mining company (or any industry) is inexorably linked to its ability to engage in activities within the following realities:

(1) Community involvement in the decision-making process at city, county, and state levels
(2) Dedication to operational excellence
(3) Commitment to environmental responsibility prior to, during, and following operations
(4) Belief in the value of historic preservation
(5) Commitment to community improvement
(6) Dedication to being a good neighbor

As in the first century of production, innovation in applying improved technologies to fit the demands of dynamic financial markets, responding positively to challenges in the realm of mining and processing, and meeting increased social expectations will be the keys to successfully extending the Cripple Creek mining district's activities across three consecutive centuries. Those activities will allow mining to continue its progress into an active, positive, contributing reality for the future. Today the CC&V Gold Mining Company is once again prolonging the production history of the Cripple Creek Mining District. The new district milestone of 22 million troy ounces of gold is scheduled to be reached during the first quarter of 2000. CC&V's ability to apply improved technologies, combined with its dedication to historic and environmental preservation, is positioning Anglogold (Colorado) Corporation as the premier gold-mining company in the United States.

Notes

1. Karen Kelley, "Origin and Timing of Magmatism and Associated Gold-Telluride Mineralization at Cripple Creek, Colorado" (Ph.D. diss., Colorado School of Mines, 1996).

2. Albert. H. Koschmann, *Structural Control of the Gold Deposits of the Cripple Creek District, Colorado*. U.S. Geological Survey Bulletin 955-B. (Washington: Government Printing Office, 1949), 19–58.

3. Robert M. Hutchinson and Carl E. Hedge, *Depth-zone Emplacement and Geochronology of Precambrian Plutons, Central Colorado Front Range* (Boulder, Colo.: Geological Society of America, 1968).

4. Jeff A. Pontius and others, *Geologic Map of the Cripple Creek Mining District*. Colorado Geological Survey, Map Series 31. Denver: Colorado Geological Survey, 1996.

5. Scott D. Birmingham, "The Cripple Creek Volcanic Field, Central Colorado" (bachelor's thesis, University of Texas, Austin, 1987).

6. Richard Sillitoe, letter to author, June 1998.

7. Michelle Murray, CC&V Mining Company, 1999 Structural Report (internal company document).

8. Tim Harris and Jeff Pontius, Mineral Potential Mapping for Gold Deposits in the Cripple Creek Mining District, Colorado, USA, Field Guide Tour, 1996 (internal company document).

9. Larry Smith, *Cripple Creek & Victor Gold Mining Company, Review of Exploration Strategic Plan Cripple Creek Mining District*, 1999 (internal company document).

10. Whitman Cross and Richard A. F. Penrose Jr., "Geology and Mining Industries of the Cripple Creek District, Colorado," in *Report of the Secretary of the Interior . . .* , 5 vols. (Washington: Government Printing Office, 1895), 4, pt. 2:119.

11. Tim P. McNulty, "A Metallurgical History of Gold," in *Mining Technology, Economics and Policy 1989 Session Papers* (Washington, D.C.: American Mining Congress, 1989).

A Review of Nevada's Gold-Mining History and the Discovery of the Carlin Deposit

J. Alan Coope

J. Alan Coope earned his doctorate in applied geochemistry at the Royal School of Mines, Imperial College, London. He spent thirty-three years with Newmont Mining, retiring as director of geochemistry. He was a co-discoverer in 1961–1962 of the Carlin gold deposit in Nevada, the first major development along the enormously productive Carlin trend. He also explored for minerals in Australia, Botswana, Canada, Namibia, the Philippines, Scotland, South Africa, and Tanzania and was the principal organizer and founding president of the Association of Exploration Geochemists. Dr. Coope passed away August 5, 2001, in Tucson, Arizona, at the age of sixty-six, following an extended illness.

Introduction

Nevada is the third largest producer of gold in the world, behind only South Africa and Australia. Gold production from the state reached 8,860,000 troy ounces in 1998 from thirty-six mining operations (Tingley and LaPointe, 1999). The majority of those operations are located along clearly defined deposit trends in the northern and northeastern parts of the state (figure 1). Of them, the northwesterly Carlin and Battle Mountain-Eureka trends are the longest and most prominent. The north-northeasterly Getchell and Independence trends are shorter in length but still economically and geologically very important. Developments along the trends have been predominantly open-pit operations, but in recent years—beginning approximately six years ago—underground operations have become much more common. Barrick Goldstrike Corporation's Betze-Post open pit, the largest gold mine in the United States, has produced in excess of 1.4 million ounces of gold each year since 1994—including 2.03 million ounces in 1995 and 1.9 million in 1996. The nearby Meikle mine, also owned by Barrick Goldstrike, produced 847,313 ounces in 1998 and is the nation's largest underground gold mine.

A Brief History of Precious-Metal Mining in Nevada

The first discovery of gold in Nevada came in July 1849 at the mouth of Gold Canyon near the present site of Dayton, approximately twelve miles northeast of Carson City. Mormon forty-niners made that discovery of placer gold as they made their way westward to the Sutter's Mill area of California, where gold had been discovered the prior year (Lincoln, 1924). The Gold Canyon placers did not generate great excitement, but in later years knowledge of the discovery attracted other prospectors, who followed the placer indications upstream into what is now known as the Virginia Range. It was there that the surface indications of the fabulous Comstock Lode were discovered in 1859 (figures 1 and 2). There was little outside interest until a heavy, black mineral, which interfered with gold-placer separations,

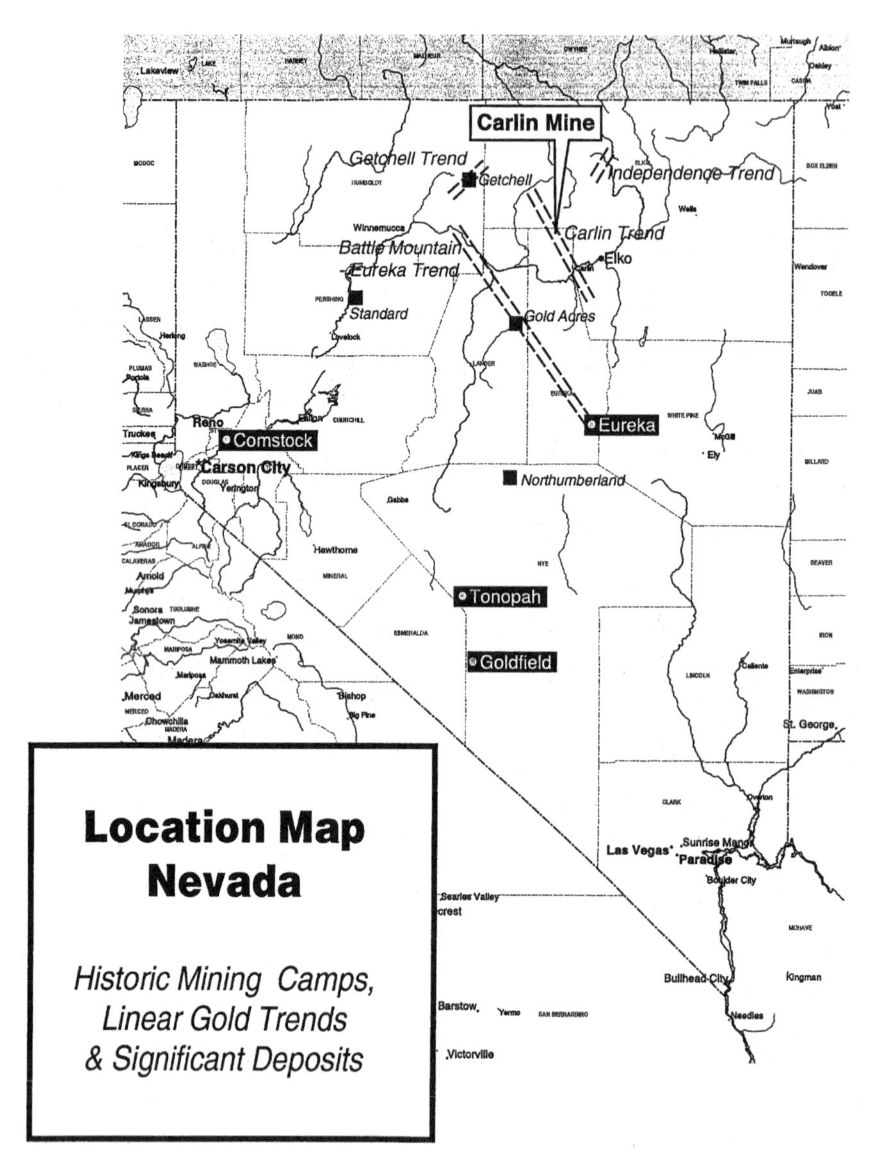

Figure 1: Location map—state of Nevada. Historic mining camps, linear gold trends, and significant deposits. *Map supplied by the author.*

was assayed in Grass Valley and found to be silver sulfide (argentite). The first nationally significant silver mine in the United States had been discovered.

Hundreds of miners crossed the Sierras from California in late 1859, and, very soon, approximately twenty thousand of them descended on Virginia City. Many

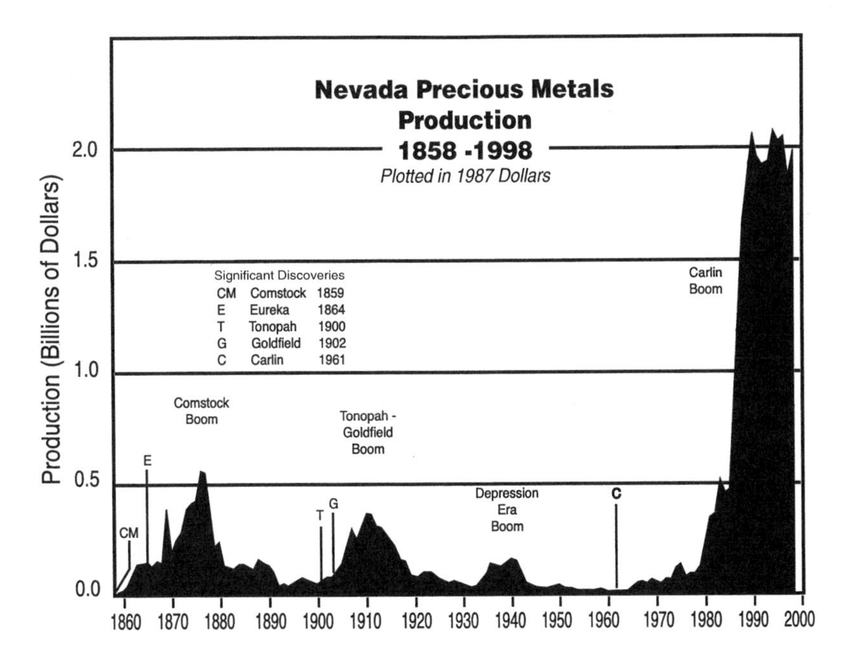

Figure 2: Nevada precious-metals production, 1859–1998, in 1987 dollars. *Chart adapted in part from data in Nevada Bureau of Mines and Geology, Special Publication 15, 1993.*

stayed on the Comstock; others spread out over the state to search for other deposits or join later rushes. It was the influx of people created by the Comstock discovery that enabled Abraham Lincoln to successfully seek the admission of Nevada as a state in 1864 (Lincoln, 1924). The Comstock ores were dominated by silver, with a production total of more than 192 million ounces. Gold was not insignificant, however, with a total output of approximately 8.4 million ounces. Another major mining camp—at Eureka—was discovered in 1864. It was the first great silver-lead district in the United States, producing $20 million worth of gold, $40 million worth of silver, and 225,000 tons of lead between 1869 and 1882 (Lincoln, 1924).

The mining industry in Nevada suffered a period of decline and depression during the last two decades of the nineteenth century. Discovery rates and production in the established mining districts declined (figure 2). Another important economic factor was the gradual fall in the price of silver to about 60 cents per ounce at the end of the century, compared with $1.33 per ounce price prior to demonetization in 1873. That unfruitful period ended with the discovery of the rich gold and silver ores at Tonopah in 1900. Completion of a railroad in 1904 led to significant increases in production, and the output was further enhanced by the operation of cyanide mills beginning in 1906. Developments at Tonopah led to increased prospecting activity and the discovery of the Goldfield district, approximately thirty miles to the south, in 1902. The railroad reached Goldfield in 1905, and for many years thereafter the district was the largest producer of gold in Nevada (Lincoln, 1924).

World War I generated an increased demand for Nevada metals production, and silver and base-metal prices moved higher. At the end of the war the demand for base metals fell, but the Pittman Act of 1918 authorized the overseas sale of silver bullion from the U.S. Treasury and the purchase of replacement bullion from domestic producers at a price of one dollar per ounce. As a result, silver mining flourished in Nevada until 1923 (Lincoln, 1924). The total output from Tonopah between 1900 and 1930 amounted to more than $35 million worth of gold (approximately 1.8 million ounces) and almost 162 million ounces of silver (Lindgren, 1933). Gold production from the Goldfield camp between 1903 and 1921 totaled $85 million (Lincoln, 1924) (approximately 4.3 million ounces).

In 1924 Nevada's precious-metal industry entered another period of decline that, except for a few years of recovery related to the increase in the gold price to thirty-five dollars per ounce in 1933, saw production fall to a near-record low in 1961. According to Tingley (1993), from the time the Comstock boom began in 1859, only the year 1894 recorded less production of gold and silver than the amount of each produced in 1961. New discoveries made in the 1930s, however, proved very important. They included the Gold Standard mine in Pershing County, the Getchell mine in Humboldt County, the Gold Acres mine in Lander County, and the Northumberland mine in Nye County (Horton, 1964). These deposits are all characterized by "invisible gold" (i.e., extremely fine micron [one-millionth of a meter]- to submicron-sized free-gold particles not visible to the naked eye), a feature of the Carlin-type deposits responsible for the enormous boom in Nevada gold mining that has occurred over the last third of the twentieth century (Tingley, 1993).

The Carlin Era

The Carlin boom began with the discovery of the Carlin gold deposit in the Tuscarora Range in 1961. That success eventually led to the discovery and development of more than thirty other deposits forming a northwesterly-trending belt (the "Carlin trend") that extends over fifty miles through northern Eureka County and Elko County (figure 3). By the end of 1998 almost forty million ounces of gold had been produced on the Carlin trend, and the historical production plus proven and probable reserves and additional mineral inventory exceeded one hundred million ounces (Teal and Jackson, 1997). The Carlin discovery was a culmination of a series of critical observations and recordings, probably the most important of which was the recognition and appreciation of the significance of invisible gold.

In the preceding brief historical review, it was noted that invisible gold characterized the Getchell, Gold Standard, Gold Acres, and Northumberland mines, which were discovered and developed during the depression years of the 1930s. It was during those years that W. O. Vanderburg, a mining engineer with the U.S. Bureau of Mines, visited both the hardrock gold and placer mines in north-central Nevada and other locales throughout the state. Vanderburg's observations were published in three very important publications (Vanderburg, 1936, 1938, 1939) in the years before World War II. In his Lander County report, Vanderburg (1939) reported that at Gold

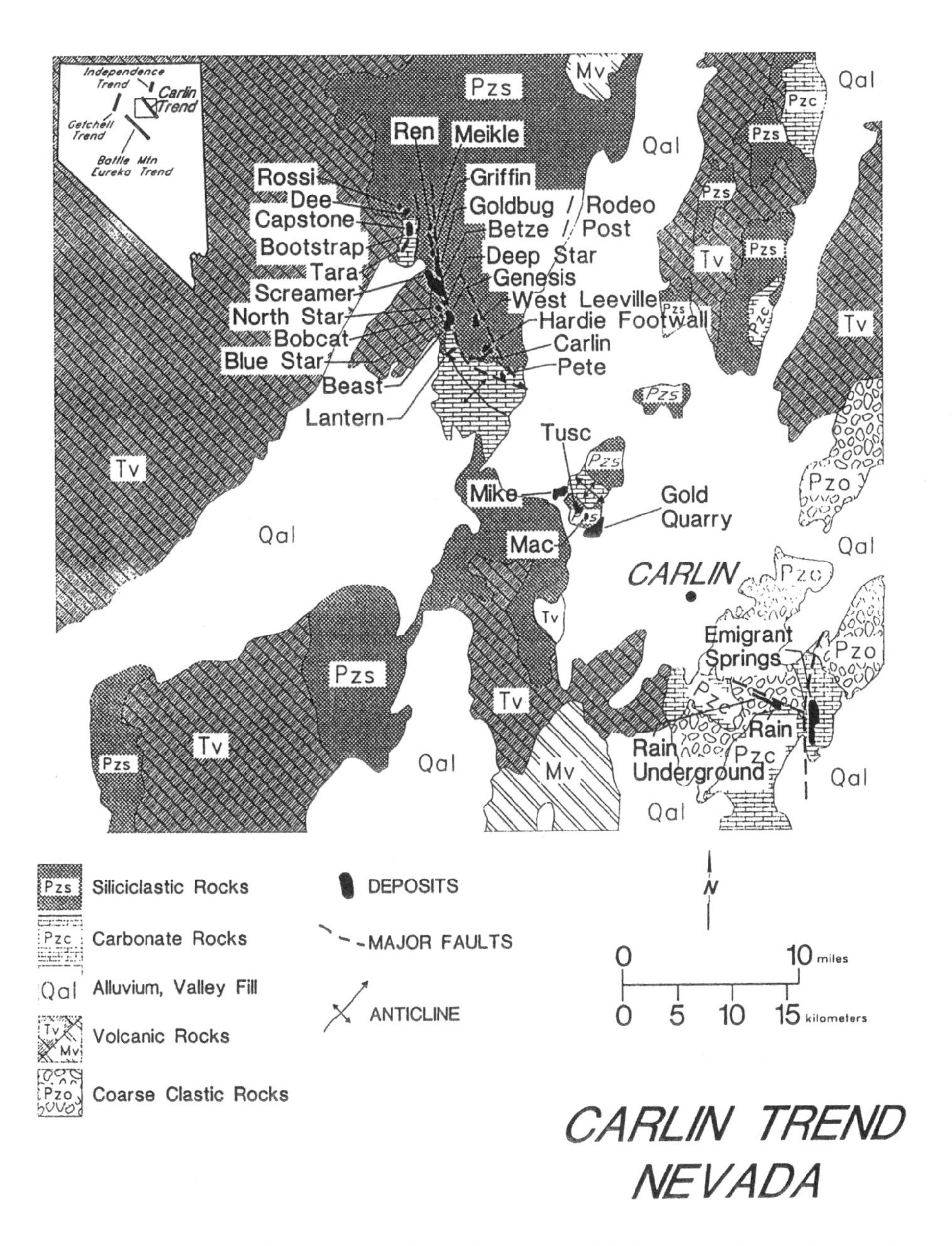

Figure 3: Location map showing general distribution of gold deposits of the Carlin Trend, as well as major structures, including faults and anticlinal folds. Inset map shows other trends defined by the distribution of major gold deposits in northern Nevada. *Map from Teal and Jackson, 1997; reproduced courtesy Society of Economic Geologists.*

Acres it was "impossible to distinguish between ore and waste except by assay" and that "the gold is present in such a state that it is impossible to obtain a single color by panning." In the final paragraph of the same report he observed that "sedimentary gold deposits do not possess easily recognizable indications" and would have been passed over in former years by prospectors who depended largely on panning. He predicted that other deposits similar to Gold Acres remained "to be discovered in Lander County and in other areas of the state, where sedimentary formations, like shale and limestone lying in proximity to acid intrusives, are common." Vanderburg and others in Nevada were apparently unaware of the discovery of very fine gold at the Mercur mine in Utah in the 1880s. There is an unofficial record of a legal action against an assaying company in Salt Lake City that detected gold by fire assay—gold that could not be confirmed by Mercur miners using a gold pan (Kornze, personal communication).

About the same time that Vanderburg was completing his reports, the United States Geological Survey (USGS) embarked on regional geological mapping studies in the northern and central parts of Nevada. Ralph J. Roberts and other USGS geologists began that work in 1939. Early in the project, Merriam and Anderson (1942) identified the Roberts Mountains thrust in the Roberts Mountains northwest of Eureka. Roberts (1951) recognized and described the Antler orogeny after extensive work in the Antler Peak district shortly afterward. By the 1950s the USGS mapping program had extended over the greater part of central and northeastern Nevada, and the Roberts Mountains thrust had been traced from Eureka on the east to Manhattan on the south and to Mountain City on the north (Roberts, 1986). At a meeting of the Geological Society of America in 1955, Roberts and Lehner (1955) described their mapping of the Carlin, Lynn, and Bootstrap windows and, in a more regional synthesis, identified the northwestern alignment of lower-plate carbonate windows exposed through the Roberts Mountains thrust along what are now known as the Carlin and Battle Mountain-Eureka trends. A comprehensive summary of the USGS mapping that describes the extent of the Roberts Mountains thrust, the distribution of Paleozoic rocks, and the alignment of the lower plate windows was published in 1958 (Roberts and others, 1958). Those mapping results are illustrated in figure 4. Building on this basic geologic framework, Roberts (1960) published a short note in 1960 that describes the close association of base-metal and precious-metal mining districts with the aligned lower-plate windows (figure 5).

Newmont Mining Company's interest in gold in northeastern Nevada developed in a very logical manner. John S. Livermore, a geology graduate out of Stanford, first read the Vanderburg reports in the late 1940s. He was impressed by the exploration potential for invisible gold that could have been missed by the old-timers using the gold pan as their principal prospecting tool. In 1949 Livermore worked at the inactive Gold Standard property near Lovelock. The owners were hoping to find higher-grade veins associated with the more extensive but lower-grade sediments containing invisible gold. The program ended when richer ores were not discovered.

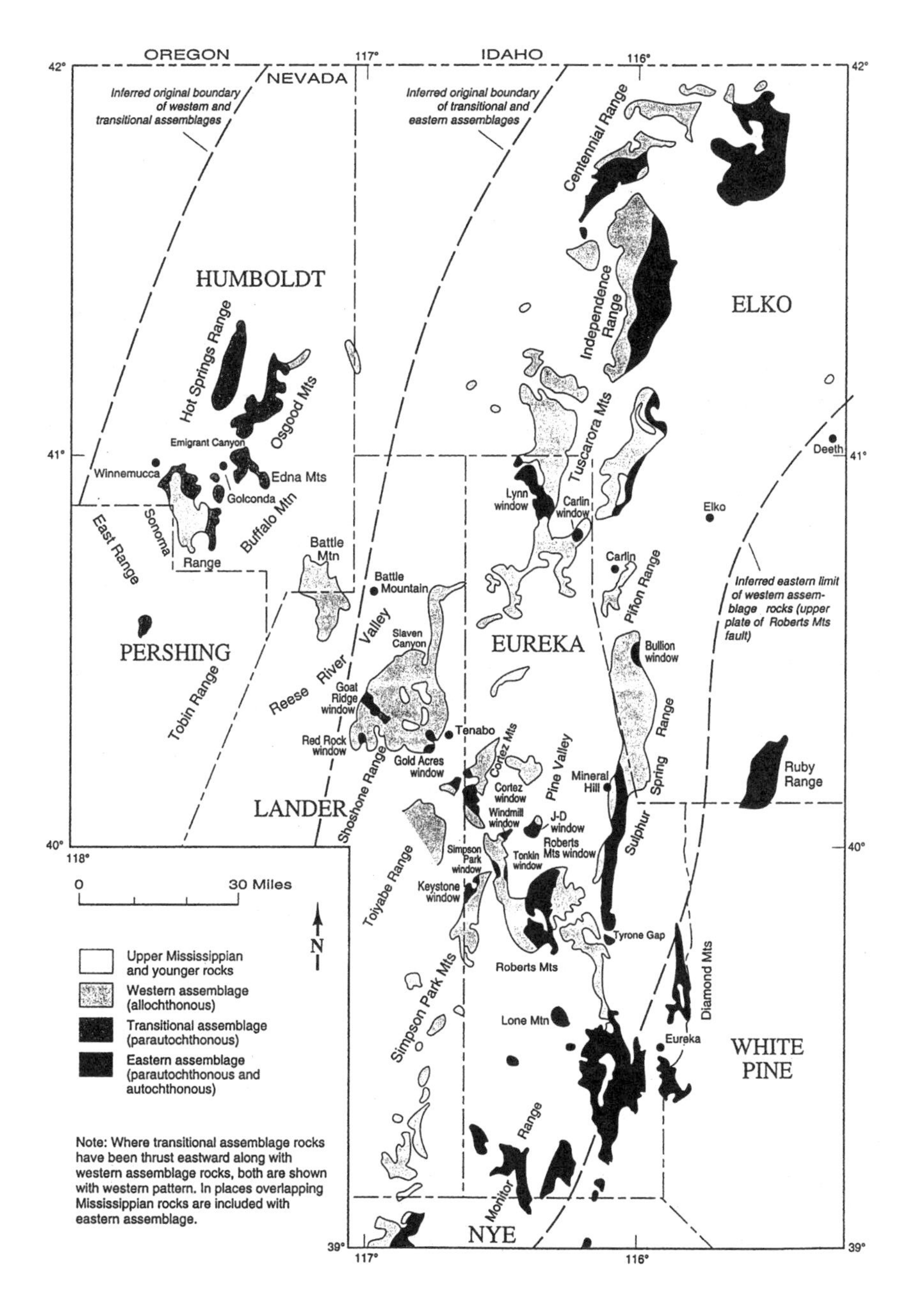

Figure 4: Map of north-central Nevada showing outcrop areas of western assemblage (upper plate), eastern assemblage (lower plate), and transitional assemblage rocks and the extent of the Roberts Mountains thrust. *Map adapted from Roberts and others, 1958; reproduced courtesy Nevada Bureau of Mines and Geology from Coope, 1991.*

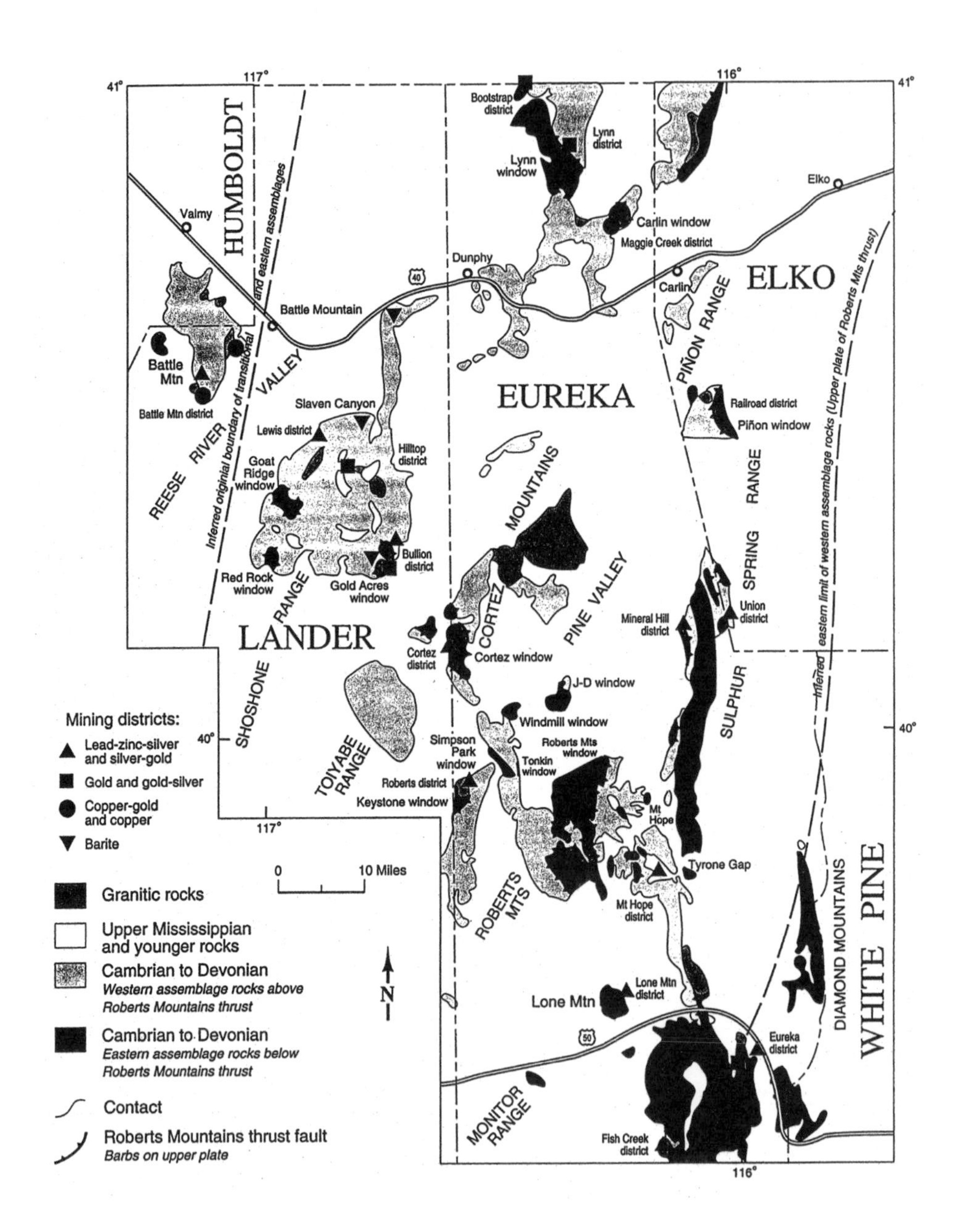

Figure 5: Distribution and alignment of Paleozoic facies, granitic rocks, and principal mining districts in Eureka County, Nevada, and adjacent areas. *Map adapted from Roberts, 1960; reproduced courtesy Nevada Bureau of Mines and Geology from Coope, 1991.*

Livermore's experience at the Gold Standard site further stimulated his interest in invisible gold, and in the early 1950s he visited both the Getchell and Gold Acres mines—two other locations of invisible gold identified by Vanderburg in the 1930s. After becoming a Newmont employee, Livermore returned to Nevada in 1960 to manage a drilling program on the silver/base-metal deposit at Ruby Hill in Eureka. From records in the Eureka County Courthouse and through numerous contacts, he was able to monitor prospecting and mining activities, particularly those in the north-central part of the state. He made contact with the USGS personnel active in the mapping programs, and, with his interest in invisible gold still at a peak, he visited gold-mining operations in Lander, Eureka, and Humboldt Counties. A significant contact was Harry Bishop of Battle Mountain, who was the manager of the London Extension Mining Company's Gold Acres mine in Lander County. Bishop believed that the most favorable area in which to prospect for other deposits of invisible gold was in the northern part of Eureka County, some fifty miles to the north-northeast. Positive results from prospecting and development activity in northern Eureka County and neighboring parts of Elko County over several decades—but particularly at the Blue Star and Bootstrap prospects in the 1940s and 1950s—were the reasons for Bishop's recommendation.

The Bootstrap property had first become the focus of interest in 1918 with the discovery of antimony mineralization. There, almost thirty years later, in 1946, Frank Maloney discovered gold when he sent out samples of mineralized rock for fire assay. The gold could not be panned, however. First production of gold was achieved in 1948, and after several periods of development the total production by 1960 was estimated at between 10,000 and 11,000 ounces of gold from 40,000 tons of ore. That output was the largest production from a lode deposit along the Carlin trend to that time (Coope, 1991).

Initial development at the Blue Star property was for turquoise in the 1920s. Marion Fisher of Battle Mountain sampled the turquoise workings and identified gold in 1957. Three years later a small, locally owned company, M M and S Mining, drilled several holes in the area of the workings and obtained values up to .44 ounces per ton over ten-foot lengths. Higher values were obtained from channel sampling in the deeper workings. None of the gold was pannable. By 1961, M M & S had leased the property to Combined Production Associates of Salt Lake City, and a two-hundred-ton-per-day cyanide mill had been established (Coope, 1991).

Early in 1961, in a series of geological talks at various locations in Nevada, Ralph Roberts described the Paleozoic stratigraphy and the alignment of mineral deposits associated with the lower-plate windows. John Livermore was present at one of Roberts's talks in Ely and later met with Roberts to discuss the geological relationships in more detail. Over the ensuing few months, Livermore prepared an exploration proposal to Newmont that highlighted the geological and economic potential for invisible gold deposits in northern Eureka County and western Elko County. Livermore's proposal eventually gained the support of Newmont's key officers and directors, who visualized a potential for low-cost open-pit gold mining supported by relatively cheap but efficient cyanide extraction. By that time, Newmont had commenced a gold-exploration program in the Battle Mountain area, where geologist Alan Coope was investigating the potential of the Marigold and Buffalo Valley properties near Valmy. It was concluded that neither property was economically viable with gold at thirty-five-dollars per ounce. Those decisions were reached at the same time the drilling program at Ruby Hill was completed, and Livermore and Coope were instructed to meet in Carlin in June 1961 to proceed with Livermore's proposed program in Eureka and Elko Counties. As indicated previously, 1961 was the year of the second lowest precious-metal production in Nevada since the Comstock discovery.

Newmont's Carlin Program

John Livermore had visited the Blue Star property soon after his meeting with Harry Bishop. The cyanide mill was experiencing operational difficulties because of slimes in the ore. After a brief examination of the workings, Livermore recommended to the chairman of Newmont that the company investigate the property more carefully, and the necessary arrangements were made. At about the same time (May 1961), Newmont received verbal permission to examine the Maggie Creek prospect, located adjacent to the southern boundary of the Carlin window, eight miles northwest of the town of Carlin. Although the Maggie Creek prospect was not ignored, the initial exploration focused on the Blue Star mine. Three weeks of geological mapping and sampling produced encouraging results, and Livermore and Coope recommended to Newmont's chairman that Newmont conclude a deal on the property. Negotiations with Combined Production Associates were unsuccessful (Coope, 1991).

Attention then turned to Maggie Creek, but the acquired knowledge obtained through the Blue Star examination became the basis for a regional evaluation of the Lynn and Carlin windows. The working geological model developed at Blue Star was that of upwardly migrating hydrothermal fluids carrying mineralization becoming trapped beneath a low-dipping thrust structure, leading to the precipitation of fine-grained gold. The trace of the Roberts Mountains thrust bounding the Lynn window could be readily followed because of the contrasting lithologies above and below the thrust, and that boundary was carefully mapped and prospected. The results from that reconnaissance are summarized in figure 6. Anomalous gold values (in excess of .03 ounces of gold per ton) were discovered in highly silicified and barite-veined exposures approximately 2 3/4 miles southeast of Blue Star after just a few weeks of work. Those values were confirmed by resampling, and several grab and channel samples that assayed in the range of .03 to .20 ounces per ton were collected.

The Lynn mining district (and, subsequently, the Lynn window) was named after Fred Lynn, who discovered placer gold in Lynn Creek in 1907 (Vanderburg, 1936). In the rush that followed, placer gold was likewise discovered in neighboring Sheep, Rodeo, and Simon Creeks (figure 7). The new indications discovered by Newmont were located approximately 1.5 miles south of Lynn Creek. Most of the

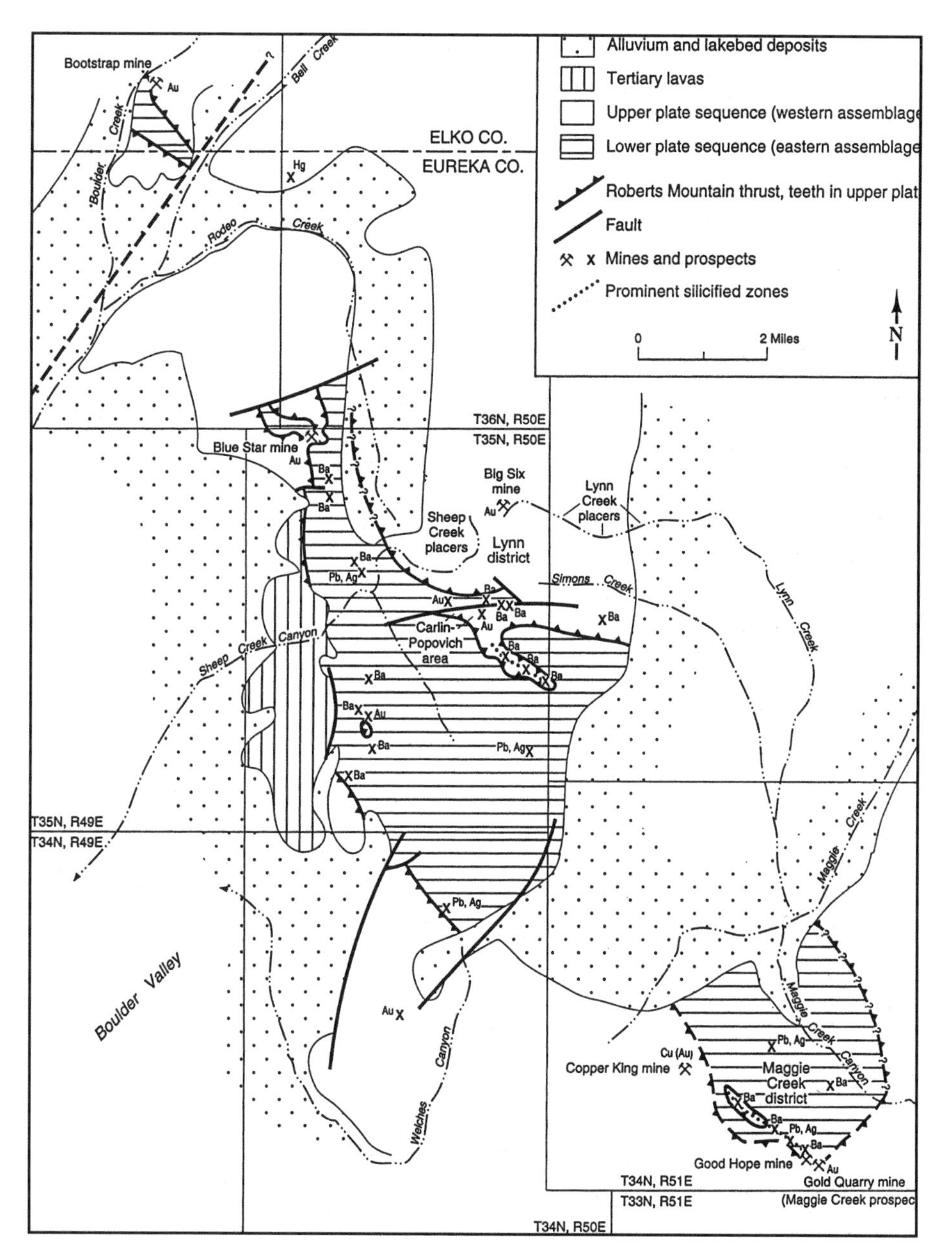

Figure 6: Reconnaissance geological map of the Bootstrap, Lynn, and Carlin windows, including the locations of active and dormant mines and prospects. *Map (prepared by J. S. Livermore and J. A. Coope, October 1961) from Coope, 1991; reproduced courtesy Nevada Bureau of Mines and Geology.*

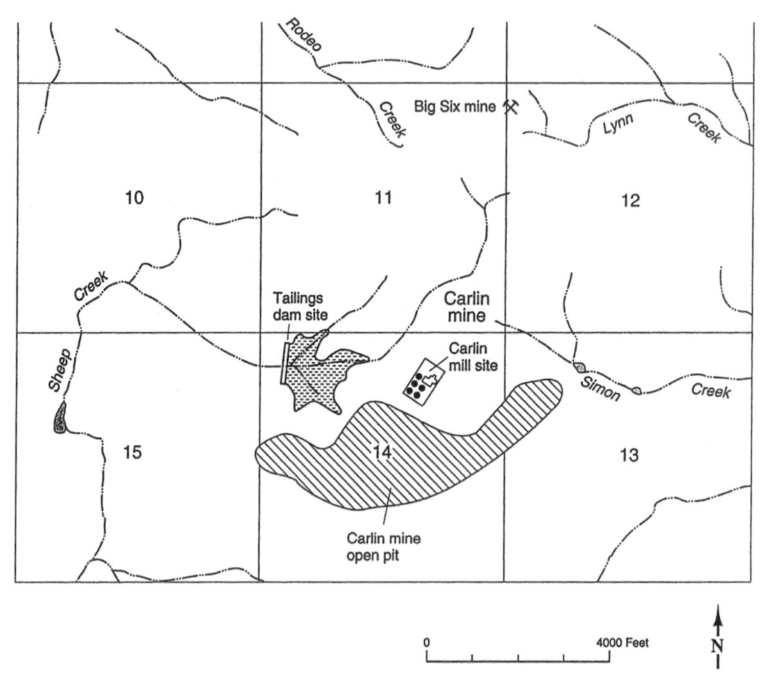

Figure 7: Present location of the Carlin mine and Lynn, Rodeo, Sheep, and Simon Creek mines, which were worked for placer gold prior to discovery of the Carlin deposit. *Map from Coope, 1991; reproduced courtesy Nevada Bureau of Mines and Geology.*

production from the placers was achieved prior to World War II, although placer miners continued to work in the district as recently as the 1960s (figure 8). The bedrock source of the placer gold consisted of quartz veins and stringers in upper-plate (Vinini) rocks similar to the occurrences at the Big Six mine. Total gold production from the placers is estimated at 7,500 to 10,000 ounces (Vanderburg, 1936; Roberts, Montgomery, and Lehner, 1967), but only very limited production came from the lode occurrences. The Lynn placer area was covered by the Newmont reconnaissance, but the occurrence of coarse gold in upper-plate rocks several hundred feet above the Roberts Mountains thrust did not fit the working model developed at Blue Star and was rated positive but secondary in importance.

Detailed geological mapping using the pace-and-compass method was extended to the discovery area to the south and over a larger area surrounding the initial indications. USGS geologists Ralph Roberts and Hal Masursky were invited to visit the area to help in identifying the stratigraphic units and locally abundant graptolite fossils. In late September, during a visit to the area by officials of Newmont Mining, approval was given to stake some claims. Seventeen twenty-acre lode claims, positioned over what eventually became the location of the initial Carlin open-pit mine, were staked in October 1961. Discovery work on the claims began almost immediately, and soon a continuous zone of mineralization averaging .20 ounces of gold per

Figure 8: Two men loosening gravel with spring-tooth harrow, Sheep Creek, Lynn district, 1935. *Photograph from Vanderburg, 1936; reproduced courtesy Nevada Bureau of Mines and Geology.*

ton over eighty feet was exposed in a bulldozer trench in altered sediments adjacent to a quartz porphyry dike. Gold was anticipated in the dike, which proved to contain only low values; nevertheless, the eighty-foot intersection yielded the first discovery of significant gold in the Carlin mine area (figure 9).

A snowy winter followed, and active exploration on the property was discontinued until the following April. Exploratory bulldozer trenching and road building commenced in advance of planned drilling programs on both the Carlin and Maggie Creek properties. While that work was getting under way, Bob Morris, one of the principals of M M & S Mining, advised Newmont that he had negotiated an agreement on the Popovich homestead property, an eighty-acre holding contiguous with the Newmont claim block (figure 9). Morris offered the Popovich eighty acres to Newmont, and an agreement was quickly negotiated. For the drilling program, Pete Loncar, a longtime Newmont employee, contracted an Ingersoll-Rand down-the-hole hammer drill being used on a highway project near Reno. That device operated solely with compressed air and had a depth capacity of only 130 feet. Sample recovery below the water table was poor, but under the local conditions that was not a serious problem. Drilling began at Maggie Creek. Low-grade values were encountered, and, although results were generally disappointing, discontinuous zones five and ten feet in length averaging .15 ounces of gold per ton were intersected. (It is important to note for the record, however, that follow-up drilling in the late 1970s, when the price of gold was considerably higher, proved those intersections to occur on the northeastern margin of the large Gold Quarry ore body.) Bulldozer trenching and road building on the seventeen Newmont claims at Carlin were based on the

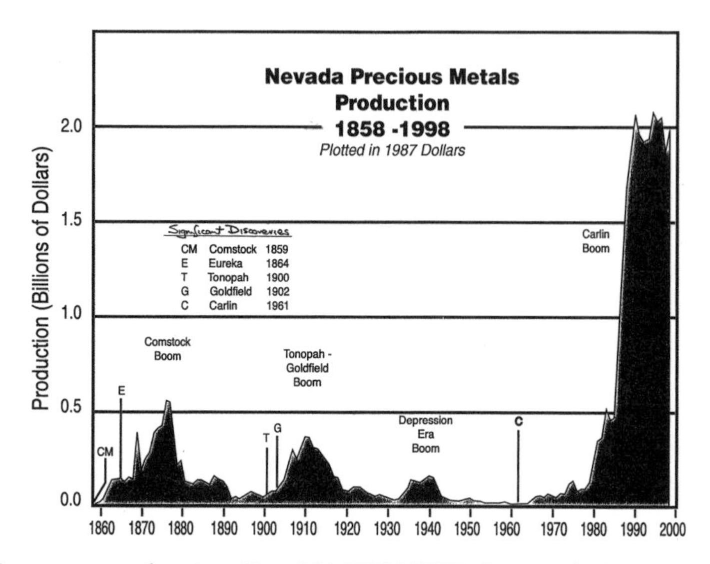

Figure 9: Property map of sections 13 and 14, T35N, R50E, showing the location of the original seventeen claims, the Popovich eighty acres, and the sites of the significant gold discoveries of the Carlin deposit. *Map from Coope, 1991; reproduced courtesy Nevada Bureau of Mines and Geology.*

results of the 1961 exploration program. The resulting exposures were mapped and sampled in detail, with assay results as high as .74 ounces of gold per ton. Those results led to additional staking.

Because of the complexity of the altered stratigraphy of the Popovich property, Alan Coope remapped the eighty acres in detail. Careful sampling revealed that less spectacularly altered (hydrothermally bleached and leached) silty-limestone float contained more gold than the more prominently altered silicified and iron-stained units that formed bold exposures. An unexpected value of .22 ounces of gold per ton was obtained in float from the immediate footwall of an exposed, slickensided, iron-stained fault plane, which assayed only .07 to .08 ounces of gold per ton. Because cash payments were due six months from the commencement of the Popovich agreement, the eighty acres was the first area investigated once the drill had been moved from Maggie Creek to the Carlin area late in the summer of 1962. A drill traverse of eleven holes was laid out just fifty feet north of and parallel to the boundary separating the Popovich property and the TS Ranch land to the south. A central hole was located in the footwall of the slickensided fault. Logistics determined that this footwall hole was the third hole drilled. The hole intersected approximately one hundred feet averaging 1.03 ounces of gold per ton, which, at that time, had a value of thirty-six dollars per ton. The spectacular discovery (figure 9) confirmed the exciting potential of the Carlin area, and the results from the ongoing drilling program considerably advanced the understanding of geological controls on mineralization. Additional sampling of less spectacularly altered float in the area revealed values of up to 2.0 ounces of gold per ton, and systematic work quickly outlined the sub-outcrop of the ore body. By the time the Carlin deposit was brought to production (at the rate of two thousand tons per day) in early 1965, eleven million tons of open-pittable ore averaging .32 ounces of gold per ton had been proven. Project costs through to production totaled ten million dollars, and the mine paid its first dividend after eight months in operation.

Subsequent Developments

The Carlin gold discovery was one of the most significant events in Nevada mining history and may rival the discovery of the Comstock in overall historical and economical importance. Following the discovery, exploration for gold increased dramatically. The industry received a major incentive in 1972 when the price of gold was released from government control and allowed to seek its own level in the world market (Tingley, 1993). In subsequent years, many important discoveries were made, and sediment-hosted gold deposits are broadly included within the Carlin, Battle Mountain-Eureka, and Getchell trends. Additional deposits, including several important ones associated with volcanic-host rocks, have been located in the western and southern parts of the state. Gold deposits discovered in the northern part of the Carlin Trend through 1997 are shown in figure 10.

Developments through the mid-1980s were oriented toward open-pit operations. In late 1986 Western States Minerals drilled a deep hole below the oxidized Post deposit and discovered the high-grade but refractory Deep Post deposit. In early 1987 Barrick purchased the Western States-Pancana (Goldstrike) property, in part because of the results at greater depth at Deep Post. Barrick's follow-up exploration led to the discovery of the twelve-million-ounce Betze deposit about one thousand feet below surface. That success ushered in deep-drilling programs by virtually all property holders along the Carlin and other trends, resulting in the discovery of several important, high-grade deposits. In 1989 Barrick intersected the margins of the Deep Star deposit, and late that year it drilled a discovery hole on the Meikle deposit: 580 feet averaging .41 ounces of gold per ton beginning at a depth of 1,400 feet. A 1989 hole drilled by Newmont through the core of the Deep Star intersected a body of ore 230 feet thick assaying 2.06 ounces of gold per ton (Teal and Jackson, 1997). Development of those deep discoveries, which are characterized by refractory ores, has required the construction of more complex and expensive extraction facilities such as autoclaves and roasters. The first ores processed from underground came from workings driven from the lower levels of the Carlin pit in 1993.

The Geology of the Carlin Trend Deposits

Since the 1960s, the geological relationships of many gold deposits along the Carlin trend have received intense scrutiny. Those deposits occur in a structurally complex sequence of Paleozoic sediments of Ordovician to Lower Mississippian age, as illustrated in the idealized figure 11. These sediments are intruded by stocks and dikes of late Triassic to late Tertiary age. Teal and Jackson (1997) note that the age of

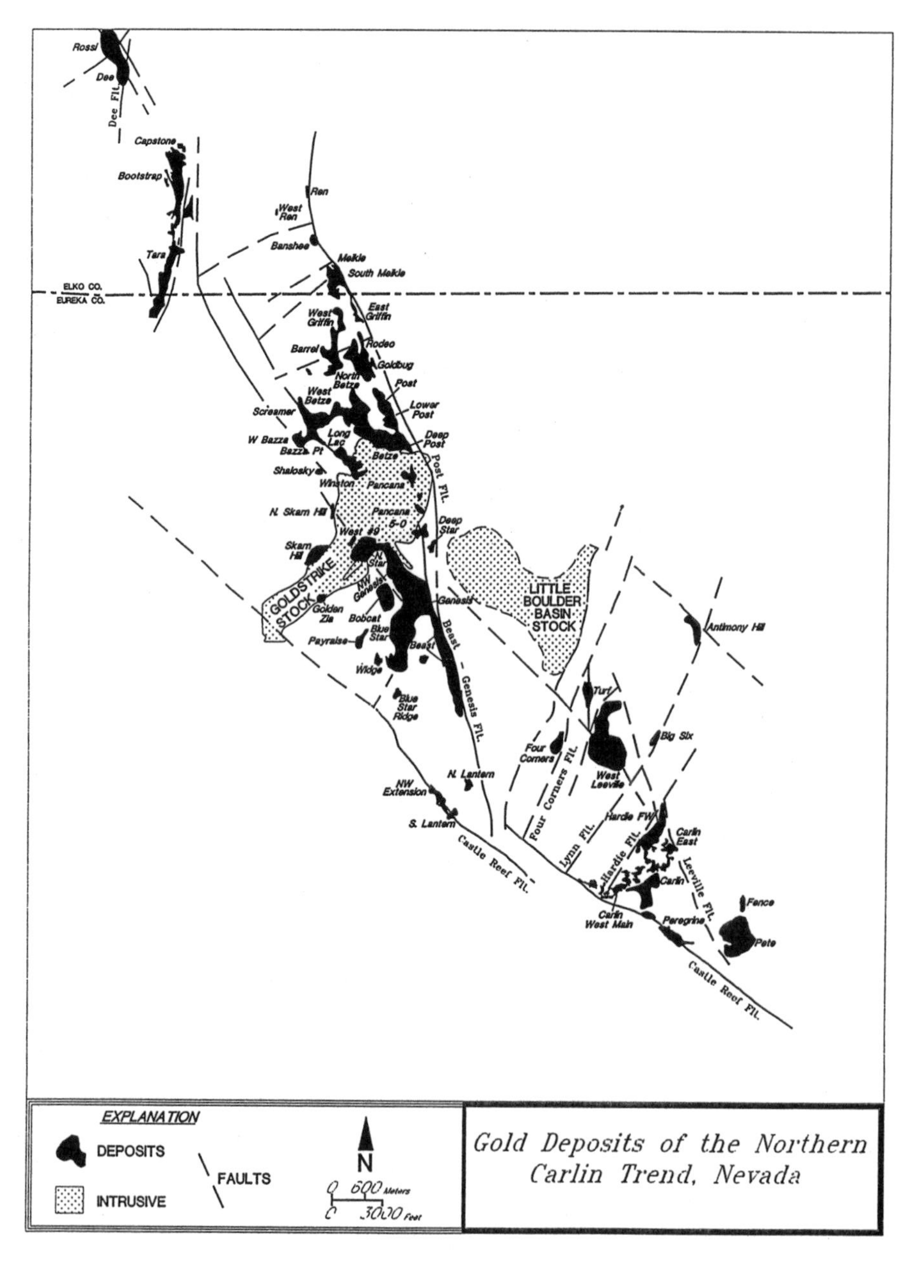

Figure 10: Gold deposits, northern part of the Carlin Trend. *Map from Teal and Jackson, 1997; reproduced courtesy Society of Economic Geologists.*

the gold mineralization is bracketed by premineral granodiorite stocks such as the Goldstrike intrusive dated at approximately 158 million years and mineralized Tertiary dikes in the Betze-Post deposit at approximately 39 million years. Kuehn (1989), in his study of the Carlin deposit, suggests a most probable age of between 37 million and 14 million years bracketed by the intrusion of early Oligocene dikes and the extrusion of Miocene volcanics.

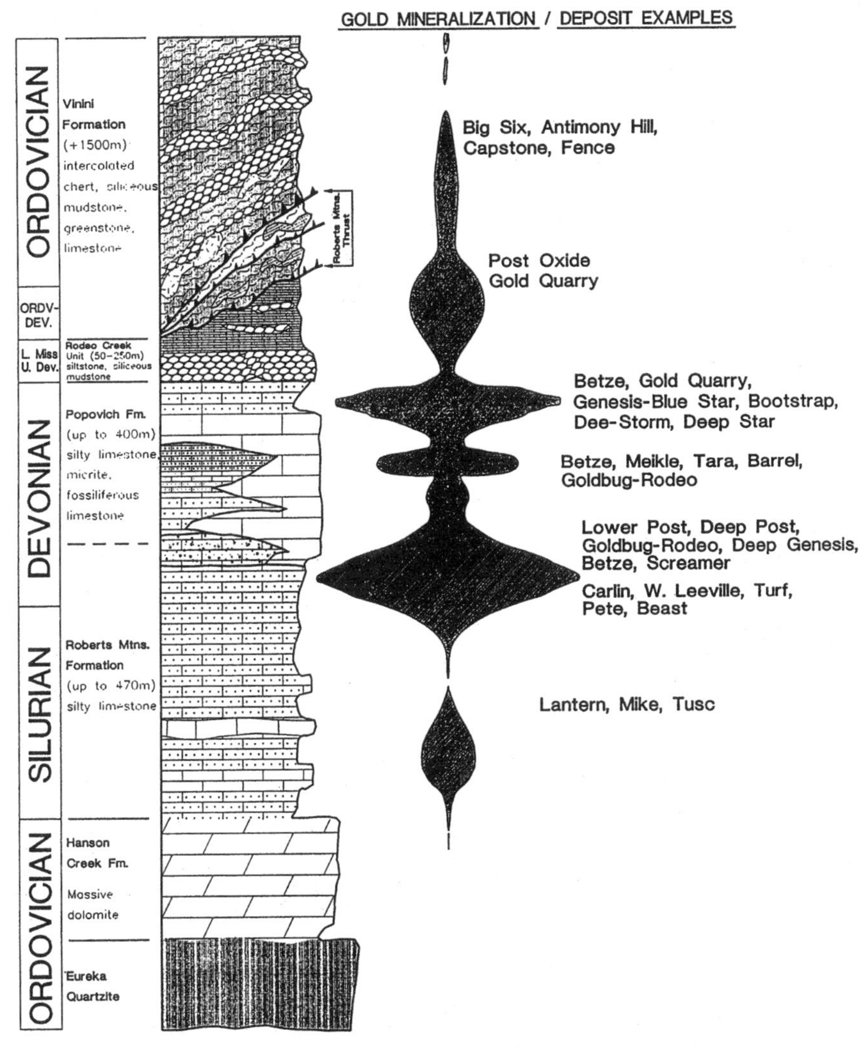

Figure 11: Idealized stratigraphic column and gold mineralization, Carlin Trend. *Diagram from Teal and Jackson, 1997; reproduced courtesy Society of Economic Geologists.*

Individual gold occurrences along the Carlin trend constitute a variety of deposit types. Teal, Jackson, Bettles, and Schutz (in Teal and Jackson, 1997) have compiled a spectrum diagram for the Carlin trend deposits that relates mineralization controls to mineralization styles (figure 12) in a valiant attempt to quantify the relative importance of those factors in deposit formation. Discussion of the complexities of those mineral deposits is beyond the scope of this paper. However, it is pertinent to note that the Carlin deposit is predominantly stratigraphically controlled, in contrast to many other structurally controlled deposits discovered in subsequent years. The reader is referred to Teal and Jackson (1997) and to published papers referenced by those authors and others (Coope, 1991) for in-depth discussion of the geological relationships.

The Impact of Carlin

Tingley (1993), in recording developments since the Carlin discovery, notes that "Nevada was the leading gold producing state in the nation in 1972, 1975, and 1977 but in the intervening years dropped to second and, in some years, third place. In 1980, however, Nevada moved into first place and stayed there. . . . In 1983, the value of precious metals production for that one year exceeded the total recorded production value of the Comstock Lode." Precious-metal production in 1987 exceeded the $1 billion figure and only two years later rose above $2 billion with a production record of five million ounces of gold. Newmont's Gold Quarry mine in 1991 became the first mine in North America to produce one million ounces of gold in a single year, and Barrick Goldstrike's Betze-Post deposit produced more than two million ounces in 1995. In 1998 Nevada's gold production climbed once more—to 8,860,000 ounces valued at more than $2.65 billion (figure 13). That annual output for the state far exceeds the total recorded production of gold from the Comstock lode since 1859.

Those developments have stimulated many improvements in mining and metallurgical technology and set new standards for gold mining throughout the world. Large-scale mining, heap leaching, and automation at various levels in the mining, milling, and assaying processes have cut overall costs and allowed lower and lower grades to be mined (Price, 1991; Coope, 1991). In addition, metallurgical (including biometallurgical) advancements in the treatment of a variety of deeper, refractory ores have served to expand the economically recoverable reserves. As noted by Price (1991), "discoveries of Carlin-type deposits have helped to diversify the Nevada economy. Gold mining directly provides thousands of jobs for Nevadans, indirectly provides thousands more, helps build and maintain infrastructure in rural parts of the state, and broadens the tax base for education and other government programs." Moreover, in an attempt to relieve added pressures on schools, police and fire departments, and other public services that have resulted from new mining developments, several companies have made monetary donations to local authorities.

Despite increased regulation and, at times, floundering gold prices, the significance and importance of the Carlin trend continues to grow. By the end of 1998,

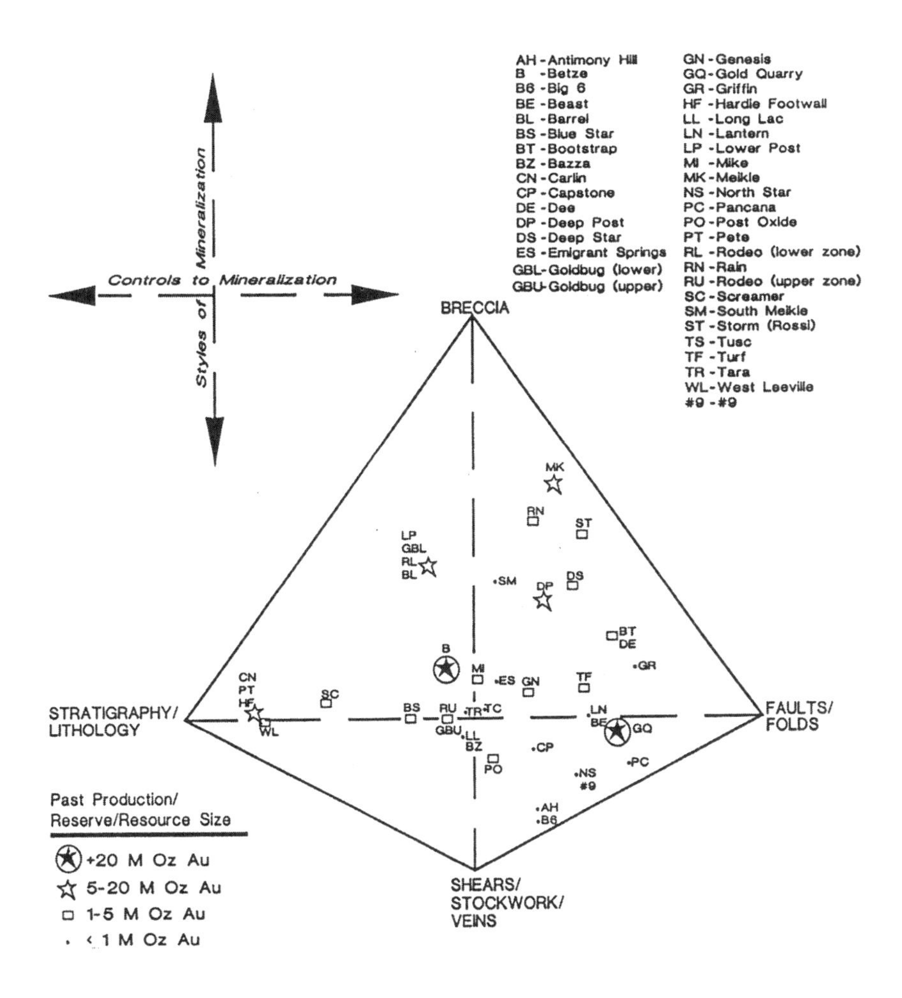

Figure 12: Quaternary diagram showing the relative importance of lithological and structural characteristics of Carlin Trend gold deposits. *Diagram from Teal and Jackson, 1997; reproduced courtesy Society of Economic Geologists.*

approximately 39.5 million ounces of gold had been extracted, and the combined production and proven and probable reserves totaled approximately 80 million ounces. The latter figure far exceeds the estimated 60 to 65 million ounces of gold produced from the Porcupine Mining District in Ontario, which is currently the largest gold-producing district in North America. The combined production from, proven and probable reserves of, and additional indicated mineral inventory in

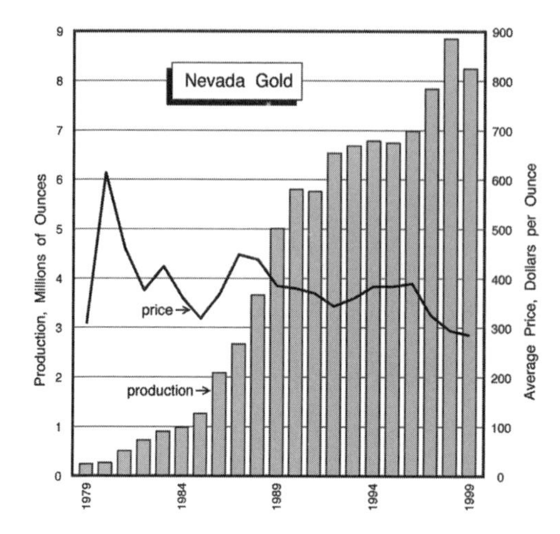

Figure 13: Gold production (in ounces), state of Nevada, 1979–1999. *Graph courtesy Nevada Bureau of Mines and Geology.*

deposits on the Carlin Trend exceed 100 million ounces of gold. That total is sensitive to changes in the price of gold and other economic factors. There is potential for the discovery of more ore, and one can reasonably anticipate that the combined scientific and technical capabilities and ingenuity of the skilled individuals in the gold-mining industry will eventually lead to the winning of an even greater total prize.

Acknowledgments

Significant amounts of the information and data contained in this review have been gathered from previously compiled records and publications. Notable among them are the publications of the Nevada Bureau of Mines and Geology (NBMG). The author is grateful to Joseph V. Tingley and other NBMG staff members for answering inquiries and providing sources for both historic and recent information. NBMG permission to reproduce maps and diagrams published in several of its publications is gratefully acknowledged. Geological understanding of the Carlin Trend deposits has advanced considerably over the years and more so with the development of the underground mines. The publication by Teal and Jackson, 1997, provides a recent, concise reference to current thinking on the geologic controls on the numerous deposit types, and the authors and the Society of Economic Geologists graciously allowed reproduction of illustrations from that paper. David Groves, Owen Lavin, and Charlene Gendill of Newmont are thanked for their help in assembling information. Mr. Lee Brumbaugh is thanked for providing access to the extensive photographic records of the Nevada Historical Society and for his permission to use the photographs that illustrate this text.

References

Coope, J. Alan. *Carlin Trend Exploration History: Discovery of the Carlin Deposit*. Special Publication 13. Reno: Nevada Bureau of Mines and Geology, 1991.

Horton, Robert C. "An Outline of the Mining History of Nevada, 1924–1964" (1964). In *Outline of Nevada Mining History*. Special Publication 15. Reno: Nevada Bureau of Mines and Geology, 1993.

Kuehn, Carl A. "Studies of Disseminated Gold Deposits near Carlin, Nevada: Evidence for a Deep Geologic Setting of Ore Formation." Ph.D. diss., Pennsylvania State University, 1989.

Lincoln, Francis C. "An Outline of the Mining History of the State of Nevada, 1855–1923" (1924). In *Outline of Nevada Mining History*. Special Publication 15. Reno: Nevada Bureau of Mines and Geology, 1993.

Lindgren, Waldemar. *Mineral Deposits*. New York: McGraw-Hill Book Company, 1933.

Merriam, Charles W., and Charles A. Anderson. "Reconnaissance Survey of the Roberts Mountains, Nevada," *Geological Society of America Bulletin* 53 (1942). Pp. 1675–1728.

Price, Jonathan G. Foreword to J. Alan Coope, *Carlin Trend Exploration History: Discovery of the Carlin Deposit*. Special Publication 13. Reno: Nevada Bureau of Mines and Geology, 1991.

Roberts, Ralph J. "Alinements of Mining Districts in North-central Nevada." In *Short Papers in the Geological Sciences: Geological Survey Research, 1960*. Geological Survey Professional Paper 400-B. Washington: U.S. Government Printing Office, 1960. Pp. B17–B19.

______. "The Carlin Story," in Sediment-hosted Precious-metal Deposits of Northern Nevada. *Nevada Bureau of Mines and Geology Report* 40 (1986). Pp. 71–80.

______. "Geology of the Antler Peak Quadrangle, Nevada." *U.S. Geological Survey Geological Quadrangle Map GQ-10* (1951).

Roberts, Ralph J., and Robert E. Lehner. "Additional Data on the Age and Extent of the Roberts Mountains Thrust Fault, North-central Nevada." *Geological Society of America Bulletin* 66 (1955). Pp. 1661.

Roberts, Ralph J., Kathleen M. Montgomery, and Robert E. Lehner. "Geology and Mineral Resources of Eureka County, Nevada." *Nevada Bureau of Mines Bulletin* 64 (1967).

Roberts, Ralph J., and others. "Paleozoic Rocks of North-central Nevada." *Bulletin of the American Association of Petroleum Geologists* 42 (December 1958). Pp. 2813–2857.

Teal, Lewis, and Mac Jackson. "Geologic Overview of the Carlin Trend Gold Deposits and Descriptions of Recent Deep Discoveries." *SEG* [Society of Economic Geologists] *Newsletter* 31 (October 1997). Pp. 1, 13–25.

Tingley, Joseph V. "An Outline of the Mining History of Nevada, 1965–1992" (1993). In *Outline of Nevada Mining History*. Special Publication 15. Reno: Nevada Bureau of Mines and Geology, 1993.

______. "Outline of Nevada Mining History." In *Nevada Geology* [newsletter of Nevada Bureau of Mines and Geology] 20 (fall 1993). Pp. 1–3.

Tingley, Joseph V., and Daphne D. LaPointe. "Annual Review 1998." *Mining Engineering* 51 (May 1999). Pp. 80–84.

Vanderburg, William O. *Placer Mining in Nevada*. University of Nevada Bulletin, Bulletin of Nevada State Bureau of Mines and Mackay School of Mines 30 [Nevada Bureau of Mines and Geology Bulletin 27]. Reno: Nevada State Bureau of Mines, 1936.

_______. *Reconnaissance of Mining Districts in Eureka County, Nevada*. U.S. Bureau of Mines Information Circular 7022 (typewritten), 1938.

_______. *Reconnaissance of Mining Districts in Lander County, Nevada*. U.S. Bureau of Mines Information Circular 7043 (typewritten), 1939.

Financing the Gold Rush

Peter Maciulaitis

Peter Maciulaitis, a professional geologist, earned a degree in geological engineering at the Colorado School of Mines in 1967. In three decades as a geologist, he has searched for silver, uranium, molybdenum, antimony, barite, and gold in Mexico, Colombia, France, Scotland, Ireland, and Canada. More than half of his career has involved generative work in gold exploration, particularly in Nevada. He has served as a consultant for and employee of two junior companies that went from start-up ventures to major gold-mining concerns. One of his other interests is mining history.

Introduction

This paper is an informal look at gold discovery and mining finance in the American Far West. It is not a comprehensive study but rather an attempt to illustrate gold-rush dynamics and provide examples of some fund-raising methods and business structures used in selected gold-discovery cycles. The focus is on four rushes that occurred between 1848 and the present. The first is the California gold rush, which to many Americans is *the* gold rush. The other rush cycles discussed occurred in neighboring Nevada. The Nevada story starts with the rush to the Comstock Lode in 1859, moves to Tonopah in 1900, and ends with the current gold-discovery cycle.

First, a little statistical information on gold resources and production. The California-Nevada region is one of the world's major gold provinces. Those two states have produced approximately 235 million ounces of gold, which is 52 percent of all the gold mined to date in the United States. The region's gold resources are in excess of 125 million ounces, of which 93 percent is in Nevada. The 1998 production for Nevada (the current gold-discovery cycle) was 8.86 million ounces of gold. By comparison, it took almost the first five years of the California gold rush to produce a similar amount.

The California Gold Rush:

From the Miner with a Gold Pan to the Corporation

The discovery of gold in North Carolina in 1799 marked the introduction of gold mining as part of the American experience. However, gold had already been found in the northern parts of Spanish Mexico, where small amounts of the metal were mined as early as the 1770s. Production in Spanish Mexico was from placers along the Colorado River and from lode deposits in the Cargo Muchacho Mountains in what is presently the state of California (south of the area in which the California gold rush was spawned). During the 1830s, as gold mining spread through the Carolinas and Georgia, Mexicans were mining gold from placers northwest of Los Angeles and south of Santa Fe. The production was small and, far from major population centers, those activities remained largely unknown to the outside world.

The first half of the nineteenth century was marked by the continual westward expansion of the borders of the United States, which led to war with Mexico in 1845.

On January 24, 1848, James Marshall discovered placer gold in a tributary of the American River in northern California, and on February 2, 1848, the treaty of Guadalupe Hidalgo was signed, ending the war between Mexico and the United States. The treaty gave the United States possession of what has become the states of Arizona, California, Colorado, Nevada, Utah, and New Mexico. At the time, an estimated one thousand non-Spanish white men lived in California. Marshall's discovery would help populate the region and graft it firmly to the United States.

An important element of gold-rush dynamics and mining finance is the use of publicity and promotion. The *Californian* and the *California Star*, newspapers in the then small town of San Francisco, first reported Marshall's gold discovery in March 1848, but the news attracted little attention. Other newspaper reports followed, gradually starting a flow of men to the foothills of the Sierra Nevada. California was distant from eastern population centers, but ships passing through the ports of San Francisco, Monterey, and Los Angeles carried the news to other ports around the world. The crowning promotional touch came in December 1848. In a speech delivered before Congress, President James Polk stated that significant amounts of gold had been discovered in California and that the gold was free for the taking. The president's speech, as well as an increasing number of reports coming out of California, soon appeared in newspapers throughout the United States and Europe. The rush was on. Between 1849 and 1854, three hundred thousand people traveled to California to make their fortunes.

When thinking about the California gold rush, many envision the individual miner or prospector working alone. However, from the very beginning of the rush, it was common for men to form partnerships, associations, and joint-stock companies to finance travel to the gold fields. The partnership and the joint-stock company provided a degree of organization, sharing of expenses, and protection. Some of these business arrangements extended into placer mining once would-be miners arrived in the California foothills, while others lasted only as long as the westward trip.

The logistics of mining also affected the roles of individuals, partnerships, associations, and joint-stock companies and eventually led to the formation of corporations. Placer mining was not easy work. An individual could shovel only so much gravel, bar aside only so many rocks, and wash only so many pans in a day. He might have to move frequently from place to place and build his own accommodations, and he certainly had to find a way to feed himself. In a partnership with one or more other people, he could take advantage of larger, more effective placer mining equipment such as a sluice, long tom, or rocker. He could pool finances with others and apply more manpower to any task required. From daguerreotypes and journals from the period, it appears that partnerships of ten or twenty people were not uncommon. Placer miners reinvested the gold they recovered to buy supplies and continue mining. Partnerships and associations could be formed or dissolved quickly.

As more people came to the gold fields, the easily accessed gold was quickly depleted. The average yield per miner declined steadily from twenty dollars per day in 1848 to eight dollars or less per day in 1851. To make a profit, many turned to river mining, which required more equipment and an even larger labor force. River mining replaced the gold pan and rocker with waterwheels, dams, ditches, flumes, and long sluice networks. Large-scale river-mining operations could be financed by direct investment. In some river-mining associations, a man could invest by providing labor in lieu of money; but individuals not directly engaged in mining were increasingly becoming investors in river-mining operations by providing only cash, and many miners were becoming paid laborers. By 1851 there were fewer miner-owners than paid mining laborers in California.

Miners followed the alluvial gold upstream, which led to the discovery of two additional types of gold deposits: paleo-placers and lodes. In paleo-placer deposits, the gold occurs in ancient, consolidated stream beds. Mining such deposits required that force be applied to break up cemented sediments to free the gold. In 1851 it was discovered that if a large nozzle (monitor) was fitted to a hose into which pressurized water was directed, the resulting water flow could blast apart the paleo-placer beds, freeing the gold. That technique is known as hydraulic mining. The technology quickly spread along the western edge of the Sierra Nevada Mountains. In 1852, as a direct result of hydraulic mining, California reached its all-time annual gold production peak of approximately 3.9 million ounces. Hydraulic mining accounted for the lion's share of California gold production until lawsuits connected with the disposal of debris forced many operations to close during the 1880s. Hydraulic mining required capital to build flumes, dig ditches, and construct recovery works, as well as to purchase monitors and hoses. Capital was raised through direct investment or through the sale of shares in joint-stock companies or, later, corporations.

As early as 1849, quartz veins, referred to as lode deposits, were beginning to attract the attention of prospectors. But after an early flurry of lode claim staking, miners quickly discovered the difficulties involved in developing vein mines. Underground workings had to be driven, timber was needed for support, machinery was required for crushing, and then the gold had to be recovered somehow. Added to these considerations was the high cost of transportation. Many lode claims were simply dropped, while some owners switched their sights to selling or trading claims.

Hard-rock mining required large amounts of capital, and the investment outcome was highly uncertain. The corporation, a type of business organization just beginning to gain acceptance in the United States, was soon seen as a way to tackle the development of quartz mines. Corporations had two significant advantages over partnerships, associations, and joint-stock companies. The corporation provided a way to raise substantial amounts of capital through the sale of shares to a large number of investors, and it limited liability to creditors to the corporation's assets, should the venture fail.

In 1850 the California State Mining & Smelting Company became the first corporation to be formed in California. Later that year the Mariposa Mining Company was organized. It issued one million dollars' worth of capital stock, and its shares were traded on the London and Paris stock exchanges. Over the decade of the 1850s, more than 576 corporations were formed in California, and about three-quarters of them were mining companies. Other corporations, formed to work in California, were organized in eastern cities such as New York and Philadelphia. Initially, investors showed considerable interest in mining shares, but most such companies failed to show profits or provide dividends. By the latter part of the decade, little capital could be attracted to California mining companies.

Before moving on to the gold rushes of Nevada, something should be said about the influence of British and French investment markets. News of the discovery of gold in California attracted people to go to the gold fields from around the world and drew foreign investment capital as well. British and French companies were quickly formed to participate in California mining ventures, and California mining stocks soon were traded in London and Paris. A flurry of British promotion and investment occurred between 1850 and 1855. Between October 1851 and January 1853, thirty-two British mining companies were formed to operate in the United States, principally in California. Col. John C. Frémont, the well-known explorer, owned a large land grant known as Rancho Las Mariposas. Lode gold was found at Mariposa in 1850, and Frémont actively promoted leases on his property. Several British companies attempted to develop quartz mines— several on the Frémont land grant—but none were known to have made a profit.

In France, news of California gold discoveries came during a period of widespread unemployment and social unrest. California seems to have captured the French imagination and offered a sense of hope. Expedition Française por les Mines d'Or de Sacramento, organized in 1849, was the first French company formed to operate in California. Its purpose appears to have been to transport and sell merchandise. At least one company, La Mariposa Mine D'Or (formed in August 1850), was organized to operate a concession on Colonel Frémont's land grant. More than a dozen other French companies were formed to transport miners to the gold fields and to engage in mining. Persons wishing to go as miners were encouraged to double their chances of obtaining wealth by also buying shares. Stock sold quickly. Late in 1850 the investment spurt ended when newspapers reported that most emigrants deserted the companies on reaching California and that gold was becoming difficult to find.

The Comstock:

The Spread of the Use of the Corporation and Stock Exchanges

As outside investment interest in California waned, the discovery of the Comstock Lode in Nevada revitalized western mining and popularized the use of the corporation as a tool for mining development. In 1849 a party traveling to the California gold fields discovered placer gold in Gold Creek, northeast of present-day Carson City. The amount of gold was small, and the discovery attracted little initial attention. Over the next several years a few placer miners traced the gold upstream to outcrops that were part of the Comstock Lode. The first lode claims were located in early 1859. Initially, the miners recovered only the gold that occurred within a dark metallic material that clogged their sluice boxes. Finally ore samples were brought to an assayer in Grass Valley, California, who determined that they were rich in silver sulfides. That news lured to the Comstock several hundred Californians, including George Hearst, a small lode mine operator from Nevada City. Along with a few other men, Hearst bought several claims and organized a partnership, the Ophir Mining Company. Assays from surface samples reportedly ran $1,595 in gold and $4,791 in silver per ton. The partners soon began systematic mine development.

News about the Ophir Mining Company caused thousands of other miners, prospectors, and speculators to rush to the Comstock region. Between sixteen thousand and seventeen thousand claims were staked. Next, attention shifted to speculation in mining claims. A partial interest in a mining claim was usually expressed in terms of lineal feet owned along a ledge. More often than not, feet on one lode were bartered for feet on other lodes without any money changing hands. Gradually the claims were converted into corporations, for which the former owners were given shares. Thus, initially the shares in many corporations were also denoted in feet or even inches, but it was not long before companies began issuing more shares than there were feet or inches along the ledge. As a result, by the late 1860s corporations no longer linked shares and assets. Instead, shares were said to have a designated par value.

On April 28, 1860, Hearst and his partners incorporated their mine in California, calling it the "Ophir Silver Mining Company." The Ophir was the first corporation to be formed on the Comstock Lode. It issued $5.04 million worth of capital stock, making it the largest corporation in the West. In 1860-1861, eighty-six Comstock mining companies were incorporated in California with combined capital stock of more than sixty million dollars. In the 1860s, thousands of companies were incorporated for mining in California and Nevada. Although many were formed in California, others were organized in eastern financial centers. It was the largest wave of incorporation yet seen in the United States.

To create a market for those companies, promoters engaged in press campaigns and the use of elaborate prospectuses. Mining towns sprang up in the wake of new discoveries, and local newspapers, which printed glowing reports of local discoveries and mine developments, were quickly established. According to Mark Twain (who wrote articles for one of the Comstock newspapers), it was common for promoters of newly formed corporations to offer reporters shares of stock in exchange for mentioning their "mines" in the newspapers. Twain said that such articles usually emphasized the "promising" nature of the property and the excellence of the workings without commenting directly on the rock, which oftentimes included not a trace of ore.

To collect brokerage fees and handle the large number of stock transactions now occurring, the San Francisco Stock and Exchange Board was created in 1862 as the first stock exchange in the West. It immediately helped create a market for mining stocks, facilitate the formation of corporations, and attract outside investment capital. By 1864 6 new stock exchanges were formed in San Francisco, 9 in the vicinity of the Comstock Lode, 1 in Sacramento, 1 in Stockton, and 2 in Portland, Oregon. Moreover, the organization of mining-stock exchanges to help create a market for local mining stocks became common in western boomtowns. Of the almost three hundred stock exchanges formed in the United States between 1860 and 1930, about two-thirds were mining-stock exchanges.

As Comstock production rose, promoters whipped the public into a stock-buying frenzy. Fortunes were made and lost on the market, which was heavily influenced by insider trading. At times, an investor could make as much on a well-promoted but worthless stock as on one with a profitably operating mine. Sometimes an operator would pay out dividends from the treasury in one quarter and issue assessments in the next; such assessments could be used both as a source of development funds and for manipulative purposes.

The Comstock Lode produced approximately 8.6 million ounces of gold and 192 million ounces of silver and had a major impact on the development of mining in the United States. In terms of exploration, prospectors were now looking beyond California. The result was the discovery of new mining districts throughout Nevada and the adjoining region. A new outcrop type was added to the prospectors' "rock memory" of the observed characteristics existent in the landscape of known gold deposits, opening up additional possibilities for discovery. The high value of the Comstock ores attracted the large amounts of capital necessary to develop those hard-rock deposits. The district became a proving ground for the development of mining and metallurgy techniques that were then brought back to California and dispersed throughout the West as new districts were discovered. The use of corporate stock offerings to finance the exploration and development of mines gained acceptance, became widely practiced, and spread to other industries.

Early-Twentieth-Century Nevada:

Bootstrap Financing Provides a Market Assist

Nevada's second major rush period began in 1900. As in the case of the Comstock Lode rush, once discoveries started to be made, wild stock speculation followed, and the stock speculation itself became a major force that drove the discovery process. Initially, however, prospective miners found themselves cash poor. One of the characteristics of this rush was the use of what might be called "bootstrap finance." Jim Butler, a rancher, made the initial discovery that sparked the boom. In May 1900 he discovered in a ledge near a spring called Tonopah a quantity of black quartz containing siliceous silver ore. He did not know what the rock was but suspected that it contained metals. Since he had no money for assaying, he offered a local assayer an interest in the property in exchange for an analysis of some samples he had collected. The assayer refused. Butler finally convinced two friends to join him in searching for claims if his samples contained ore. One of the partners found a schoolteacher and part-time assayer who agreed to analyze the samples for an interest in any claims that were staked. The assays revealed gold and silver values ranging from eighteen dollars to six hundred dollars per ton. The partners returned to Tonopah to locate their claims. In proving up the claims, they mined about two tons of ore, which they shipped to Austin, Nevada, for processing. Near the end of the year, they received a check in the amount of six hundred dollars.

News of the ore shipment spread locally, and a claim-staking rush resulted. Butler and his partners had what appeared to be the best ground, but they still lacked money to develop their claims. To get development started and build a treasury to finance operations on their own, they initiated a leasing system. They made their first lease on December 1, 1900. At least 120 leases were subsequently given out. All of them specified that a 25 percent royalty on production would go to the partnership. Production began quickly, and by June 1901 a California promoter was attracted to the property. He purchased the partnership's holdings for $336,000. The Tonopah Mining Company was incorporated to develop the property. The leases were allowed to run to the end of December 1901, then the corporation took over. Capitalization was one million shares at one dollar par value. Systematic development of the property commenced.

The town of Tonopah grew to three thousand people in 1902 and to about six thousand by the end of 1903, becoming a mining center that attracted the familiar hordes of prospectors and promoters. Some men switched back and forth between mining and prospecting. Often short of funds, prospectors commonly worked on grubstakes from locals or from businessmen from cities in the East or West. Grubstakes ranged from as little as some food supplies and the use of a horse and wagon or mule to money advanced against an interest in any potential property. Although grubstake agreements were seldom in writing, local courts upheld various grantors' rights in a number of famous cases. While some prospectors were grubstaked several times over a period of years, most grubstakes probably were for single prospecting trips.

Eventually, the efforts of a number of grubstaked prospectors resulted in the discovery of most of the major mines at Goldfield, a district that ultimately produced more than 4.2 million ounces of gold. The Goldfield district is located about twenty-five miles south of Tonopah. The initial discovery occurred in December 1902 when two local prospectors named Harry Stimler and William Marsh investigated an area of siliceous outcrops there. The two men followed traces of gold detected by panning to a ledge, which they suspected was the source. Although only low-grade gold values were obtained from assays, they were able to get sufficient grubstake backing to prove up several claims. During the ensuing year, hundreds of prospectors visited the area and panned around the numerous siliceous outcrops that occurred over several square miles. Only a few prospectors, also working on grubstakes, decided to locate claims. In two of those newer locations—the Combination and Red Top claims—shipping-grade ore was found. That discovery created a stampede of claim-staking that eventually resulted in claims covering an area of approximately thirty square miles. Eight hundred and thirty-eight claims were eventually taken to patent.

Florence mine, Goldfield, Nevada, July 1906. *Photograph courtesy Nevada Historical Society.*

Many people did little more than stake claims and look for backers to advance money to develop them. Others sought out promoters. The promoter would form a mining corporation that would buy the claims, then sell stock. The original claim owner might be paid in cash, but commonly he received shares in the corporation. If there were traces of gold on the property, the promoter would give out leases in order to get mining initiated. The lessees' development work was then reported and used to increase the company's stock price. However, since 97 percent of the district production came from mines situated within a half a mile square, the ground held by most companies proved to be worthless.

As in the case of Tonopah, the leasing system was initially used to develop most of the mines at Goldfield. The Combination mine was the only one of Goldfield's major producers that was not initially developed this way, although leasing was used for later development. Goldfield leases typically were for one-hundred-foot lengths along a ledge and the full width of the claim. If the ground was in a less desirable claim, the lease area might be larger. Leases typically ran for eight to twelve months, and a 25 percent royalty on production was retained by the lessor. Sometimes lessees would sublease part of their ground to obtain additional operating funds. Some leases paid out handsomely. The Hayes-Monnette lease became the most famous lease block in the district. It measured 373 by 700 feet and produced ore valued at $4.6 million. The Frances-Mohawk lease, a sublease that went for $4,000, produced $2,275,000. If a mining company enjoyed actual ore production on a lease, its treasury grew through royalty payments, and its stock value increased as well, generating additional funds as more shares were issued. Another result of the spectacular success of some leases was the creation of leasing corporations. Like the mining companies, leasing corporations raised money by simply selling stock. Approximately 25 percent of Goldfield's total gold production was obtained from leases operated between 1904 and 1908.

Advertising agencies such as the L. M. Sullivan Trust Company, run by the notorious stock manipulator George Graham Rice, promoted district stocks through the publication of investment sheets and advertisements in major newspapers. Business came largely from brokers and mine promoters selling new stock issues. Enthusiastic advertisements combined with initial stock offerings at ten, twenty, and thirty cents a share sparked wild speculation, and the success of the Hayes-Monnette and Francis-Mohawk leases drove the stock prices higher across the board. According to Rice, Mohawk Company of Goldfield (the lessor of the famous lease blocks) went from ten cents to twenty dollars, and many other companies with little or no production increased in stock value by multiples of ten and fifteen. The speculative bubble in Goldfield stocks burst at the end of 1906 with the financial collapse of the Sullivan Trust Company. Within two months, listed Goldfield securities, which were valued at more than $150 million during the boom, had dropped by $60 million. In late 1906 and early 1907, Goldfield Consolidated Mines Company gained control of all the major producers except two. The company, which had an authorized capital of $50 million, took over production and dominated the district.

100 SHARES
N.Y.5845
SHARES 100
INCORPORATED UNDER THE LAWS OF THE STATE OF WYOMING
CAPITAL $50,000,000.00
The Goldfield Consolidated Mines Company
GOLDFIELD, NEVADA.
FIVE MILLION SHARES OF TEN DOLLARS EACH
This Certifies that Walter E. Beer
is the owner of ONE HUNDRED Shares of the
Capital Stock of The Goldfield Consolidated Mines Company full paid and non-assessable,
SEP 29 1910

Certificate for one hundred shares of the Goldfield Consolidated Mines Company, 1910. *Photograph supplied by the author.*

The success of Tonopah and Goldfield as producing mining camps, as well as the profits to be made on speculative ventures, stimulated prospectors to scour the state in search of new discoveries. New districts such as National, Jarbridge, Gold Circle (Midas), Seven Troughs, Rawhide, and Bullfrog were found. Development and promotion often followed the pattern established at Tonopah and Goldfield. Although 1907 probably marked the end of the main discovery period in this rush cycle, other, smaller booms erupted occasionally. The last rush was to the Divide District, located just outside of Tonopah. The Divide boom started in 1918. More than two hundred companies were quickly formed, over fifty of which were well funded. The boom collapsed in 1919 with little significant production.

The Modern Era:
Science, Technology, and Higher Gold Prices Result in a Long Rush Period

The third major gold boom to hit Nevada began in late 1961 when Newmont Mining Corporation staked the first of its Carlin claims. Three years later, Newmont announced that it had a minable gold deposit. The announcement marked the start of the greatest gold discovery period in U.S. history. More than two hundred million ounces of gold have been discovered in Nevada since that time, and more gold continues to be found. Although some major gold discoveries have been made in volcanic rocks, most of the precious metal has been found in sedimentary rocks. Such gold does not occur in veins or ledges. Those deposits, which contain very fine, microscopic gold, are commonly referred to as sediment-hosted or Carlin-type gold deposits.

To understand how those gold deposits were missed during the earlier periods of exploration, it is important to remember that early prospectors used the gold pan as their main prospecting tool. They panned the streams for traces of alluvial gold. To test for lode gold, they crushed rocks and panned the pulverized material. If they found traces of gold in the pan, they could then send the rock for assay. Because the gold in Carlin-type deposits is very fine and often encapsulated in silica, little or no placer gold is formed, and the gold cannot be panned from crushed rock. The new indicator rock for prospectors and geologists became silica-replaced carbonate sediments called jasperoids, and whether or not a jasperoid was mineralized had to be determined by assay.

Carlin-type gold had in fact been found earlier. In the late 1800s, "invisible" gold was found and mined at Mercur, Utah, after an assayer discovered that there was gold in rock that was being mined for mercury. In the 1930s, at least three Carlin-type gold deposits—the Gold Standard, Gold Acres, and Getchell—were mined in Nevada. By the 1950s, three occurrences were known north of the town of Carlin, Nevada. From one, the Bootstrap mine, about ten or eleven thousand ounces had been mined. But most of the known deposits were small and generally of too low a grade to attract much interest.

The framework that led modern prospectors and geologists on a systematic search that resulted in the enormous new gold discoveries now being mined in Nevada took years of geologic mapping by the U.S. Geological Survey and the Nevada Bureau of Mines and Geology. Dr. Ralph J. Roberts, a career USGS geologist who began work in Nevada in the late 1930s, was instrumental in inaugurating the modern gold-discovery cycle. Roberts's short essay titled "Alignments of Mining Districts in North-central Nevada," which appeared in *USGS Professional Paper 400-B*, published in 1960, was the first of several papers that linked mineralization to the geology that was being revealed by ongoing mapping. But even though science and prospecting combined to start this gold rush, progress was not without its ups and downs. Ore deposit models were formed and altered as gold discoveries were continually made. Over time, the search for gold has been stimulated by increases in the price of gold, economies of scale achieved through the use of large-scale mining equipment, and advances in recovery techniques such as heap leaching, carbon-in-pulp recovery, and autoclaves. Exploration targeted increasingly larger-sized deposits. Starting at deposits of 100,000 ounces, expectations grew to 500,000, one million, and five million.

In a long gold-rush cycle such as this one, many changes occur over time, both in the number of companies involved and in the methods of fund raising. This paper highlights only a few such changes. Initially, metal-mining companies—including some of the majors, as well as smaller companies listed on local stock exchanges such as Salt Lake City and Spokane—followed Newmont's lead into Nevada. At that time, very few companies in the United States were engaged strictly in mining for gold. Mining-company geologists were not the only people to start looking around. Many small Nevada towns were populated by scores of individuals who worked intermittently at small-scale mining operations, concentrating on whatever mineral commodity was in demand. Some joined the hunt by staking known gold occurrences that had been abandoned by previous owners. Those local entrepreneurs peddled their properties to the mining companies, usually receiving in return annual rental payments, property-work commitments, and a royalty on any production. Some individuals put together quite a stable of properties, and properties could often be turned over several times as exploration strategies continually changed. The royalty was commonly 4 or 5 percent of net receipts. Most such properties never produced any ore, but the rental payment was sometimes worth the effort. Companies came and went. In the 1970s, oil companies joined the hunt but then withdrew as gold prices turned downward in the mid-1980s. Of more significance was the arrival of junior and mid-sized companies during those two decades. They remain an influence to this day, and a few have grown considerably.

Mining is a business that can experience quick up-and-down turns. To obtain funding, a start-up company needs a buoyant market. In earlier U.S. gold rushes, companies were quickly formed, listed, and promoted on a myriad of mining-stock exchanges, but in this modern rush, things were done differently. For one thing, the U.S. had tightened its regulation of the securities market. The time required for formation and listing of companies had increased. The mining-stock exchanges so common in earlier times had largely disappeared. In Canada, things could be done more quickly. Many people, Canadians and Americans alike, chose to form mining and exploration companies in Canada. Such companies were listed on the Toronto, Montreal, Vancouver, and Alberta exchanges. Most chose Vancouver, which has sometimes been referred to as the largest venture-capital market in the world. This is not to say that there were not start-up companies formed in the U.S.; there were, and some of them went on to make significant gold discoveries.

Mining exploration is a high-risk business, and most prospects do not develop into ore bodies. Smaller companies have made major contributions through aggressive exploration and by tackling marginal projects that the larger mining companies have turned down. Their diligence and willingness to take a greater risk or pay a higher price have led to some major discoveries. This is exemplified by Barrick Gold Corporation's purchase of the Goldstrike property, which contained a few hundred thousand ounces in reserves at the time of acquisition and is now the largest gold mine in the U.S. From a start-up company in the early 1980s, Barrick has become the world's largest gold company in terms of market capitalization.

In general, smaller companies use a combination of staff and consulting geologists to conduct "grass-roots" exploration. Promising areas are staked or leased. A company usually acquires several properties. Systematic mapping and sampling follow in an effort to identify drill targets, and several rounds of drilling are required to take a property from a promising prospect to a mine. Most properties do not make the whole trip, but a lot of money can be spent drilling a tantalizing prospect. At any point, the company might elect to lease or joint-venture the property to a larger company, with the requirement that the larger company conduct the higher-cost drilling and reserve development. A small company often engages in a mixture of leases, joint ventures, and direct participation, and publicity from such activities is commonly used to maintain investor interest. Additional funds are usually raised through the sale of stock.

The modern rush has seen financing innovations generated by larger mining companies as well. As interest in investing in exploration for gold heated up in the mid-1980s, a trend developed that carried into the 1990s. Major mining companies began to spin off companies that were pure plays in gold, sometimes retaining large equity interests in their offspring. Among the companies formed were Newmont Gold, Freeport Gold, Amax Gold, and St. Joe Gold. Nonmining companies joined in by spinning off their gold divisions. These included Battle Mountain Gold, formed out of Pennzoil; Santa Fe Gold, spun off from the Santa Fe Pacific Railroad; and First Miss Gold (later Getchell Gold), formed out of First Mississippi Corporation. By the late 1990s, most of these companies had been re-absorbed by or been merged into other companies.

Another recent innovation in gold-mining finance has been the introduction of companies that purchase gold royalties. Royalty companies had existed for many years in the iron ore and oil industries, but it was not until 1986 that a company developed a strategy to purchase royalties on gold properties. Franco-Nevada Mining Corporation, then a junior company, led the way by purchasing the royalty on the Goldstrike mine. Franco-Nevada has since become the fifth largest gold company in the world, as measured by market capitalization, and is by far the world's leading mineral royalty company. Franco-Nevada has gone on to find and develop a major underground property, the Ken Snyder mine in northern Nevada. As a result of the success of royalty companies, several other companies now employ strategies that mix royalty purchasing, exploration, and mine development.

For companies with mines or properties undergoing mine development, a variety of hedging techniques—a strategy initially developed by the Australian mining industry—has evolved over the past fifteen years. Those techniques include puts and calls on gold futures, spot deferred contracts, and gold loans. Use of such financial instruments reduces risk associated with price volatility and can provide development funds. Barrick reported that over the past ten years it has realized an additional $1.1 billion in revenues through hedging and has developed one of the world's major gold-mine complexes in the process. Several mid-sized companies have attracted investment through the use of gold bonds and gold warrants, which give speculators the opportunity to take a part of future production.

As gold prices fell in the late 1990s, those few companies able to show positive earnings could usually attribute their profits to an aggressive hedging portfolio. A surprise came in late September 1999 when European central banks announced that they would cap their gold bullion over the next five years. The result was an immediate sharp upward surge in gold prices. That surge caused huge losses for some gold-mining companies with overextended hedge-book positions. While mining companies operating in Nevada were not severely affected, investors began to look more closely at company hedging practices. As a result, some mining companies announced that they were reducing or even curtailing new additions to their gold-hedge books.

Traditionally, many investors have purchased shares of emerging and major gold-mining companies as growth stocks representing an investment in value. At the same time, investors willing to take greater risks have been attracted to the shares of junior gold-mining companies, for which news of a few good drill holes might result in a rapid increase in price. During the latter part of 1999 and early 2000, spectacular increases in the share value of many Internet and related high-technology stocks led many investors to remove their money from stocks in mining and other traditional industries and to reinvest it in the surging Internet sector. Indeed, the newly formed Internet stocks have filled the niche commonly occupied by junior mining companies—that of the high-risk, speculative investment. Gold-mining companies are adapting to the new market conditions. Three trends are emerging: mergers between gold-mining companies, broadening of mining operations to include metals other than gold, and private-placement investments by the major gold companies in aggressive junior mining companies with good property positions in emerging gold regions.

Conclusion

James Marshall's discovery of gold in 1848 has led to a series of gold rushes that has uncovered one of the world's great gold provinces. As discoveries moved eastward, business structures and fund-raising techniques have evolved to tackle the problems associated with discovery and mine development. Direct financing by individuals, partnerships, and associations has given way to corporate financing as the main vehicle for developing mines. Mining corporations have adopted a mixture of financing strategies, including stock offerings, royalty purchases, forward sales, and gold loans. U.S. mining-stock exchanges, which were developed to handle transactions, create markets, and facilitate the creation of new companies, have gradually disappeared, and modern-day mining entrepreneurs and investors alike have often sought out Canadian stock exchanges as sources for capital for investment in mining. In an industry in which development of a major mine can cost hundreds of millions of dollars, individuals and junior companies continue to play an important role by making discoveries and occasionally going on to develop a major mine of their own.

Bibliography

Coope, J. Alan. *Carlin Trend Exploration History: Discovery of the Carlin Deposit*. Nevada Bureau of Mines and Geology, Special Publication 13. Reno: University of Nevada, 1991.

"French Companies in the California Gold Rush." *International Bond & Share Society Journal* (March 1993). Pp. 32–34.

"French Companies in the California Gold Rush (Update)." *International Bond & Share Society Journal* (September 1993). Pp. 32–33.

"French Companies in the California Gold Rush." *International Bond & Share Society Journal* (November 1995). P. 21.

"The Gold-Rush 'La Bretonne' Appears at Last." *International Bond & Share Society Journal* (June 1995). P. 11.

Hammelbacher, Frank W. *A Treasury of Mining Stocks from Nevada Territory, 1861–1864*. Flushing, N.Y.: Norrico, Inc., 1996.

Horton, Robert C., Francis C. Lincoln, and Joseph V. Tingley. *Outline of Nevada Mining History*. Nevada Bureau of Mines and Geology, Special Publication 15. Reno: University of Nevada, 1993.

Jung, Maureen A. "Capitalism Comes to the Diggings: From Gold-Rush Adventure to Corporate Enterprise." *California History* 77 (1998). Pp. 51–77.

_______. "The Comstocks and the California Mining Economy, 1848–1900: The Stock Market and the Modern Corporation." Ph.D. diss., University of Michigan, Ann Arbor, 1988.

Koschman, Albert H., and Maximilian H. Bergendahl. *Principal Gold-Producing Districts of the United States*. U.S. Geological Survey, Professional Paper 610. Washington, D.C.: Department of the Interior, 1968.

Lassonde, Pierre. *The Gold Book: The Complete Investment Guide to Precious Metals*. Toronto: Penguin Books, 1994.

MacDonald, Douglas, and Gina MacDonald. *Mines of the West, 1863*. Helena, Mont.: Gypsyfoot Enterprises, 1996.

"*Mark Twain in Virginia City, Nevada.*" Excerpted from Mark Twain (Samuel L. Clemens). *Roughing It*. Hartford: American Publishing Company, 1888; excerpt, Las Vegas: Nevada Publications, 1985.

Rice, George Graham. *My Adventures with Your Money*. Las Vegas: Nevada Publications, 1986.

Roberts, J. Ralph. *Alignment of Mining Districts in North-Central Nevada*. U.S. Geological Survey, Professional Paper 400-B. Washington, D.C.: Government Printing Office, 1960.

Sears, Marian V. *Mining Stock Exchanges, 1860–1930: An Historical Survey*. Missoula: University of Montana Press, 1973.

Shakespeare, Howard. "British Mining Companies in Australia and the U.S.A." *International Bond & Share Society Journal* 16 (spring 1993). Pp. 10.

Shamberger, Hugh A. *The Story of Goldfield*. Sparks, Nev.: Western Printing and Publishing Company, 1982.

von Arx, Rolf. "Ave Maria Gold Quartz Mine." *International Bond & Share Society Journal* 11 (autumn 1988). Pp. 11–12.

The Last Great Gold Rush: Gold in the Klondike

Ken Coates

Ken Coates is dean of the faculty of arts at the University of New Brunswick at Saint John in Canada. His lengthy list of publications includes Canada's Colonies: A History of the Yukon and Northwest Territories *(1985),* Best Left as Indians: Native-White Relations and the Yukon Territory *(1991),* North to Alaska: Fifty Years on the World's Most Remarkable Highway *(1991), and (with W. R. Morrison)* The Forgotten North: A History of Canada's Provincial Norths *(1992).*

Two of the most enduring images in North American history come from the Klondike gold rush. The first, a widely recognized shot of stampeders carrying supplies up the Chilkoot Pass, captures the spirit, grit, determination, and hardship of the Klondike experience. The second, far more crucial to Canadian than American audiences, shows a small contingent of officers of the North West Mounted Police (NWMP) in their detachment at the summit of the Chilkoot Pass. There the Mounties stood "on guard" for Canada, protecting the British traditions of the Canadian frontier from the banditry and vigilantism that Canadians long associated with America's "wild west." The Klondike has been called, in Pierre Berton's phrase, "the last great gold rush," an apt description that captures the continent-wide and international excitement about an event that took place in one of the most isolated, hard-to-reach corners of North America.

One hundred years ago, the Klondike was probably the most famous mining camp in the world. News of the northern "Eldorado" was on the lips of would-be prospectors from Britain to New Zealand, Japan to Brazil. Newspapers in countries around the globe provided regular, breathless accounts of the fabulous wealth that had been uncovered in the far northwest, and hundreds of thousands of men, and hundreds of women, laid plans to make their way to the Yukon River valley. The discovery of gold on Rabbit Creek struck an unusually responsive cord. Few events in world history—and no other episode in the history of Canada—have so captured the public's imagination.

The Klondike gold rush coincided with many key elements in the history of late nineteenth-century North America: tense relations between Canada and the United States over the Alaska-Canada boundary; a continent-wide depression that sapped the vitality of the economy and dampened prospects for personal advancement; the "closing" of the frontier that had been so crucial to America's self-identity; the emergence of "yellow journalism"—the tendency of newspaper owners to embellish reality in the interest of improving circulation; international fascination with the exoticism of the far north; and the Canadian government's preoccupation with expanding its control across the massive, thinly populated, politically unstable, and rather poor nation. And, of course, there was the continuing and enduring fascination with gold. The juxtaposition of those influences, forces, and sentiments helped to set the

Packers ascending summit of Chilkoot Pass on their way to the Yukon Territory. *Photograph supplied by the author.*

stage for what followed: the discovery of a shotgun shell full of gold nuggets and dust on a tributary of the Klondike River in August 1896 would set in motion, in short order, one of the most chaotic, lively, and colorful stampedes in world history. (P. Berton, 1987)

The luxury of hindsight allows us to see the unfolding of events in a logical and consistent fashion. There is, in retrospect, a certain inevitability to the Klondike gold rush, although the general public thought little about the Yukon basin in the mid-1890s. Prospectors and miners had moved steadily across the American West in search of gold and other minerals, and the California gold rush had sparked intense national interest in the rich resources in the western territories. And as one field was staked, the unfortunate and the adventurous pushed on—to new discoveries in Oregon and Washington and then, in the 1850s, to British Columbia, where the flood of American miners into British territory tested again the friendship and diplomacy of U.S. and British officials alike. That gold strike, too, petered out, for the death of a field becomes virtually inevitable as soon as the discovery is made, and the hardiest miners pushed on still further north.

The vanguard of the nineteenth-century mining frontier was populated with a strange breed of men—eternal optimists, grudgingly determined, and socially marginalized. Many of them took First Nations (Native American) wives, a personal decision that often cut them off from families and friends in the south—although the discovery of gold frequently resulted in the quick abandonment of the native partner. They operated with little government supervision, both in the United States and in Canada, at least until paying claims were staked and sizable communities sprang into existence. Their grubstakes were often funded by borrowed money, for it was not difficult to find friends, family members, and investors who shared the belief that the proverbial "mother lode" lay only a little further to the north. They lived with few luxuries and endured backbreaking and often unproductive work. They embodied the best of the North American free spirit and internalized the continent-wide optimism that the wealth of the land was inexhaustible. (Zaslow, 1991)

By the 1870s, however, the prospectors were nearing their limits. Central British Columbia, with its mountainous snowfalls and bitter winter temperatures, had tested their resolve, but the salve of gold soothes many wounds and ills. The men pushed into the Cassiar district, where a small strike excited the miners for a short time before dashing their hopes. In 1872 the first prospectors crossed into the upper Yukon River basin by way of mountain passes from the head of Lynn Canal. While other miners tried their luck along the Alaskan coast—gold was discovered near Juneau—a small but steady stream of prospectors made their way into the interior. Conditions appeared ideal, with wide river valleys and small signs of gold along almost every creek and river. To those who ventured north, the big discovery seemed to be but a matter of time.

But the North extracted its toll as well. Conditions were extremely harsh, with temperatures plunging to minus 50 degrees Fahrenheit during a winter that lasted from October to April. The extreme cold of the sub-Arctic resulted in a thick layer of permafrost, permanently frozen ground lying only a few inches underground, often covered only by small shrubs and a thin layer of dirt and ice-cold surface water. Given that the richest prospects lay at bedrock, the prospectors' realization that there were many feet of frozen soil between them and the potential deposits caused great chagrin. Miners were initially restricted to working on the shifting sandbars along the major rivers and creeks, typically finding promising signs and often digging up enough gold to convince them to press on.

Staying in the Yukon was no mean feat. The Yukon River basin was virtually an island, cut off from the rest of the continent by the formidable mountains along the Alaska coast. It was possible to reach the area overland from the Mackenzie River, the route favored by the Hudson's Bay Company traders, the first Europeans to reach the region, or along the long and winding Yukon River from the Bering Strait. The latter route opened up after the United States purchased Alaska in 1867, but the high costs of steamboat travel and the tiny population in the new territory restricted service dramatically. Securing supplies was difficult and costly. Most of the merchants in the Yukon initially came north to work in the fur trade but soon shifted to the more lucrative opportunities in the mining sector. Many, in turn, gambled on the prospectors' efforts, providing a grubstake in return for a share of whatever gold was produced.

The miners tried many areas, working for a time along the Stewart River and then abandoning that area when the small deposits of gold played out. By the early 1880s, a sizable group had congregated near Fortymile, on the Canadian side of the border, and Circle, in Alaska. Although neither camp was fabulously wealthy—neither offered even a taste of the mother lode—those with good claims generated a living return. A small, professional mining community developed in the region, which was generally unfettered by government officials and official regulations but autonomously managed through highly democratic miners' meetings. The prospectors introduced the concept of such meetings, which had been developed on the American frontier, and relied upon collective control to manage and protect claims and to regulate and punish unwelcome behavior. The miners could be vindictive in their assemblies; their decisions occasionally reflected personal pique more than carefully reasoned judicial judgment, but they managed to contain and regulate the small society. (Stone, 1988)

William Ogilvie of the Geological Survey of Canada, sent to the Yukon in the late 1880s to survey the area for the Canadian government, solved the conundrum of the permafrost. Taking a lesson from urban work crews, who thawed pipes and culverts in the winter by lighting fires, Ogilvie realized that miners would have to melt the permafrost if they were to progress toward the gold-bearing ground at bedrock. The simple solution worked easily, if not pleasantly. Miners built a fire on top of the permafrost and, once the fire had gone out, scraped away a few inches of now-pliant dirt, then repeated the process. Laboring in that fashion, and scrunched over in small, smoke-filled shafts, the miners worked their way slowly down toward bedrock. Once they reached the prime ground, they worked their way along the hard rock, stockpiling what they hoped would be mounds of gold-bearing gravel on the surface. Of course, during the winter months there was no surface water with which to work the gravel through the sluice boxes (they could and did take test pans in their cabins), so the miners had to work an entire season without knowing if their efforts were being rewarded with substantial wealth or paltry returns. For most of the miners working the Fortymile field, annual returns were barely enough to justify the winter's work, and many left the area after one or two disappointing years. (Ogilvie, 1913; Gates, 1994)

The Canadian government worried a little about the presence of American miners in their territory. In 1894, prodded by Anglican bishop William Carpenter Bompas (whose affection for the First Nations was not matched by appreciation for the prospectors), they sent two members of the North West Mounted Police to investigate conditions at Fortymile, near the Canada-Alaska border. Inspector Charles Constantine reported back that although the miners were not debauching the First Nations, as the bishop alleged, they did pose a small threat to Canadian sovereignty in the area. The federal government responded by sending Constantine back to the Yukon the following year, in one of the best examples of fortuitous timing in Canadian history. The small detachment of twenty officers had no difficulty controlling the already well-mannered community, and Constantine moved quickly to eliminate the miners' meeting as any kind of regional authority. Canadian supremacy had been well and firmly reestablished. (Morrison, 1985)

A short distance away from Fortymile, a Nova Scotia miner named Robert Henderson had been working several of the creeks in the Klondike River area. When he met George Carmack, an American, and Carmack's two companions, Skookum Jim and Dawson Charlie, Henderson felt duty bound to pass on word of his small discovery of paying quantities of gold. While Henderson had little regard for Carmack, whom he shunned as a "squaw man" and generally regarded as a blowhard of little reliability, he had even less respect for the two First Nations men. Carmack chose to ignore Henderson's invitation (not extended to Jim and Charlie), and the trio instead decided to explore along Rabbit Creek. Much debate has ensued about what happened next. Regardless of the specifics of the discovery—some say Skookum Jim made the first find, others say Carmack, and a few point to George's wife Kate—the outcome was that on August 16, 1896, George Carmack left the Klondike River bound for Fortymile with a sizable quantity of gold in his hand. In Fortymile, he convinced an initially suspicious audience of prospectors and miners of the truth of his story by ceremoniously dumping a shotgun shell full of gold on a counter. The three men, Carmack claimed, had found huge quantities of gold right on the surface, suggesting the presence of other riches to be discovered at bedrock. (Carmack declared himself to be the discoverer and took two claims on Discovery Creek; Skookum Jim and Dawson Charlie staked one claim each.) (Wright, 1976)

The evidence before the people of Fortymile was overwhelming, and, in the time-honored tradition of gold rushes the world over, they threw down their tools, jumped into boats, and poled their way upstream to the Klondike. Within a matter of days, Fortymile was a virtual ghost town, and several miles of gold-bearing ground had been staked. Word passed quickly throughout the upper Yukon River basin, and miners descended on the Klondike and the newly established town of Dawson City, at the junction of the Klondike and Yukon Rivers. Robert Henderson, incidentally, was several valleys away from the excitement and was not told about the Carmack-Jim-Charlie discovery. By the time he learned of the find, all of the good gold-bearing ground had been taken up. The North West Mounted Police were on the scene within days to ensure that the staking process was orderly and that there was an official registration system in place. The claim-jumping and violence that often accompanied initial strikes were kept in check by the presence of the Mounties.

By the time the new creeks—Discovery, Eldorado, Hunker, and others—had been staked and the strike made by the discoverers confirmed a hundred times over, it was too late to get word to the outside. Through the winter of 1896-1897, the miners in the Yukon worked in a frenzied pace, desperate to get to bedrock, determined to stack up mounds of gold-bearing dirt, and anxious for the spring wash-up to prove their newfound wealth. Tests throughout the winter confirmed that the Klondike strike was exceptionally rich. A few miners gambled that their claims, staked in haste, would not be worth a great deal and sold out to others. Most of them lost out on the opportunity to make hundreds of thousands of dollars. Euphoria

Street scene, Dawson City, 1898. *Photograph supplied by the author.*

filled the camp, forcing up prices for goods and labor and adding a celebratory giddiness to life in the region. When spring finally came and the sluice boxes could be operated, miners converted piles of hard-won dirt into bags full of gold dust and nuggets. Many could hardly wait to get outside to share the news of their good fortune and to capitalize on the newfound wealth. Again, a frenzied buying and selling of properties ensued, accompanied by further struggles over fragments of claims, all occurring under the watchful eye of the police.

In the summer of 1897, with no forewarning, the first miners arrived by steamship in Seattle and San Francisco. The continent gasped in collective amazement. Individual miners, transformed from dirt-pushing poverty into dashing wealth over the course of a winter, carried thousands of dollars in gold. The miners were greeted like conquering heroes along the West Coast, as news of the discovery spread like wildfire. Few across North America knew where the Yukon was before the summer of 1897; the place was one everyone's lips and in everyone's imagination within weeks. The Klondike strike occurred at a crucial time—when Americans were wondering about the future of their nation in light of the closing of the frontier and while they were locked in the grip of a severe depression. Canadians, less dramatically, had wondered where the newly acquired northland fit into national priorities and worried that the vast western and northern expanse would be but a continuous drain on the country's limited resources. Newspapers fanned the flames of sensationalism, releasing countless stories, many of them grossly inflated, about the vast wealth that had been uncovered in the far Northwest.

The Klondike consumed the continent, providing fantasies of creeks paved with gold in the far distant and frozen Yukon River valley. Through 1897 and 1898, tens of thousands of would-be prospectors prepared themselves for the journey to the Yukon. They descended on Seattle, Vancouver, and Victoria in droves. Others, falling victim to the scandalous hucksterism of Edmonton promoters, opted to try for the Klondike by way of the Mackenzie River. Still others followed various trails, real and imagined, through Alaska. Some came prepared for the worst the North might throw at them. Others, counting on their good fortune and demonstrating their profound ignorance, carried little more than the clothes on their backs. Companies across North America engaged in shameless entrepreneurship, attaching the Klondike name to anything imaginable, promoting wild schemes for carrying prospectors to the North, selling a wide variety of Klondike guidebooks (some dangerously misleading), and rushing to supply the stampeders with the clothing, food, and equipment that they required. In the Yukon, the Mounties and the small Klondike society braced themselves for a massive assault, with many worrying that the arrival of thousands of ill-prepared stampeders would cause chaos, starvation, and anarchy in the region.

The Canadian federal government rushed additional men to the Klondike as soon as it learned of the strike, but the task those men faced was formidable. More to the point, the ill-paid Mounties found themselves responsible for enforcing the law while all around them grand fortunes were being made and lost. Miners bought and sold claims containing hundreds of thousands of dollars' worth of gold. The Mounties earned 50 cents a day, with an additional 50 cents a day offered as a special Yukon bonus. The small group of men—there were no women in the NWMP until the 1970s—were not allowed to marry or, obviously, bring wife and family with them to the North. They were, as well, on call around the clock and had few personal freedoms or rights. They lived in near-military conditions and were burdened with a wide range of police and civil responsibilities. They may have joined the NWMP to serve as officers of the law, but they found themselves performing duties relating to the recording of mining claims, customs information, vital statistics, and myriad other responsibilities. And all around them, as they tackled the banality of nineteenth-century federal administration, long-suffering miners struck it rich, lost their fortunes, made thousands in quirky business ventures, and "hooted it up" in the creeks and in the dozens of bars that sprang up in Dawson City. (Morrison, 1985)

The problem, beginning in 1897, lay not with the miners in the Klondike but with the massed migratory herd of stampeders working its way up the coast to Skagway and Dyea with designs on making it to Dawson City either that summer or the following spring. When the stampeders reached the border, the police turned back "suspicious" characters or those with a known criminal past. Ignoring legal restrictions on their authority, the NWMP imposed some special measures. Would-be miners were not allowed into the Yukon unless they carried a year's provisions—roughly interpreted to mean one thousand pounds of food and supplies. That requirement alone forced many of the ill-prepared travelers to return to the south.

For the rest, it turned the grueling climb over the Chilkoot Pass into a tortuous endurance test. But—and here the particular genius of the police shone through—it also meant that there was no starvation, as feared, in the Yukon during the winters of 1897-1898 and 1898-1899. (Neufeld and Norris, 1996) Concerning the river journey from Lake Bennett to Dawson City, the police worried about the mounting death toll at the infamous Miles Canyon and Whitehorse Rapids. The fatal combination of poorly built boats and untrained boatsmen resulted in numerous accidents and more than a few tragedies. To head off further problems (and quite outside their formal legal authority), the Mounted Police simply decreed that all stampeders had to hire experienced guides to pilot their craft through the difficult waters. From the moment the stampeders crossed into Canadian territory, and long before, by reputation, they knew that they would be working and living under the watchful eye of the North West Mounted Police. (Bennett, 1978)

The Klondike gold rush attracted stampeders from around the world, although the majority came from the western regions of Canada and the United States. It is difficult to get a precise handle on the nationality of the gold seekers, for many who came north were recent migrants to North America. According to the 1901 census—taken after the peak of the boom, when many had moved on to smaller strikes in Alaska—40 percent of the Klondikers were American, 27 percent were Canadian, and 13 percent were British (most having come via Canada). An additional 15 percent came from continental or northern Europe. The popular literature made much of the "exotics" who joined the rush—the few Japanese or Chinese who attempted to come north (most were turned back), and the Russians or New Zealanders who traveled thousands of miles, only to be disappointed in the sub-Arctic. But the cultural mix in the Klondike, judged by birthplace as per the 1901 census, was considerable, with more than 300 German-born miners, 61 from Japan, 177 Norwegians, 202 Scots, 2 each from Turkey and Egypt, 7 from Hungary, and so on. The more than 3,000 Americans and 2,200 Canadians dominated the gold fields, however, and many of those born in other countries had in fact been reared in the United States. The Klondike was largely, but not exclusively, male. Many women joined the rush, and most of them were not prostitutes or dance-hall girls. The largest number accompanied their husbands on the trip north; others came for the adventure or to open restaurants, boardinghouses, or laundries to service the miners. At the height of the rush, in 1898, more than 90 percent of the regional population was male, with all of the masculine tensions and realities that that kind of demographic composition produces. (Porsild, 1998)

There were concerns that the rapid influx of miners might disrupt the local First Nations population and, as happened further south, result in bloodshed and violence. In the south, First Nations people found well-paying jobs as packers, helping wealthier stampeders carry their one thousand pounds of supplies over the mountain passes. Most First Nations people withdrew from the Klondike corridor (the river from Lake Bennett to Dawson City and the small mining area) during the heyday of the rush. (McClellan, 1987) Small NWMP detachments were opened in outlying areas, partially to bring Canadian law and order to the First Nations but, even more importantly, to protect the First Nations from the stampeders. The police reacted swiftly to the first—and only—outbreak of serious violence, an attack on two prospectors by the Nantuck brothers in the southern Yukon in 1898. Two of the brothers were hanged in 1899, the third having died in custody—a swift example of the power of Canadian law. The actions of the police also reflected the social and cultural values of the age, for they worked to keep First Nations people out of major communities, save for short visits, and did little to protect the indigenous population from numerous acts of racial discrimination. (Coates, 1993)

The Klondike evolved in a uniquely Canadian fashion. The government accepted gambling and prostitution as necessary evils, believing that the miners needed an occasional "release." But at the same time, they enforced quite rigidly the rules of Canada's Sabbatarian laws, which prohibited work and commerce on Sunday. Amusing stories emerged of men arrested for working on the Sabbath—who were then sentenced to cut wood on the police woodpile as punishment. The police also ordered that guns could not be carried in Dawson City, a decision that significantly reduced the likelihood of armed confrontations and helped make the Klondike experience markedly less violent than other gold rushes. They used the famed northern "blue ticket" (an unofficial order issued by the police that compelled an individual to depart the district or face prosecution) to evict from the Yukon anyone they felt was likely to cause a disturbance—crooked gamblers were among the many tossed out of the territory without the benefit of trial. The enforcement of Canadian laws did not diminish the dominant American character of the stampede; the Fourth of July celebrations were the largest annual events in the territory. (Porsild, 1998; Riley, 1997)

Popular mythology has cultivated a variety of fascinating images about the Yukon. One of the most prominent is of a handful of NWMP officers staring down a violent mob of (largely American) prospectors and camp followers, imposing order in the face of daunting odds. While the accomplishment was considerable, the imagery is flawed. Perhaps most importantly, it is vital to recall that the stampeders and would-be prospectors were not a slathering mob of vicious frontiersmen. Instead, they represented a broad cross section of North American society, with a substantial number of other nationalities thrown into the mix. They were, in the main, a peaceable lot, heading north to pursue a dream; and they feared disorder, social chaos, and violence as much as the police did. There were a few ruffians, to be sure, but the police and other Canadians loved to highlight the contrast between crime-ridden Skagway, where the notorious outlaw Soapy Smith held sway, and the tranquil communities in the Yukon. But the miners themselves deserve a fair measure of the credit for the calmness of the Yukon experience. Furthermore, the federal government had left little to chance. In an era when municipal police forces were relatively small and the hand of government lay very gently on the shoulders of Canadians, the Yukon enjoyed a massive, almost omnipresent, police presence. To ensure that no part of the Yukon was out of the vigilant gaze of the authorities, the federal government ordered the opening of a string of police posts throughout the region, dispatched more than three hundred officers to the district, and buttressed the NWMP presence with two hundred members of the Yukon Field Force of the Canadian Army. The Yukon was, effectively, blanketed with a determined, fast-acting, rigid, and highly formal police presence. (Morrison, 1985)

The Canadian government, determined that the Klondike gold rush should remain a Canadian event, controlled it through the strict enforcement of Canadian laws. It created a safe, nonviolent environment, one much appreciated by the vast majority of stampeders. Though rigid in their adherence to most laws and regulations, the police were not overly moralistic. They tolerated the "inevitable" evils of gambling and prostitution and saw to it that the gamblers, dance-hall girls, and prostitutes operated in an appropriate fashion. (Backhouse, 1995) They imparted a distinctly Canadian flair to an otherwise "American" episode and received the approbation and support of the entire gold-rush community. It is probably too much to say, as some have argued, that the presence of the North West Mounted Police protected the Yukon from an American takeover. After all, the Klondike miners operated with few restrictions based on nationality (Americans could not vote in the Yukon's first elections) and paid only a tiny price for removing their gold to the United States when they were either rich enough to leave or, as in many cases, too poor to stay. At one point, there were rumors that the prospect of mass starvation in the Yukon might lead to American intervention, but the presence of the Yukon Field Force and American understanding of Canadian sensitivities prevented such an overt action. (Hall, 1984-1985)

Not all of the police efforts at "Canadizing" the Klondike worked as well. The major access routes to the Yukon—the Chilkoot Pass and the White Pass—both passed through the Alaska Panhandle, thus placing the Yukon Territory at the mercy of American customs—and politics. With the two countries embroiled in a bitter war of words over the Alaska-Canada boundary, the Canadian government was anxious to secure an alternate, all-Canadian, route to the far Northwest. It turned, once again, to the police, who were asked to chart a workable trail from northern Alberta through northern British Columbia to the Yukon, a task that they approached with their usual relish. The expedition, conducted in 1897-1898, was a logistical nightmare. The mountainous route, carved up by raging rivers, defied easy conquest. The police made it through to the Yukon, but only at considerable personal costs. The idea of encouraging subsequent travelers to reach the Klondike via the all-Canadian trail was quickly abandoned. (Morrison, 1985; MacGregor, 1970)

The North West Mounted Police emerged as one of the defining symbols of the Klondike gold rush. They were celebrated at the time for their ability to bring order to a potentially chaotic situation. Their swift, decisive, and effective actions endeared them to Canadians and the Canadian government and ensured the continued operation of the national police force. During and immediately after the gold rush, the works of poet Robert Service, novelist Jack London, and literally hundreds of writers and journalists immortalized the NWMP by lavishing praise upon it. A mystique soon engulfed the Klondike experience of the police, who emerged as stalwart defenders of the British Empire and Canadian jurisprudence, men of integrity and grit who stared down would-be villainous desperadoes and shaped the unruly mob that swarmed over the Chilkoot Pass and White Pass between 1897 and 1900 into a calm, controlled, and competent group of miners, shopkeepers, and community members.

But the Klondike produced many other images as well—of amiable prostitutes servicing their clients in Klondike City (Lousetown), of the many business people who climbed the Chilkoot Pass determined to "mine the miners," and of the vicissitudes of the northern winter that dug so deeply into the minds and souls of the stampeders. There are countless stories of fortunes made and lost, and of stampeders who crossed the Chilkoot Pass with a grand piano or a parcel of eggs, both of which "turned to gold" in isolated Dawson City. The Klondike engendered numerous figures who were bigger than life: Swiftwater Bill Gates, Gussie Lamore, Diamond Tooth Gertie, the Oregon Mare, NWMP officer Sam Steele, Martha Black, Bishop Bompas, and Gather Judge. It was, at one stroke, a grand adventure and a majestic tragedy. The loss of human life along the Klondike trail, although small, caused great pain. The urgent rush to carry supplies across the White Pass left a graveyard of abandoned mules and horses at the bottom of the pass. First Nations people were pushed out of their traditional territories and marginalized in their homelands. Almost all the stampeders who ventured north discovered no gold and made no money, save enough as a laborer to pay for the passage home. A few lingered and participated in "echo booms," including the much smaller gold strikes at Atlin in British Columbia and at Fairbanks, Rudy, Iditarod, and Nome in Alaska. Another few hundred, believing that the "real" mother lode lay somewhere in the far Northwest, persisted, shifting back to the lonely and uncertain life of the prospector. But the vast majority of those who participated, including the ones who went home empty-handed, called it the greatest time of their life and took great satisfaction in having completed the journey to Dawson City. (Coates and Morrison, 1990)

The general impression of the Klondike gold rush is that the gold soon petered out and the creeks were quickly abandoned. The reality was more complex. The Canadian government did not believe, from the first, that the Klondike strike could be the basis of a permanent northern society. Clifford Sifton, minister of the Interior, assumed that the Klondike camp would die an early death and instituted policies that hastened its demise. The government favored granting water concessions to large companies associated with the Liberal Party and encouraged those firms to consolidate their holdings in the Klondike region. There had been discussion of building an all-Canadian railway to the Yukon, a land link that would have encouraged regional development, but that plan, too, was abandoned, although private entrepreneurs did construct the White Pass and Yukon Route railway from Skagway, Alaska, to Whitehorse, Yukon. The railway, combined with the summertime sternwheelers that plied the Yukon River and the telegraph line constructed to connect the Klondike with the South, provided some measure of access and contact to the isolated Northwest. By 1900 only the poorest would-be prospectors among the hundreds who continued to follow the route to the Klondike in pursuit of gold were obliged to walk over the Chilkoot Pass and paddle their way down to Dawson City. (Hall, 1984-1985; Minter, 1987)

By the early 1900s, concessionaires were taking over the gold fields, moving in monster dredges to scour the worked-over claims for additional gold and to claw away at the surrounding hillsides. Individual prospectors were gradually bought out, or, when returns failed, they simply moved on. Dawson City, once a boisterous frontier city, soon became an overbuilt government center and company town. The creeks that once buzzed with the activity of hundreds of independent claims were now worked systematically by company laborers who manned massive and voracious dredges and lived in company towns. The era produced its share of conflicts, largely over concessions, and grandiose characters, especially Joe Boyle, king of the Klondike and confidant of eastern European royalty. (Green, 1972; Rodney, 1974)

Two tragic events effectively killed off the Klondike era in the North. World War I arrived in 1914 on the heels of years of continuous bad economic news and steady decline. Faced with the challenge of doing their share for the war effort, the Yukoners decided to use the occasion to demonstrate their commitment to the nation. The territory had one of Canada's highest regional enlistment rates and donated thousands of dollars to wartime causes; in the process, its work force fell so low that its mining companies had trouble running the dredges. By the time the war ended in 1918, the Yukon gold-mining economy was on its last legs. As if to signal the death knell of the Yukon, the Canadian Pacific steamship *Princess Sophia*, with more than two hundred northerners aboard, including many key citizens of Dawson society, ran aground off Juneau, Alaska, in October 1918. Rescue efforts failed, all on board perished in the frigid waters of the north Pacific, and the Yukon joined with Alaska in mourning the loss of many crucial members of the northern community. (Coates and Morrison, 1990)

The Yukon's dredges continued to operate through the 1920s and 1930s. Following the Great Depression, the business ran into difficult times again in the 1940s, when labor shortages and high costs turned the once profitable ventures into marginal operations, but rebounded under the Emergency Gold Mining Assistance Program in the 1950s. The Yukon Consolidated Gold Corporation, which dominated the mining industry in the province, finally ceased dredging in 1966. For the first time in seventy years, there was no gold-mining in the Klondike. But the hiatus proved short-lived. The advent of new sluicing technologies, coupled with gold prices that shot up from $200 (U.S.) an ounce in 1978 to $650 (U.S.) two years later, sparked a return to the gold fields. Although the activity level peaked when prices collapsed from their inflated levels of $875 (U.S.) an ounce, many small dredging and hydraulic operations continue in the Klondike basin and surrounding area. Dawson City was reborn in the 1970s and 1980s as a living tourist attraction and presently draws thousands of visitors every summer to the refurbished buildings, Klondike-era attractions, and Diamond Tooth Gertie's "gay '90s" emporium. One of the highlights, played to the full by locals, is the site of modern-era gold miners coming in from the creeks to gamble at Gertie's low-limit tables. (Coates and Morrison, 1988; Zaslow, 1988)

Perhaps the richest lode to come out of the Yukon and the Klondike was in the form of stories. Memoirs and travel accounts began to appear within months after the news of the Klondike reached the South, and an outpouring of books about the gold rush commenced immediately and continued for years. (For an aboriginal perspective on the Klondike, see Cruikshank, 1990). Novelist Jack London joined the migration, and the short stories that he penned captured much of the essence of the Klondike experience. Poet Robert Service arrived after the peak of the rush, and his view of the north from behind the bank counter offered a restricted view of life in the creeks. But he nonetheless wrote dozens of poems that immortalized the Klondike gold rush and which retain their popularity to the present day. At the time, and now long forgotten, journalists and novelists produced hundreds of stories, more fanciful than real, as well as hundreds of penny-dreadful novels based on fictitious Yukon adventures. The pattern continued, through Charlie Chaplin's movie *The Gold Rush*, the 1950s television series *Sergeant Preston of the Yukon*, and, most intriguingly, the Quaker Puffed Wheat promotion in the early 1960s that offered one square inch of a Klondike claim to any purchaser of the cereal. The glorification of the Klondike continues, through government-promoted tourist sites in Dawson City

The reality of Klondike mining—long, hard days working the gold-bearing soil at bedrock—differed greatly from the image of the gold-panning prospector of Yukon legend. *Photograph courtesy National Archives of Canada, No. 11270.*

and the continued popularity of tours by cruise ship and bus that bring thousands (mostly Americans) into the Yukon every summer. (Three of the best firsthand accounts are Adney, 1994; Berton, 1975; and Black, 1980.)

The work of academic historians, northern community groups, and First Nations people is gradually producing a more balanced view of the Klondike gold rush, but the reality is that most North Americans like the Klondike best as a myth and a fantasy. There is a desire to maintain the sense of awe that gripped the continent in 1897, when news about the massive discovery first spread southward. The allure of gold through history is well known, but the combination of social, economic, and political forces that coincided with the Klondike gold rush has seldom if ever been matched. That a huge gold discovery had been made in the frozen lands of the far Northwest, a land rich in imagery and clouded in misunderstanding, only added to the excitement. The Klondike gold rush was, perhaps more than anything, the ultimate democratic experience, available to all willing, against all reasonable odds, to test themselves against the sub-Arctic. Oddly, and perhaps because the event is such a continental, if not international, experience, the Klondike phenomenon has not figured prominently in general histories of either Canada or the United States. It is regarded, if at all, as a brief, unimportant occurrence in the far Northwest, a minor distraction from the economic and social realities of late-nineteenth-century Canada and the United States. But the Klondike was far more important than this relative neglect suggests.

The Yukon appears, in recent years, to have recaptured its gold-rush past (although it has removed the long-famous gold miner from its license plates). The Klondike figures prominently in tourist-promotion material and continues to attract thousands to the territory each year. Even more striking, gold has emerged as the single most significant element in the Yukon's overall mineral production in five of the past eight years. (That new prominence reflects the closure of the Yukon's largest mine, a lead-zinc operation at Faro, as well as an improvement in the efficiency of gold production.) The Klondike of 1999 is not the same place as the Klondike of 1899. But while the gold will eventually run out, the Klondike legend will persist. For the story of the last great gold rush—a tale of human resilience, entrepreneurial determination, Canadian resolve, American expansionism, and the quirks and characters of nineteenth-century North American life—seems destined to live forever in our memory and imagination.

References

Adney, Tappan. *The Klondike Stampede*. Vancouver: UBC Press, 1994.

Backhouse, Francis. *Women of the Klondike*. Vancouver: Whitecap, 1995.

Bennett, Gordon. *Yukon Transportation: A History*. Ottawa: Parks Canada, 1978.

Berton, Laura. *I Married the Klondike*. Toronto: McClelland and Stewart, 1975.

Berton, Pierre. *Klondike: The Last Great Gold Rush, 1896–1899*. Toronto: McClelland and Stewart, 1987.

Black, Martha Louise. *My Ninety Years*. Edmonds, Wash.: Alaska Northwest Publishing, 1980.

Coates, Ken. *Best Left as Indians: Native-White Relations in the Yukon Territory, 1840–1973*. Montreal: McGill-Queen's, 1993.

Coates, Ken, and William R. Morrison. *The Land of the Midnight Sun: A History of the Yukon*. Edmonton: Hurtig, 1988.

_______. *The Sinking of the* PRINCESS SOPHIA: *Taking the North Down with Her*. Toronto: Oxford, 1990.

Cruikshank, Julie. *Life Lived Like a Story: Life Stories of Three Yukon Elders*. Vancouver: UBC Press, 1990.

Gates, Michael. *Gold at Fortymile Creek: Early Days in the Yukon*. Vancouver: UBC Press, 1994.

Green, Lewis. *The Gold Hustlers*. Vancouver: J. J. Douglas, 1972.

Hall, David. *Clifford Sifton*, 2 vols. Vancouver: UBC Press, 1984–1985.

McClellan, Catherine. *Part of the Land, Part of the Water: A History of Yukon Indians*. Vancouver: Douglas and McIntyre, 1987.

MacGregor, James G. *The Klondike Rush through Edmonton, 1897–1898*. Toronto: McClelland and Stewart, 1970.

Minter, Roy. *The White Pass: Gateway to the Klondike*. Toronto: McClelland and Stewart, 1987.

Morrison, William R. *Showing the Flag: The Mounted Police and Canadian Sovereignty in the North, 1894–1925*. Vancouver: UBC Press, 1985.

Neufeld, David, and Frank Norris. *Chilkoot Trail: Heritage Trail to the Klondike*. Whitehorse: Lost Moose, 1996.

Ogilvie, William. *Early Days on the Yukon*. Toronto: Bell and Cockburn, 1913.

Porsild, Charlene. *Gamblers and Dreamers: Women, Men and Community in the Klondike*. Vancouver: UBC Press, 1998.

Riley, Bay. *Gold Diggers of the Klondike*. Winnipeg: Watson and Dwyer, 1997.

Rodney, William. *Joe Boyle: King of the Klondike*. Toronto: McGraw-Hill Ryerson, 1974.

Stone, Thomas. *Miners' Justice: Migration, Law and Order on the Alaska-Yukon Frontier, 1873–1902*. New York: Peter Lang, 1988.

Webb, Melody. *The Last Frontier: A History of the Yukon Basin of Canada and Alaska*. Albuquerque: University of New Mexico Press, 1985.

Wright, Allen. *Prelude to Bonanza: The Discovery and Exploration of the Yukon*. Sidney, B.C.: Gray's, 1976.

Zaslow, Morris. *The Northward Extension of Canada, 1914–1967*. Toronto: McClelland and Stewart, 1988.

_______. *The Opening of the Canadian North, 1870–1914*. Toronto: McClelland and Stewart, 1971.

Gold in Alaska

Joan M. Antonson

Joan M. Antonson, deputy state historic preservation officer and state historian for Alaska, as well as coordinator for Alaska's Gold Rush Centennial task force, documents historic properties and promotes Alaska history. She also teaches Alaska history, is associate editor of Alaska History, *and has written* Alaska's Heritage, *a popular high school textbook, and numerous articles.*

Thanks to singer Johnny Horton, I knew about "big nuggets they're finding . . . just a little northeast of Nome" in Alaska before I knew that Walt Disney's Scrooge McDuck made his fortune in Canada's Klondike. This is unusual. While students will find the Klondike mentioned in their American history textbook, few will find Alaska's gold rushes mentioned.[1] Only those who search will find that before the Klondike discovery, mines around southeast Alaska had produced more than twenty-five million dollars' worth of gold. And it is not common knowledge that Alaska had gold mines not only at Juneau in southeast Alaska but at more than sixty other places around the territory, among them Nome in the northwest, Circle City and Fairbanks in the interior, and Willow Creek and the Kenai Peninsula in south-central Alaska.

In 1848, the same year gold was discovered in California, a Russian geologist named Petr Doroshin explored Alaska's Kenai Peninsula for minerals. He reported traces of gold in the mouths of streams emptying into Kenai Bay and after a second trip in 1850 reported finding gold elsewhere in the area. For the remainder of his five-year tour of duty in Alaska, however, Doroshin was directed to investigate coal deposits around Cook Inlet.[2] Other Russians reported deposits of gold and copper in streams in southern Alaska, but, with the exception of coal, the Russians did not continue mineral investigations, encourage mining, or attempt to develop mines. Nevertheless, rumors of gold in Alaska circulated through California mining camps. In 1856 California's United States senator William Gwin introduced a bill to buy Russian Alaska for five million dollars, pointing to the area's furs, whales, fish, and vast mineral resources.[3] The *Dred Scott* decision, an election, and other more pressing issues prevented action on the bill.

California miners believed there were gold deposits throughout the Pacific Coast and Rocky Mountains. Working their way north, prospectors in 1861 found gold on Telegraph Creek, a tributary of the Stikine River, which has its headwaters in British Columbia and flows to the Pacific Ocean through southeast Alaska. After Russia sold Alaska to the United States in 1867, more prospectors searched streams in southeast Alaska and British Columbia. In 1872, following another discovery in a Stikine River tributary, there was a rush to what became known as the Cassiar gold district. The rush was large enough to cause the U.S. Army to station troops at the fur-trading post near the mouth of the Stikine to maintain order among the gold seekers. More than three thousand people passed through Fort Wrangell on their way to the Cassiar gold fields.

Small gold discoveries by prospectors who headquartered at Alaska's Sitka, the only place that could be called a town along the coast north of Victoria, British Columbia, before 1880, led to the organization of a mining district on May 10, 1879. Prospectors sought a way to have their claims legally recognized, inasmuch as Congress had failed to provide civil government for the territory. This action proved well timed: the big strike was made the following year. Chief Kowee, a local native, led grubstaked prospectors Richard Harris and Joe Juneau to Silver Bow Basin on the mainland west of Sitka and north of Fort Wrangell. Upon hearing of the strike, several thousand prospectors and entrepreneurs rushed north from Portland and San Francisco and that year founded the town of Juneau.[4] Two years later, miner John Treadwell located lode-gold deposits on Douglas Island across Gastineau Channel from Juneau. In 1887 he formed the Alaska Treadwell Gold Mining Company. Before the mines flooded in 1917, the company extracted $67 million worth of gold and at its peak operated 960 stamps and employed two thousand people. At the same time, the Alaska Gastineau Mining Company consolidated claims in Silver Bow Basin and mined lode gold there until 1921. The Alaska-Juneau

Richard Harris and Joe Juneau's discovery of gold in Silver Bow Basin in 1880 occurred sixteen years before the Klondike discovery in Canada. Even then, it was the second rush north. The discovery led to the founding of the first American settlement following Alaska's purchase from Russia. Many prospectors, Juneau and Harris among them, continued searching for gold across the Coast Mountains in the upper Yukon River basin, where the Klondike is located. *Photograph courtesy Alaska State Library, Historical Collections, Juneau.*

Prospectors found lode, or hard-rock, gold deposits in the Juneau vicinity two years after Harris and Juneau's discovery. The Treadwell mines started operating in 1887. The company had modern equipment, shown here inside the main mill building. It employed up to two thousand workers and was one of the first to practice "corporate socialism," paying high wages and providing amenities, including a natatorium. Workers went on strike, however, when the company brought in a boatload of Chinese workers. The mines flooded in 1917 and closed in 1922 but produced sixty-seven million dollars' worth of gold. *Photograph courtesy Alaska State Library, Historical Collections.*

Company, a third large-scale lode-gold-mining firm, was incorporated in 1897 and produced more than $80 million worth of gold before closing during World War II. During the 1920s and 1930s, the A-J was the largest low-grade lode-gold operation in the world. Together, the three operations produced $158 million worth of gold.[5]

Prospectors Jack McQuesten, Arthur Harper, and Alfred Mayo began looking for gold in the upper Yukon River area in 1873. Although they became fur traders to support themselves, they grubstaked hundreds of men to keep searching for the gold they believed was there. After the Juneau discovery in 1880, more gold seekers wanted to search on the other side of the Coast Mountains, i.e., in the upper Yukon River area. They sought easier access to the region and enlisted the help of the U.S. Navy to negotiate with the native Tlingits to allow prospectors to use the native peoples' jealously guarded Chilkoot Trail. By 1896, when the Klondike discovery was made, an estimated sixteen hundred prospectors and miners were in the upper Yukon River basin, as many as half of them grubstaked by traders.[6] Anticipating a significant gold discovery in the upper Yukon, Capt. William Moore, who had been a member of a Canadian survey party, built a cabin and wharf at the head of a

There are only a few photographs of the upper Yukon area before 1898. Jack McQuesten (on the left) operated the largest store at Circle City. He extended credit to hundreds of prospectors. The Alaska Commercial Company's principal competitor, the North American Transportation and Trading Company, dealt only in gold, furs, and cash. *Photograph courtesy Alaska State Library, Historical Collections.*

GOLD DISCOVERIES, 1880-1914

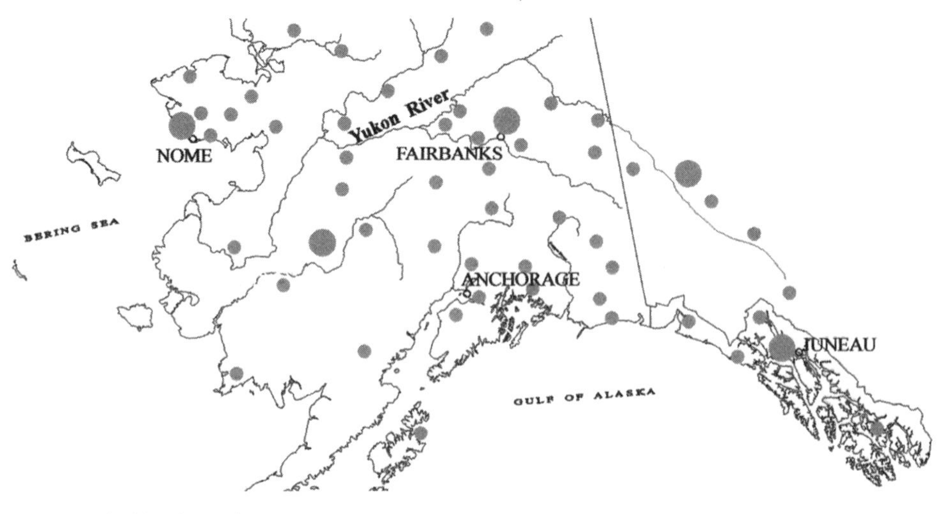

Map supplied by the author.

deepwater bay called "Skagua," east of where the Chilkoot Trail started. Moore began constructing a trail over the White Pass that was lower and not as steep as the Chilkoot Pass.[7]

Several significant gold discoveries were made in tributaries of the upper Yukon on the Alaska side of the 141st meridian before the 1896 Klondike discovery. Prospectors Howard Franklin and Henry Madison found gold on the Fortymile River in 1886, and Pitka Pavaloff and Sergi Gologoff Cherosky discovered it on Birch Creek in 1892. News of the latter discovery lured a thousand stampeders to cross the Chilkoot Pass and travel nine hundred miles down the Yukon River in 1895. Circle City, the supply town for the Birch Creek district, boasted a population of seven hundred and achieved renown as the largest log-cabin town in North America.[8] The town had theaters, saloons, dance halls, two major trading companies, and a fleet of steamboats that brought in supplies during the summer months. By 1896 miners had taken nearly $1.5 million in gold from the Birch Creek area.[9] Still other prospectors in the 1880s and early 1890s sought to find the gold deposits reported by the Russian Doroshin on the Kenai Peninsula. Although there is some dispute about who, when, and where the gold discovery was made, prospectors staked placer claims in the late 1880s at several places on the peninsula.[10] An estimated three thousand people rushed to the northern Kenai Peninsula in 1896 and founded the towns of Hope and Sunrise.

Of course, on August 16, 1896, George Washington Carmack, Tagish (Dawson) Charlie, and Skookum Jim made the discovery in Canada that started the great

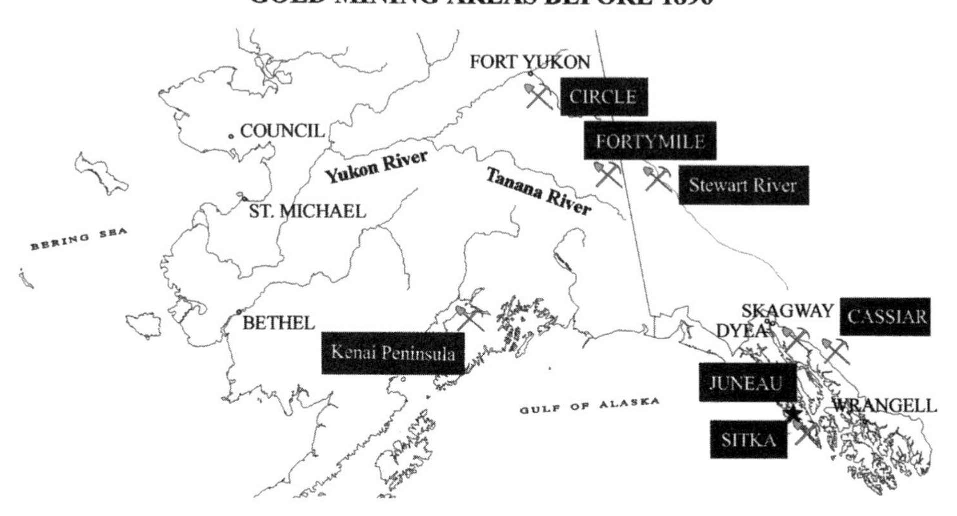

Map supplied by the author.

TOP: Most of the people who headed for the Klondike in 1897 and 1898 took the Chilkoot Trail. Southeast Alaska native peoples established the trail and jealously guarded it until 1880. That year a U.S. Navy ship aimed cannons at the village, and the captain negotiated with the natives to allow non-natives to use the trail to cross the Coast Mountains to search for gold in the upper Yukon River basin. BOTTOM: The Chilkoot Trail did not have a deepwater harbor at its head, and the trail was too steep for horses. Enterprising men developed a route that started about four miles east at a deepwater harbor and featured a more gradual climb over White Pass (shown here). When a railroad was completed over the new route, the Chilkoot Trail was abandoned virtually overnight. *Photographs courtesy Alaska State Library, Historical Collections.*

Klondike gold rush. Word did not reach the outside world until July 1897, almost a year later. Most of the estimated eighty thousand people who participated in the rush took routes through Alaska to the Yukon—the Chilkoot Trail, the White Pass Trail, the all-water Yukon River route, or the unproven Valdez Glacier Trail. At least one third of the people who embarked for the Klondike turned around. Others stopped along their route to prospect for gold, and a few made discoveries—Rampart on the Yukon, Porcupine in southeast Alaska, and Chistochina in the Copper River basin among them. Those who reached the Klondike in 1898 were too late to stake claims. Most were also too late to leave on the steamboats that year.

News of a gold strike in Alaska that fall near the mouth of the Yukon would not reach the Klondike until late the following spring. Known as the "Three Lucky Swedes" (although one was a Norwegian, and the men had mining experience), Jafet Lindeberg, Eric Lindblom, and John Brynteson discovered gold in the Snake River and its tributaries on the Seward Peninsula in September 1898.[11] The men had come to prospect on the Seward Peninsula after David Libbey, who had been a surveyor with the Western Union Telegraph Expedition in the area in the early 1860s, returned in 1896 and discovered gold inland on the peninsula. After hearing the news, people raced for the area named Nome. In a single week eight thousand people left Dawson for the coast. Thousands came from the Pacific Northwest and California. Fortuitously, in July 1899 placer gold was found in the beach sands at Nome, some say by a soldier digging a well. A judge determined that the beach could not be staked, so more than two thousand people camped along thirty miles of the beach and mined the sands twenty-four hours a day, recovering in excess of two million dollars' worth of gold that summer. Workers from mines along the creeks, making good wages of as much as eleven dollars a day, left their jobs and went to the beach. Ironically, that same year the first gold dredge started to operate on the Snake River. Between 1899 and 1910, miners took more than forty-six million dollars' worth of gold from the Nome mining district.[12]

If there was gold at the headwaters and gold near the mouth of the Yukon River, prospectors reasoned that tributaries through the middle reaches of the fourteen-hundred-mile-long river contained gold. They were right. On July 22, 1902, the Italian immigrant Felix Pedro found gold in the Tanana hills, and the discovery led to the founding of Fairbanks. The gold in that district was buried deeply, and miners often had to dig two hundred feet to reach bedrock and find the ore. It was worth the effort on Cleary, Ester, and Fairbanks Creeks. Almost two-thirds of the gold mined in the Tanana valley before 1910 came from those three streams, for a total value of more than thirty million dollars. Over the years, the total production of the Tanana gold fields exceeded that of other northern gold fields, even the Klondike.[13]

The Nome and Fairbanks gold fields proved to be Alaska's largest, but prospectors found gold in many other places around the territory by 1914. Of those discoveries, the rush to the Innoko-Iditarod district in 1909 and 1910 is the best known because of the annual dogsled race started in the 1970s that is run over old mail and freight trails that served the area. Within a year or two, large companies supplanted

TOP: Those who rushed to Nome on the Seward Peninsula thought their trip would be a pleasant ride by ocean steamer, in contrast to the ordeals of the Klondikers, who crossed the Chilkoot, White Pass, and Valdez Glacier trails. The Bering Sea is shallow, with few deepwater harbors, a tidal variation of at most a foot, and wild storms. Ships had to anchor several miles offshore. Many gold seekers wrote about wild barge rides to the beach. BOTTOM: Those who arrived at Nome in 1899 learned, as the '98ers had at Dawson, that every inch of every stream had been staked the previous fall. The thousands of disappointed gold seekers got a second chance when gold was discovered in the beach sands in July. A judge ruled that claims could not be staked on the beach. People camped on the beach, panned, rocked, and sluiced for enough gold to buy a steamship ticket home. That summer, they took more than two million dollars' worth of gold from the beach. *Top photograph courtesy Carrie McLain Museum, Nome; bottom photo courtesy National Maritime Museum, San Francisco.*

individual miners and small groups of partners. With capital from the sale of stock, the companies brought in hydraulic hoses, draglines, and small dredges for placer mining and built mills at lode mines. After the rush years, Alaska's population declined almost 15 percent—from 63,000 in 1900 to 55,036 in 1920.[14]

The completion of the government-financed and -built Alaska Railroad in 1923 cut the cost of freight to interior Alaska and tapped coal mines to provide a cheaper source of power. The project reinvigorated gold mining in the Fairbanks area. The U.S. Smelting, Refining and Mining Company of Massachusetts bought hundreds of gold claims and brought the first of what would grow to a fleet of eight dredges to the area. Operations started in 1926. The company also operated dredges on the Seward Peninsula and the Fortymile drainage. In the 1930s, one-third of the population of Fairbanks worked for the U.S.S.R.&M.[15]

The first dredge to operate on the Seward Peninsula commenced operations in 1899, the same year as the big rush. In the 1920s the U.S. Smelting, Refining and Mining Company initiated dredging operations on the Seward Peninsula, in the Fairbanks area, and in the Fortymile drainage. The company brought in larger dredges and to operate efficiently built a seventy-two-mile-long ditch in the Fairbanks area and a system of ditches that crisscrossed the Seward Peninsula. *Photograph courtesy Candy Waugaman.*

At the same time, lode-gold deposits north of Anchorage in the Willow Creek district began to be developed. They, as did all gold mines in Alaska, received a boost in 1933 when President Franklin D. Roosevelt raised the price of gold from $21 an ounce to $35 an ounce. Mines that had not been worked for years reopened. Gold mining abruptly stopped when the United States entered World War II. Early in 1942

MAJOR GOLD DREDGING AND LODE MINING AREAS, 1926-1942

Map supplied by the author.

Alaska has extensive hard-rock, or lode-gold deposits, as well as placer gold. Miners started lode-gold operations in the Willow Creek district, sixty miles north of Anchorage, in the 1920s. The scale of operations increased in the late 1930s after President Roosevelt raised the price of gold from twenty dollars to thirty-five dollars an ounce. The price of gold remained fixed at thirty-five dollars an ounce after the war, although other costs rose and Alaska's lode-gold mines permanently shut down. The Independence mine, shown here, has been preserved and is a state historic park. *Photograph courtesy Stoll Collection, Archives and Manuscripts Department, University of Alaska, Anchorage.*

Roosevelt signed Executive Order E-208, which closed all mines not essential to the war effort.[16] After the war, the fixed price of gold at $35 an ounce was not enough to meet the increased cost of labor and to modernize the mining equipment. Mining virtually ceased in Alaska until the mid-1970s, when the government stopped fixing the price of gold. The price soared, peaking in 1980 at $850 an ounce, then settling between $250 and $400 an ounce.[17] Mines throughout Alaska reopened. Exploration work led to gold discoveries north, east, and west of Fairbanks. Advances in technology changed mining. So did new concerns about damage to the environment and water quality. Although gold mines remain in operation at present, the industry is riskier than ever. It is Alaska's history and its future.

Leaving the Klondike out of this story provided the opportunity to demonstrate that gold was discovered in Alaska at the same time as it was in California. It adds credibility to the statement that the Klondike discovery was "an otherwise routine event."[18] Arguably more so than the fur trade or salmon-canning industry, gold mining encouraged the building of an infrastructure in Alaska. Gold brought more people to the territory than did other industries. Those people in turn established towns and demanded services from the U.S. government,[19] which responded by building and maintaining a telegraph line, trails, roads, and a railroad. It also expanded the judicial system, gave Alaska nonvoting representation in Congress, and, finally, in 1912, territorial status and a legislature. It is not an understatement to say that gold changed Alaska forever.

Notes

1. Some historians assert that gold was discovered in the "Alaskan Klondike." See, for example, Paul S. Boyer et al., *The Enduring Vision: A History of the American People*, 2d ed. (Lexington, Mass.: D. C. Heath and Company, 1993), 573.

2. Richard A. Pierce, *Russian America: A Biographical Dictionary* (Kingston, Ontario, and Fairbanks, Alaska: Limestone Press, 1990), 123–124. Of interest, Doroshin with a crew of ten Russians and Tlingits dug for gold on the Yuba River in California in March and April 1849 and found enough gold to enable them to buy a ship for the Russian-American Company and purchase 39,300 rubles, which Doroshin delivered to the company office at Sitka.

3. Hallie M. McPherson, "The Interest of William McKendree Gwin in the Purchase of Alaska, 1854–1861," *Pacific Historical Review* 3 (March 1934): 28–38.

4. R. N. DeArmond, *The Founding of Juneau*, centennial ed. (Juneau: Gastineau Channel Centennial Association, 1980), 41–51.

5. David Stone and Brenda Stone, *Hard Rock Gold: The Story of the Great Mines that Were the Heartbeat of Juneau* (Juneau: Juneau Centennial Committee, 1980), 7, 10–13.

6. Alfred Hulse Brooks, *Blazing Alaska's Trails*, 2d ed. (Fairbanks: University of Alaska Press, 1973), 334.

7. David Neufeld and Frank Norris, *Chilkoot Trail: Heritage Route to the Klondike* (Whitehorse, Yukon: Lost Moose, the Yukon Publishers, 1996), 64.

8. Michael Gates, *Gold at Fortymile Creek: Early Days in the Yukon* (Vancouver: University of British Columbia Press, 1994), 39, 65–67, 115.

9. Of note, Birch Creek is one of few streams in Alaska from which gold has been extracted every year since being discovered.

10. Rolfe Buzzell, "The Turnagain Arm Gold Rush, 1896–1898," *Proceedings of the International Symposium on Mining, Fairbanks, Alaska, September 9–14, 1997* (Fairbanks: Festival Fairbanks, Inc., 1997), 231–240.

11. In 1903 the discovery claim on Anvil Creek yielded the largest gold nugget ever found in Alaska, weighing 155 troy ounces and worth about $1,500.

12. Terrence Cole, "Nome: 'City of the Golden Beaches,'" *Alaska Geographic* 11 (1984): 26.

13. Dermot Cole, *Fairbanks: A Gold Rush Town that Beat the Odds* (Fairbanks and Seattle: Epicenter Press, 1999), 13–44.

14. Until 1930, Alaska natives comprised more than half of the territory's population. By law, they were not allowed to hold mining claims, although European immigrants could.

15. Clark C. Spence, *The Northern Gold Fleet: Twentieth-Century Gold Dredging in Alaska* (Urbana and Chicago: University of Illinois Press, 1996), 13–98.

16. William M. Stoll, *Hunting for Gold in Alaska's Talkeetna Mountains, 1897–1951, with a Background Sketch of Alaska's Great Gold-Lode Camps* (Ligonier, Pa.: privately printed, 1997).

17. *Anchorage Daily News*, March 17, 1999.

18. William Bronson, *The Last Grand Adventure: The Story of the Klondike Gold Rush and the Opening of Alaska* (New York: McGraw-Hill Book Company, 1977), 1–3.

19. Alaska's non-native population increased from 4,289 in 1890 to 30,450 in 1900, equaling and then surpassing the native population as measles and influenza epidemics took larger tolls among the native people.

In the Realm of the Senses: The Aesthetic Dimension of Gold

William L. Bischoff

Dr. Bischoff holds a B.A. in history from Stanford University and a Ph.D. in history from Harvard University and has been an associate curator for the American Numismatic Society and curator of numismatics at the Newark Museum in Newark, New Jersey. He has organized several major exhibits of rare coins and edited Coinage of the Viceroyalty of El Perú *(New York: American Numismatic Society, 1989). He is the author of* The Currency of Africa *(San Francisco: Pomegranate, 1995) and has recently edited and published* The Cob Coinage of Colombia, 1622-1756 *(New York: Pertinax, 2000). Dr. Bischoff is presently a numismatic consultant and book producer in New York City.*

Introduction

This paper began as a study of gold coinage through the ages. I soon found my theme expanding into a consideration of popular attitudes concerning the precious metal and its special physical qualities—including its suitability for artistic purposes—and to the highlights of its political, religious, economic, and aesthetic history. That is an ambitious theme indeed, and I offer no pretense to an exhaustive or definitive treatment. In the context of a conference on the discovery of gold in this country some two hundred years ago, however, I think my observations can provide a useful perspective on the non-monetary importance of gold throughout history.

Popular culture has long betrayed a singular ambivalence about gold's desirability. The Greeks loved to tell of King Midas, whose lust for gold was sated when a god granted him his wish that everything he touched should turn into the precious metal. That included, in due time, his beloved daughter, who thereby lost the human qualities that made her worth loving.

But one need not go back to ancient Greece for examples of the way gold has often stood as a metaphor for the dangers of materialistic values. William Jennings Bryan captured the idea most memorably in his famous speech warning that "You shall not crucify mankind upon a cross of gold!" That rallying cry of the late nineteenth century reflected the economic hardship of the time: silver miners in the West and farmers and small businessmen in the heartland saw themselves threatened with ruin if the monetary gold standard favored by Wall Street and the industrial barons won out over the interests of humbler folk.

The same issue is addressed in a more roundabout way in Frank Baum's populist book *The Wonderful Wizard of Oz*, which recently marked its centenary. Most will remember the story from the movie, not the book, but perhaps without being aware of the contemporary references:

*Dorothy's location. Kansas was a hotbed of free-silver (inflationary) agitation and a Populist stronghold for Bryan in his presidential campaign.

*THE COWARDLY LION. Some interpreters take this character as an allusion to Bryan. Not that the statesman was cowardly—just that he had a leonine mane of hair.

*THE TIN WOODMAN is a stand-in for the dehumanized, spiritually impoverished factory worker described by Karl Marx and other nineteenth-century social critics.

*THE WICKED WITCH OF THE EAST. She stands for the forces of evil; the East was, after all, the home of Pierpont Morgan and other arch-capitalists and robber barons whose influence made support for the gold standard the touchstone of monetary propriety.

*THE YELLOW BRICK (i.e., paved with gold ingots) ROAD and DOROTHY'S *SILVER* SLIPPERS. In light of the foregoing, no exegesis is necessary here. It is worth noting, however, that whereas in Baum's original story Dorothy's slippers were of silver, when it came time to make the movie, Technicolor had just come in and the producers couldn't resist the greater impression ruby-red slippers would make on the audience.

A number of films picked up on the theme of gold and materialism. Charlie Chaplin's *The Gold Rush* (1925), for all its outrageous wit and sentimentality, shows us men who will stop at nothing in their pursuit of the sudden wealth gold can bring. Readers may have seen Erich von Stroheim's silent movie *Greed* (1924), inspired by Frank Norris's *MacTeague*. Although the film itself was badly damaged by studio editing, it contains compelling images of what the lust for wealth does to some very ordinary people.[1] Finally, there's John Huston's *Treasure of the Sierra Madre* (1948), in which a ne'er-do-well Humphrey Bogart teams up with two other prospectors to get rich quickly by prospecting for gold in Mexico's Sierra Madre.[2] Bogart, of course, scoffs at his partner and would-be mentor Walter Huston's prophetic warning, "I've seen what gold does to men's souls." And as a consequence, he soon goes to pieces before our very eyes.[3]

Such cautionary tales, however, do less than justice to the complexity of popular attitudes about gold. On the one hand, as the epitome of "filthy lucre," it's the Satanic force that leads men—and women—astray, even into crime. But to have a *heart of gold* is admirable—because in another part of our minds we recognize gold as a shorthand term for all that is valuable, alluring, and lasting. And that view was already pervasive when Hesiod lamented the bygone "Golden Age" when Saturn ruled the world, when Aristotle described his ethical standard of the "Golden Mean," or when Jesus of Nazareth put forth what we call the "Golden Rule." What is it, then, that gives this second, undeniably positive, connotation to all things reminding us of that yellow metal? The question calls for a review of gold's physical and aesthetic qualities, then a consideration of how they have manifested themselves in history.

Physical and Aesthetic Properties

Gold, hammered while cold into small objects of adornment, has been found in sites dating from as far back as the late Paleolithic era. Along with copper, gold counts as the earliest metal used by humankind. Yet unlike copper, it had no clearly utilitarian value. Why, then, has it always been prized so highly? Briefly, because of its scarcity, its virtual indestructibility, and its ease of working, exemplified by its malleability and ductility.

Scarcity

Ordinarily no one gets excited about air or water, however essential they are to our very lives, for there is usually plenty of both to go around. But gold is one of the rarest of useful substances. Its average occurrence in the earth's crust is on the order of .005 grams per metric ton. That works out to about one part gold to two billion parts of undesirable other material. For years—in South Africa at least—it has been profitable, despite enormous capital costs, to refine ore many times the bulk of Cheops's pyramid in order to produce the equivalent of a cube nine feet on a side.[4] The resultant tailings that ring Johannesburg make it look like a moonscape with a city in the background. Nearly forty years ago the renowned numismatist C. H. V. Sutherland considered it likely that all of the new gold produced worldwide in the last five hundred years could be contained in a cube fifteen feet per side weighing some fifty tons.[5] There are, of course, other metals still rarer than gold—platinum for one—but they do not exercise the same irresistible appeal. What else is involved?

Indestructibility

Once purified, gold—as innumerable finds of treasure salvaged from shipwrecks attest—is not corroded or even tarnished by immersion in seawater, underground burial, or electrolytic reaction with other metals. It is impervious to most inorganic acids—nitric, sulfuric, whatever. Only acids that produce chlorine and bromine—such as the famous aqua regia (royal water)—can dissolve the king of metals. And that can be useful in freeing it from its host medium and alloyed metals, although smelting temperatures of between 1,555 degrees and 2,500 degrees Celsius are necessary to separate gold from alloyed metals of the platinum group.[6]

Imagine what this virtual indestructibility means. For one thing, it means that the metal can be cast as bars or coins, melted down again (the very word bullion refers to this melting), then recast as rings, inlaid on other metals, or even made into royal crowns. And from that kind of personal, princely, or priestly adornment it can be transformed into coin again, or used in industrial processes—or to fill teeth. If you wear a gold ring, think of the ancient pedigree it might well have. Such durability means that the world's stock of gold has grown steadily for at least five thousand years. The demand for it, however—and its relative scarcity—has never been greater. What else enters into the equation that explains the millennial gold rush?

Aesthetic and Metallurgical Considerations

Gold possesses the metallurgical qualities of malleability and ductility and the aesthetic qualities of color and sheen. The first two can be quantified; the latter two are more subjective. Goldsmiths and their patrons have always been enamored of gold because of its ease of working and suitability for articles of adornment—princely, cultic, or personal. Although the softness of gold makes it most useful when alloyed with some other metal—usually copper—in its pure form it is extremely ductile: one ounce can be drawn out to a fine wire five miles long.[7] It is perfect, as one can imagine, for delicate filigree work.

Gold is also extremely malleable: one troy ounce can be beaten into a sheet of 350 square feet just 1/282,000th of an inch thick, or, rather, thin—about one thousand times thinner than common notepaper.[8] Now, *that* is thin: so much so that light can be seen as a vaguely green color through the unbroken surface. Nothing could be better for plating large objects—such as the statue of Athena made by Phideas in the time of Pericles.[9] When accused by Pericles's enemies of cheating on the amount of gold actually used, Phideas was able to strip the gold sheeting off the statue and weigh it, and not a mina was missing! (Unfortunately, in the final agony of the Peloponnesian War, that gold was stripped off again in 403 B.C. and used as the first and only gold coinage of classical Athens.)[10]

The luster of pure gold seems to glow from within. Moreover, since in pure form it does not tarnish or corrode, it readily evokes intimations of immortality. What could surpass it, therefore, for the adornment of palaces and princes, temples and churches? And, in reciprocal fashion, its association with power and the sacred has added to its allure. In our own age of disenchantment—when power is suspect and the sacred seems anachronistic or worse—in such an age it is hard to appreciate what gold has meant during most of history. This essay is an attempt to redress, in part, that situation.

It is appropriate to adopt a roughly chronological approach to the aesthetic manifestations of gold. After all, the times and places in which it was most readily available coincide largely with the high points of its use for artistic expression. The exception that proves the rule is the period usually called the Middle Ages, when the very scarcity of gold evoked new ways of calling attention to its use, as seen later in this essay.

Preclassical Antiquity

Egypt

Spain and Egypt/Nubia were the most important producers of gold in the ancient world. Egypt got off to a very fast start that lasted from the predynastic period well into Hellenistic and Roman times. So ample was Egypt's supply that at times its gold was traded for silver at a mere 1-to-2 ratio (1-to-2.5 under Menes, the first pharaoh), whereas it fluctuated around 1-to-10 in the rest of the ancient world.[11]

Decorative uses antedate monetary uses by at least three thousand years. The Egyptian hieroglyphic symbol for gold reflects this: it is a string of beads, presumably cold-hammered for necklaces very early on. Cowrie shells appear as items of decoration even today, and a distinctive necklace of gold cowries from about 1900 B.C. shows how the Egyptians strove to improve on nature: they created cowries that were lustrous, perfect in form, and immune to the breakage of real cowries.[12] In order to make such objects, the goldsmiths of Pharaonic times must have mastered the repoussé technique, which implies that they understood the need for repeated annealing and quenching when making any but the simplest objects of gold.

Indeed, technological progress in Egypt was continuous if not always rapid, and the technology of extracting and working gold that was developed there was never lost but simply improved upon in its long history. Writing in the second century B.C., Diodorus Siculus provides a long, detailed description of the complex division of labor involved in "deep mining"—the extraction of gold ore from granite and quartz mines so deep that some of the unfortunate workers never saw the sun. This was, of course, forced labor—and an indication that such mining was a Pharaonic monopoly (even if the Pharaohs were now named Ptolemy) and had been since the fourth millennium B.C. Another important technological invention of the Egyptians was the *cupellation* process, which involved baking the impure ore for several days at very high temperatures in clay vessels: because gold has a higher melting point than the metals with which it is commonly found alloyed, the impurities "migrated" by vaporization into the surrounding clay. Contemporary illustrations reveal how well developed the division of labor in gold refining had become.[13]

The kind of quarrying in deep shafts described by Diodorus was typical of a sixty-mile-wide stretch of rich ore that bordered the Red Sea for two hundred miles on the Egyptian side. Farther south and west, from the first to the fifth cataract of the Nile, lay the rich alluvial deposits of Nubia, where natural hydraulic processes had for millions of years freed gold from its rocky matrix and made it available as dust and nuggets to be gleaned from the banks of the Nile. In fact, the very word *Nubia* derives from the Egyptian word for gold, which was *nub*.[14] Speaking of that source, Nubian gold was probably used about 2300 B.C. to fashion a gold falcon from Hierakonopolis in upper Egypt.[15] The restraint and economy of means evident in that early work shows that the "classic" style was not unique to the Greeks nearly two thousand years later.

It is important to realize that much of Egypt's large stock of gold was regularly taken out of circulation for the ornamentation of Pharaonic and other important graves. In light of those circumstances, as British archaeologist V. Gordon Childe noted long ago, Egypt's numerous and skilled grave robbers performed a useful social function by regularly returning those riches to circulation, after first melting down their loot. The bad news in that regard, however, is that we can scarcely imagine the sophisticated use made of gold for cultic and personal adornment in ancient Egypt. In fact, we would have almost no idea of it except for the survival of one tomb virtually untouched by the customary pillage. The celebrated tomb of Tutankhamen (fourteenth century B.C.), discovered in 1922, was not even one of the grandest of royal burials. It does, however, give us some idea of the lavish but tasteful use of gold by Egyptian goldsmiths of the time.[16] The famous gold mask of "King Tut," adorned with enamel and precious stones, reflects the hieratic, as opposed to the more humanistic, aspect of this era, during which Tutankhamen's father-in-law Amenophis (Akhenaten) attempted to establish a religious reform bordering on monotheism.[17]

Egyptian goldsmithing techniques maintained a high level of competence throughout the country's existence: burial goods dated to near the end of the indigenous rule of Egypt show no sign of decadence.[18] The commercial utility of gold was discovered quite early. Menes, the founder of the First Dynasty of unified upper and lower Egypt (about 3100 B.C.), is known to have issued pure gold ingots of fourteen grams inscribed with his name. The Tell-al-Amarna archives contain correspondence between Mesopotamian rulers and the Pharaohs concerning the exchange of Egyptian gold for products of Sumeria. In one case a local ruler complains that the gold shipped to him was far from being as pure as represented, which at the very least demonstrates the sophistication of Mesopotamian assay techniques.[19]

Mesopotamia

The Fertile Crescent region possessed no native gold supplies, despite being a river-valley civilization like Egypt. What gold the states that developed there did have was obtained in trade from Egypt and from smaller deposits in the Arabian peninsula. Nevertheless, goldsmiths of the region developed a skill in working the metal that rivals that of the Egyptians, as in the case of this king's ceremonial helmet from Ur, made about 2700 B.C.[20]

Troy

The fabled riches of Troy deserve at least a footnote here: they belong to this time period and era, although we lack the kind of reliable documentation we have about other Middle Eastern kingdoms. The treasures recovered by German archaeologist Heinrich Schliemann and his successors reflect a very high standard of goldsmithing talent and good taste on the part of their patrons.[21]

Persia

Persia's conquest of the entire Near East, as well as Egypt, made it by far the richest state of its time. Persia not only gained through conquest the vast gold stocks of Egypt but also continued the Egyptian practice of extracting the metal from Nubia and Arabia. Moreover, Persia conquered, then took control of the wealth and continuing gold production of Lydia, whose King Croesus was a watchword then and now for the riches he derived from alluvial gold produced by the river Pactolus. Persia acquired from Croesus the system of minting coins in pure gold and pure silver, instead of the combination of both metals in the form of *electrum* previously used for coinage. Persia's gold *darics* and silver *sigloi* found use mainly in the western part of the empire, however—the part inhabited or strongly influenced by the Greeks. Both gold and silver coin were major contributors to Persia's strength, whether for the hiring of mercenary troops or for bribing avaricious Greek political leaders.

The Classical Civilizations

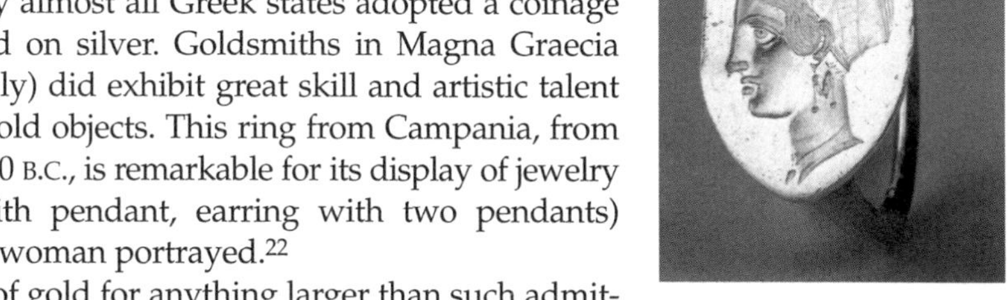

Greece

The Greeks were paupers when it came to gold—which is why almost all Greek states adopted a coinage system based on silver. Goldsmiths in Magna Graecia (southern Italy) did exhibit great skill and artistic talent with small gold objects. This ring from Campania, from about 400–350 B.C., is remarkable for its display of jewelry (necklace with pendant, earring with two pendants) worn by the woman portrayed.[22]

But use of gold for anything larger than such admittedly exquisite personal adornments was the exception, not the rule, until the advent of Macedonian hegemony in the Greek world. To begin with, Philip II of Macedon changed the coinage system radically when he began to exploit aggressively the gold deposits in his Thracian provinces. That change led to an outpouring of gold staters that featured Apollo's laureate head on the obverse and on the reverse a victory-bound biga with Philip's name in the exergue.[23] The flood of gold and silver staters produced under Philip laid the groundwork for the conquest of Persia by Philip's son, Alexander III. By the time Alexander died in 323 B.C., the center of gravity for world gold stocks had shifted profoundly to the Greek world, albeit a Greek world that soon shattered into separate kingdoms founded by Alexander's successors.

Of those successor kingdoms, Egypt, acquired by the Ptolemies, was the richest prize, for its temples were full of gold, and its mines were still producing more than any other area. The founder of the dynasty, Ptolemy I Soter ("Savior"), was the first of Alexander's successors to put his own likeness on the obverse of his coinage, thereby initiating a brilliant tradition of Hellenistic coin portraiture that lasted well into the first three centuries of the Roman Empire.

The kingdom of another successor dynasty, that of the Seleukids, centered on present-day Syria and Iraq. Seleukid rulers thus had access to gold from Arabia, which they controlled, and from the Phoenicians—either from the mines exploited by their colonies in southeastern Spain or from Phoenician traders who obtained the precious stuff from as far east as India and possibly as far south as Somalia or even the Transvaal area of Africa (the Egyptians' fabled land of "Punt"). Some gold undoubtedly also trickled in from Baktria, a former Persian satrapy centered on present-day Afghanistan and ruled in part for some two hundred years by the Greek epigones of Alexander's conquest.

Third to share in the Greek world's growing supply of gold was Macedonia, which had its own mines, as well as access to large quantities of the metal from the Black Sea and Caspian Sea regions, home of the fabled Golden Fleece in Scythia. (It is thought that the myth of the fleece itself may have originated from the practice of using sheepskin to catch flakes of alluvial gold.) Nor should one overlook Thracian settlements such as Pantikapeum as sources. It was thought, by the way, that the non-Greek Thracians themselves had to get their gold from the even wilder northern reaches of Scythia, an area famous for its ferocious warriors and, equally, for its excellent gold artifacts. Good specimens of this work are to be found in the ten bracelets unearthed in a hoard at Kul Oba, one of which (now at the Hermitage in Russia) features two friezes of griffins and stags formed with one die cutout each and soldered onto the plain background of a broad band of pale gold.[24]

Rome

If the Greeks were poor by comparison with the Persians—which explains why they coined almost exclusively in silver—the Romans during their early history had access to neither silver nor gold. Consequently, their early coinage consisted of ingots, then disks, of copper—a most inconvenient means of exchange. Initially, silver coinage was introduced only for trade with the Greek colonies of Magna Graecia in southern Italy. A silver coin for internal as well as external use—the *denarius*—was not introduced until the Second Punic War (218–201 B.C.).

With their hard-fought victory in that war, however, the Romans suddenly found themselves masters of some of the world's richest gold mines—those of southeastern Spain, formerly worked by Carthage. The Romans brought their justly famous administrative and engineering skills to bear on those mines, where they even employed hydraulic methods to separate gold from its rocky matrix; the same skills were soon applied to areas in northwest Spain (presently Asturias and Galicia) that had previously lain virtually untouched. Small wonder that with reserves such as these, Julius Caesar could initiate a regular coinage of very pure gold, establishing a tradition carried on in the West even after the fall of Rome by the Second Rome, Byzantium. These coins, down to the third century at least, are works of art in their own right, embodying the realistic conventions of Roman portraiture in the noblest, most easily worked, most beautiful of metals.

Octavian's conquest of Antony and Cleopatra's Egypt was duly celebrated on coins of bronze, silver, and gold. The first emperor's victories brought so much of the world's gold supply within the Roman Empire that the stuff was now available to even moderately wealthy Roman private citizens for hoarding and ostentation. The Roman poets and essayists of the time harped on the corrupting influence of so much

wealth. And Nero's master of entertainments, C. Petronius Arbiter, brilliantly described the lavish use of gold in his *Satyricon*, particularly the part of it that portrayed the vulgar ostentation of the nouveaux riches invited to "Trimalchio's Dinner." Perhaps these first-century earrings, with their intricate granulation and design, were worn at that same dinner or others like it.[25] When the Emperor Trajan defeated the Dacians in what is now Romania, his booty amounted to enough gold to coin 2.25 million *aureii* and enough silver to make 90 million *denarii*, as well as ownership of the Dacian mines' future production.[26]

One of the most outstanding examples of numismatic use of gold by the empire was in the form of medallions thought to have been awards to particularly important political and military figures. Only recently did this author become aware that a goodly proportion of those special tokens of esteem (or perhaps bribery) were also awarded to potentates outside the empire—rather on the lines of colonial American "Indian Peace Medals." Some of those medallions were outstanding works of art whose appearance is quite distinct from anything issued from the Mint of Rome, much less in the provinces, as demonstrated by a specimen portraying Alexander the Great and excavated at Abukir, Egypt.[27]

The contrast with a gold medallion of Constantine I could hardly be greater, although the date of the Alexander piece has not been determined so far.[28] Constantine's portrait communicates a late-imperial ideological message, for it shows the emperor in an idealized manner, eyes looking upward in prayer as if in touch with the Deity himself. The transformation of the emperor into a lofty, distant, unapproachable Oriental monarch was complete by the early fourth century, when the focus of wealth and power shifted to the "Second Rome" established by Constantine, the hybrid state we call the Byzantine Empire.

The Medieval World

Byzantium

With the collapse of Rome in the West, traditionally dated A.D. 476, coinage—and all other uses of gold—virtually disappeared there, although to the east the Byzantine Empire survived in increasingly parlous condition for another thousand years. During most of that time, Constantinopolis, the *polis* founded by and named for Constantine, survived as the only real city in Europe. (The modern name "Istanbul" reflects this: it is a corruption of the Greek for "to the City.") In the early period—down to about 1000—Byzantine emperors had ample reserves of gold to underwrite their ongoing wars with a coinage based almost entirely on that metal. The hapless Justinian II, in fact, was the first ruler to put a portrait of Christ (heavily stylized, to be sure) on his coins.[29] Besides a major production of gold trade coinage, Byzantium made lavish use of the metal in mosaics such as those found in Ravenna. Not that secular purposes were denied the use of gold, so long as it lasted. A notable example is the gilding of the magnificent classical bronze horses that once stood in the Hippodrome at Constantinople. When the Crusaders conquered the city in 1204, they carried the figures off to Venice, where they presently stand in the plaza of San Marco.

Gold found many intimate sacramental uses as well, as seen in a tempera-and-gold-on-parchment liturgical manuscript of the tenth century that depicts St. Mathew writing his gospel. This kind of artwork would have been seen only by the priesthood, not the laity. The importance of icons in religious life during the Byzantine period can hardly be exaggerated. Among other things, they emphasized the decisive difference between Orthodox Christianity and Muslim teachings. Some icons, such as a tenth-century image of St. Nicholas from a monastery in the Sinai, make an especially effective use of gold. Elements of the Classical tradition survive in the half-length format and "the soft shading of the face and the slight turn of the head," but the overall effect is ineffably, sacredly distant, in large part because of the flatness of the gold-leaf background.[30]

The Byzantines excelled at another specialty of the goldsmith's art: cloisonné. In that technique, ground glass of various colors is set into "cells" (*cloisonnes*) of gold wire, each cell intended for a different color, the whole laid into a backing of sheet gold. This composition is then fired in a kiln until the glass fuses with the gold. As a New York Metropolitan Museum handbook puts it, "The finished effect is like that of stones in a piece of jewelry."[31] It would be hard to overstate the influence of Byzantium on the development of art in Europe during the entire Dark and Middle Ages. Perhaps the best testimony for that assertion is the famous Hungarian Crown of St. Stephen, with cloisonné medallions of Michael VII Dukas, his son Constantine, and King Geza of Hungary.

Islam

Beginning in the seventh century, the spectacular rise of Islamic states in the Near East, then along the North African littoral and into Spain, deprived Byzantium of its virtual monopoly on gold. Huge treasures in gold fell to the jihad in the former Persian Empire and in the Christian churches of what became the Muslim world. In addition, the new Muslim states could now exploit the rich mining areas of Arabia, Egypt, Nubia, and Spain. In fact, Arab traders rivaled their Phoenician predecessors in the quest for gold wherever it might be found. Most astonishing, perhaps, is the thousand-mile caravan route they established across the Sahara from Marrakech in present-day Algeria to Timbuktu in modern Nigeria. Soon the Islamic dinar rivaled, and then surpassed, the Byzantine solidus (often called a *bezant*) and its subdivisions as the international means of exchange.

Medieval Europe

The "Dark" and Early-Middle Ages in Europe did see gold—but it was extremely hard to come by in the insignificant deposits and alluvial streams then accessible. Trade was also a meager source, since Europe had little in the way of luxury goods to offer the Muslims or the Byzantines. Europeans got most of their gold from trade in fur, amber, and slaves (the latter word's resemblance to *Slavs* and *Slavic* is no accident). Much of the gold that was available in the West ended up as ornaments or hoards in the burial mounds of Germanic and Nordic barbarian chieftains. (Islamic coins, for example, have been found as far afield as Iceland.) Artistically, this dearth of gold had consequences visible even today in our modern notion of "jewelry."

Churches, kings, and princes were virtually the only Westerners with access to gold, but even they could not afford to be lavish in its use. Rather than make a king's crown of pure gold, as the Hellenistic Greeks or the Romans did, medieval goldsmiths were constrained to find ways to use gold sparingly, but without any diminution in its capacity to awe and dazzle. One solution was cloisonné work, in which all sorts of bright, sparkling objects—rubies and sapphires but also garnets and even, early on, colored glass—are set in *cloisonnés*, or cells, atop the frugally used gold, whose gleam is more subdued than that of the stones and seems to come from within itself.

The Morgan Library's famous Stavelot Triptych, made for a Benedictine abbey in Belgium around 1150, is a sacral embodiment of the "jewel" idea that originated in the Middle Ages (the word itself dates to the thirteenth century). This devotional shrine, which frames a small piece of wood reputed to come from the True Cross,

uses extravagant enamel and cloisonné work to set off its metal backing, which is merely gilded copper and silver instead of gold.[32]

Another medieval treasure at the Morgan Library that utilizes small amounts of gold to great effect is the Moralized Bible created about 1235 for Blanche of Castile and her son Louis IX. Medieval "illustrated manuscripts" like this one take their generic name from the golden light (*lumen*) that seems to shine from the parchment page's illustrations and lettering. The specimen shown here employs tiny bits of gold leaf carefully overlaid on each other to create a shimmering, lustrous background for the royal sponsors and their clerical artists and scribes.[33]

In a more secular context, the Emperor Friedrich II Hohenstaufen (1197-1250) issued a glorious series of *augustalis* gold coins from his southern Italian mints.[34] The emperor is seen in profile and draped simply—a treatment inspired more by the Roman imperial *aureus* than by the full-face, opulently dressed portraits on the late Roman and Byzantine *solidus*. The *augustalis* was struck on a thicker flan than were contemporary coins in either gold or silver anywhere else, which made it possible to model the ruler's facial features more convincingly than on the wafer-thin coins of Friedrich's allies and rivals. Coins of the same beauty and substantial gold content were not seen again in the West until well into the Renaissance.

Increasing trade in the High Middle Ages—trade with both the Islamic states, approaching their apogee, and with Byzantium, approaching its nadir—made it possible for the emerging states of Europe to begin issuing gold coinage for the first time in the thirteenth century. Florence led off with the florin in 1252, followed in short order by Genoa, Sicily and Naples, France, and, at the end of the century, by Venice with its ducat in 1284.

Gold in the Modern Era

Meanwhile, the search for more and still more gold went on, whether by alchemy—which had originated in Hellenistic times—or, more promisingly, by overseas exploration. There is no need to recount again the rivalry of Portugal and Spain in this search. It bears repeating, however, that in terms of gold, the Spanish truly struck it rich in Latin America. It began in Hispaniola, where the first Spaniards settled, with a reign of terror against the indigenous Tainos. The harsh exploitation of the natives by the Spanish brought them scant reward in the form of confiscated gold trinkets or in new supplies of the metal extracted by forced labor. The oppressive regime of the Tainos' foreign masters soon led to the extinction of the entire native population.

Things were different in Mexico, at least from the Spanish point of view. Cortez and his men used every means to extract an estimated two thousand pounds of gold from the Aztecs as ransom for their emperor Moctezuma (whom they then executed—for treason).[35] Almost all this precious metal was in the form of cultic or princely display and adornment, which was immediately melted down into ingots. The result is that we have no better idea of what Aztec art in gold looked like than we would have of the same art in Egypt except for King Tut's tomb.[36]

Peru experienced similar treatment from Pizarro, where the Inca king Atahualpa paid a "ransom" by filling the room in which he was confined with gold objects fetched from throughout his enormous kingdom. When some thirteen thousand pounds had been piled up in the room, a pretense was found to murder Atahualpa under cover of law. (Pizarro had him baptized before he was killed.) It took Inca goldsmiths a month, working around the clock, to melt down the gold artwork on which they had lavished years to create.[37] In the case of South America, as of Mexico, we have only the most scanty idea of what indigenous gold artwork looked like, except for scattered grave goods discovered much later or from individual articles sent back to Charles V and added to other treasures in Hapsburg possession—articles that can now be found almost exclusively in Vienna.[38] Numismatic museums, on the other hand—not to mention private collections and the shipwreck-salvage vaults of the state of Florida—are amply supplied with specimens of the crude gold "cob" coinage issued in the Spanish New World for the next two hundred years and more.[39]

The Spanish appropriation of native gold was a not-to-be-repeated windfall, of course. And neither the native peoples nor their new masters knew how to extract more than surface deposits or pan for alluvial gold. Silver, however, was a different matter. It was both easier to mine and vastly more plentiful in the Spanish colonies, which led before long to a drastic change in the prevailing ratio of the value of gold to silver. Within a century the ratio changed from about 1-to-10 to the 1-to-16 that many of us grew up thinking of as the "normal" figure. There is insufficient space here to explore the new wealth's result in terms of the "Price Revolution of the Seventeenth Century"—only to note that on the positive side of the ledger, seventeenth- and eighteenth-century hyperinflation made it more attractive to take on debt in order to create larger-scale, more efficient, and, ultimately, steam-powered mechanical production and transport. That process in turn brought the supply of goods relative to money back into balance in the course of the Industrial Revolution. By the late eighteenth and early nineteenth century, newly productive sources of gold in Brazil and Russia merely helped maintain, not disturb, that equilibrium.

Artistically, the early modern age was not one of special distinction in gold work, unless one especially likes *objets de vertú* such as Cellini's lavishly self-praised salt cellar for the French king Francis I or elaborate silver-gilt tableware. But here, if anywhere, is the spot to include the golden nose fashioned for the brilliant Danish astronomer Tycho Brahe after his home-grown organ was mutilated in a duel or brawl (the distinction was not always clear in the sixteenth century).[40]

Although the bulk of this paper has been devoted to art and coinage in the Western tradition, it would be a shame to entirely pass over some of the innovative art forms utilizing gold developed by non-Western cultures. Pride of place in this respect would probably go to the Japanese, despite their nearly total lack of indigenous gold. A beautiful Japanese screen created in the early eighteenth century illustrates this innovative quality.[41] The six panels stretch to a maximum length of

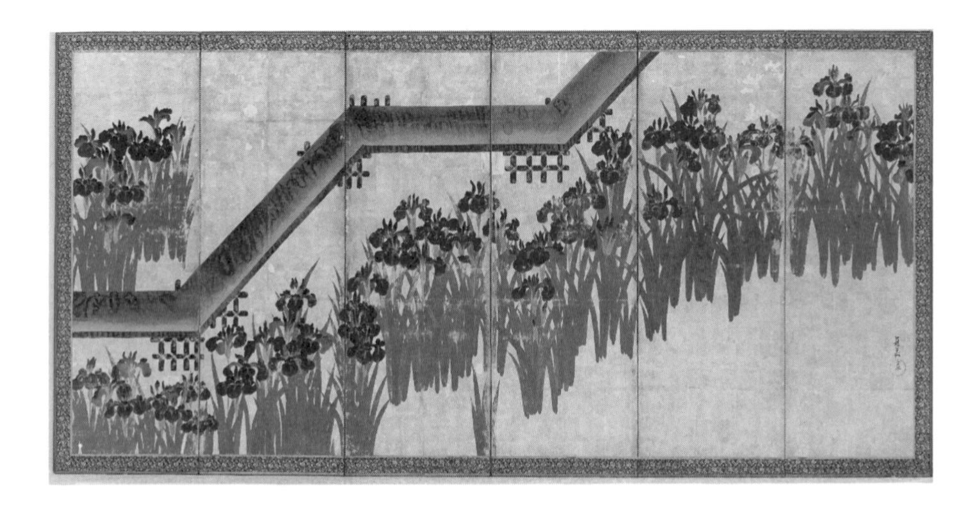

twenty-four feet, providing optimum flexibility in dividing the space in a room. The footbridge-and-irises theme is set off by a background of pure gold leaf; each square overlaps its neighbor slightly. And of course the Japanese were masters at the time-consuming process of combining gold powder with the naturally derived plastic substance of lacquer to create exquisite tableware and items of utilitarian use, at least for the wealthy.

Which brings us to the nineteenth century, the classic age of the gold rush. The United States first discovered gold in its own backyard in the Appalachian region of North Carolina and Georgia in the late eighteenth century. Those strikes provided welcome additions to the scanty American stockpile, but the total yield—estimated at about fifty thousand pounds up to 1850—was insignificant in terms of worldwide production.

The true, archetypal gold rush was, of course, the one that began near Sacramento, California, in 1848. Digging, sluicing, and panning for gold in California was, however, in two respects similar to the Appalachian discoveries and provided an intimation of where future rushes might be expected. California's ore and dust and nuggets were found in areas previously untapped for precious metal, so that gold had built up as surface or near-surface deposits for millions of years before humans began to exploit it. And, just as in the 1830s in the East, government was content to leave gold production entirely to private initiative—a major departure from past practice.

California set the pattern for the rest of the century and down to the present, at least in capitalist economies. That is, deposits are discovered large enough to inspire a mass migration. The easily reached ore and nuggets are soon exhausted, so that profitable extraction can take place only by using complex, capital-intensive methods. Companies are formed to undertake such deep mining. A few sourdoughs linger on as employees in the extractive industry or settle down to another trade, but the rest "pull up stakes" (a gold-rush term) and move on to the next fabled El Dorado, leaving mine tailings and melancholy ghost towns in their wake. This pattern repeated itself in Australia, where some of the largest masses of gold ever found were turned up. Circumstances were similar in Nevada, site of the Comstock Lode, and in Colorado, Montana, and South Dakota. And, finally, there was a last "poorman's gold rush" at the end of the century in the adjacent Canadian and Alaskan fields of the frozen Northwest.

All these "rushes," together with the increasing output of the Main Reef area of Transvaal and the Orange Free State in South Africa, meant that for the first time since the early Roman Empire, and to an even greater extent, ordinary citizens had a personal familiarity with gold. For most, that meant gold coins in denominations from $2.50 to $20. For a considerable segment of the upper-middle and upper classes, it meant that some cities, such as Newark, New Jersey, produced luxury objects previously beyond their means—gold objects of conspicuous consumption, from watch fobs for men to women's purses worked in gold mesh and citrine and worth far more than the money they might contain.[42] At the top of the scale for pure-gold objects owned by prominent citizens, consider this 1868 testimonial medallion awarded to Newark's mayor Thomas Peddie.[43]

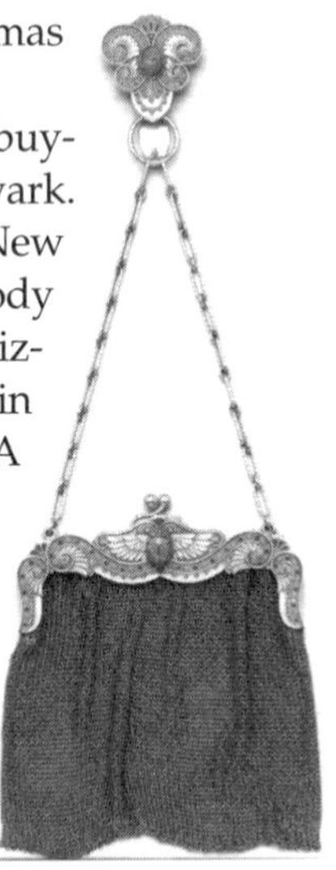

Still famous with today's jewelry buyers, Tiffany's had a workshop in Newark. But its main headquarters was in New York, where its work came to embody the highest aesthetic standards in utilizing gold and other precious materials in public as well as private works of art. A fine example of the latter is a turn-of-the-century mosaic, which stands about 8 1/2 feet tall behind a small fountain. The mosaic illustrates the inventive use of various kinds of glass, backed with gold foil, which makes the whole composition seem to shimmer from within.

No mention has been made thus far of the use of gold to enhance the luxury of priestly vestments, royal robes and hangings, and the attire of just plain rich people eager to demonstrate their status. Fabric molders or wears out more easily than do crowns, reliquaries, or rings, after all, so its material remains are rather scanty. Typically, the first meeting of the rival kings Henry VIII and Francis I took place near Calais in 1520 on the famous "Field of the Cloth of Gold," but to my knowledge none of the cloth survives. An idea of the sumptuous use of gold for private use in more recent times, however, is readily available in the Costume Institute at the Metropolitan Museum of Art in New York, where an eighteenth-century Indian chador interwoven with gold and silver thread testifies to the affluence of an upper-class woman's surroundings.[44] Included in that collection is a woman's velvet robe from Saudia Arabia that demonstrates the continued use of gold thread and sequins (the word itself comes from a small fifteenth-century gold coin) into the twentieth century.[45] This is perhaps the ideal point to close this survey of the sensual appeal of gold, for it was in the East that the techniques for incorporating it into articles of clothing were first developed.

Conclusion

From 1945 until quite recently, something like half the gold stock in the world lay entombed in America's Fort Knox, and large quantities were held under tight security by the central banks of many other nations as well. The beautiful metal that had cost so much to tear from the bowels of the earth, inaccessible in government vaults, was more difficult to get to than it was before it was panned or mined. Gold no longer enjoys legal-tender status for domestic exchange and is no longer coined anywhere for actual circulation. It still plays a residual role as last-resort backing for the fiduciary money of our age, but that function may be ending.

Over the past twenty years the price for an ounce of gold has followed a fairly steady downward trend—from a high of $851 in 1980 to a figure currently (early 2001) hovering around $265. By the end of 1998 only 23 percent of the world's refined gold stocks were held by the various central banks. Both England and Switzerland have recently decided that there are more effective ways to guard their economies against risk than by warehousing huge stocks of gold, and they plan to sell off 13.3 million ounces and 41.8 million ounces respectively in the years to come. Combined with the advances in science and mining technology that make it cheaper than ever to produce gold, the price may well fall even farther.[46] We seem to be in for a glut of gold in the years ahead.

It is unlikely, however, that we will see a new Age of Gold like that of the early Roman Empire or the period from roughly 1850 to 1914. According to the Gold Institute, "World demand for gold is now greater than ever—more than 40 percent above the total annual output of the world's gold mines."[47] But most demand comes from sources quite different from those of the past. Gold is an essential element in the burgeoning computer-driven world of technology: "every microchip in the world depends on gold in order to function efficiently."[48] In medicine, gold instruments are used to clear coronary arteries without risk of infection, microscopic gold pellets are injected to retard prostate cancer, and in some studies gold is placed in minuscule quantities on DNA to study genetic material. Larger, much larger, quantities are used to build jet turbines, to ensure the fail-safe mechanism of automobile airbags, and to protect astronautic spacecraft and astronauts themselves.[49] Thus, gold is now more pervasive in the life of humankind than it has ever been before, and it should become even more important in the future. Remarkable, however, is its general absence from all but incidental aesthetic uses and its disappearance from the monetary system.

Max Weber used the term "*Die Entzauberung der Welt*" to describe the "disenchantment" of the modern world, with its increasing predictability and prosaic pursuit of an everyday life from which all sense of magic and wonder has been banished. His term applies with special force to gold, the wonderful, alluring metal that kept mankind spellbound from the Neolithic age until the recent past. In the face of today's fiduciary, purely notional money, "all that is solid melts into air"—including gold.

The magic of gold has been vanquished by the rational, matter-of-fact nature of the modern world. But that magic can still be experienced, vicariously, without renouncing the practical and technical benefits for which gold is essential. The aim of this essay has been to communicate some of that legendary enchantment.

Notes

1. *The Complete* GREED *of Erich von Stroheim*, comp. and annotated by Herman G. Weinberg (New York: Arno Press, 1972), n.p.

2. Clifford McCarty, *The Complete Films of Humphrey Bogart* (New York: Carrol Publishing Group, 1995), 137.

3. McCarty, *Films of Humphrey Bogart*, 140.

4. C. H. V. Sutherland, *Gold: Its Beauty, Power and Allure* (New York: McGraw-Hill, 1959), 12.

5. Sutherland, *Gold*, 11. The apparent discrepancy with the nine-foot cube cited earlier arises from the "ever more elaborate and scientific mechanical methods of mining"—that is, intensive mining such as that undertaken in South Africa was simply unavailable to earlier generations. Moreover, South Africa's deep mines have been superseded in part by great open-pit mines and heap-leaching of gold, as in Nevada.

6. Heinrich Quiring, *Geschichte des Goldes: Die Goldenen Zeitalter in ihrer kulturellen und wirtschaftlichen Bedeutung* (Stuttgart: Ferdinand Enke Verlag, 1948), 4.

7. *www.goldinstitute.org/facts.html* (the Gold Institute World Wide Web site). Also in Sutherland, *Gold*, 19.

8. Sutherland, *Gold*, 19.

9. An artist's attempt to show how the ancient statue looked is found in Peter Connolly and Hazel Dodge, *The Ancient City: Life in Classical Athens and Rome* (New York: Oxford University Press, 1998), 75.

10. Photograph of coin electrotype reproduced courtesy Ray Gardner.

11. See, for example, Quiring, *Geschichte des Goldes*, 290, for a map that correlates geography, chronology, and yield in the production of gold throughout the world since the rise of Egypt.

12. Henri Stierlin, *The Gold of the Pharaohs* (Paris: Pierre Terrail/Finest S.A., 1997), 94.

13. Quiring, *Geschichte des Goldes*, 38.

14. Quiring, *Geschichte des Goldes*, 1.

15. Stierlin, *The Gold of the Pharaohs*, 83.

16. Upper part of innermost gold coffin of Tutankhamen. Photography by Egyptian Expedition, Metropolitan Museum of Art, New York.

17. Thomas Mann, *Joseph und seine Brüder* (Zürich: Buchclub Ex Libris, n.d.) includes a wealth of factually based speculation about the relationship between Akhenaten and the origins of Jewish monotheism.

18. Stierlin, *The Gold of the Pharaohs*, 191.

19. Sutherland, *Gold*, 53.

20. Illustration copyright the British Museum.

21. Beautiful photographs of this material and other visual treasures are found throughout Vladimir Tolstikov and Mikhail Treister, *The Gold of Troy: Searching for Homer's Fabled City*, trans. Christina Sever and Mila Bonnichsen (New York: Abrams, 1996).

22. Illustration copyright the British Museum.

23. Collection of Frederic G. Withington. The consensus among numismatists is that the reverse type celebrated the victories won by Philip's chariots at the Olympic Games and perhaps elsewhere. See Barclay V. Head's 1911 magisterial reference *Historia Numorum: A Manual of Greek Numismatics* (1911; reprint, New York: Sanford J. Durst, 1983), 222–224.

24. Illustration courtesy the State Hermitage Museum, St. Petersburg, Russia. The crude vigor of Scythian goldsmithing is here given a noteworthy refinement by a Greek hand.

25. Illustration from *The Good Life: Luxury Objects of the Ancient World* (New York: Antiquarian Ltd., n.d. [1999]), 24; reproduced courtesy Antiquarium Ltd. and Kerr Photography.

26. Sutherland, *Gold*, 98. This *aureus* of 114–117 appears in *Men of Rome: The Golden Military Years, Lvcivs Svlla to Severvs Alexander 82 BC–235 AD: The John Whitney Walter Collection* (New York: Stack's, 1990), lot no. 42; reproduced courtesy Stack's/Coin Galleries. Note the deliberate contrast between the aging military leader on the obverse and the youthful, idealized god Helios on the reverse. Trajan's honorary title for his Dacian victories appears at about 4:30 o'clock on the obverse. So prolific was Trajan's issue of gold coinage that even recently exquisite specimens such as this could be purchased quite reasonably: a 1990 auction estimate by Stack's was for $2,500 to $3,500.

27. Information and photograph courtesy Prof. Frank L. Holt, University of Houston.

28. *The Nelson Bunker Hunt Collection, Highly important Greek and Roman Coins* [auction catalog] (New York: Sotheby's, 1990), plate 108, © Sotheby's Inc.

29. Author's collection. Actually, Justinian II ruled at two different times (685–695 and 705–711) and used a different portrait for each reign. The illustration presented here is from the second reign and is thought by some to be closer to Christ's actual appearance than the first-reign image. A recent discussion of this coinage and the relevant numismatic literature

appears in Charles Suter, "Images of Christ on the Coins of Justinian II," *The Celator* 13 (November 1999): 18–21.

30. See the illustrations in Thomas F. Mathews, *Byzantium: From Antiquity to the Renaissance* (New York: Abrams, 1998), 24, 59, 131.

31. *Gold: A Book and a Kit* (New York: Metropolitan Museum of Art, 1998), 18–19.

32. Courtesy Pierpont Morgan Library, New York.

33. Courtesy Pierpont Morgan Library, MS-M 240, fol. 8.

34. Photograph by Sharon M. Suchma; reproduced courtesy the American Numismatic Society, New York. For an in-depth study of these coins, see H. Kowalski, "Die Augustalen Kaiser Friedrichs II," *Schweizerische Numismatische Rundschau* 55 (1976): 77–150.

35. Sutherland, *Gold*, 131–132. This was the total for the artifacts melted down on the spot; the finer pieces were set aside, at least for the time being. And, of course, this was only the first installment, as it were, of the gold to be extracted by plunder and mining in years to come.

36. Sutherland, *Gold*, 132.

37. Sutherland, *Gold*, 134–135.

38. LEFT: The ceremonial gold knife with three birds on top is from Peru and was made sometime between the twelfth and thirteenth centuries. Illustrated is a modern reproduction (gold-plated) by Galeria Cano, from the author's collection. RIGHT: Female and male figurines, roughly contemporary with Pizarro's conquest of the Inca empire, now housed in the Cali, Colombia, Gold Museum. Photograph by Stuart Franklin, in John Hemming, "Pizarro, Conqueror of the Inca," *National Geographic* 181 (February 1992): 105; reproduced courtesy Stuart Franklin.

39. Two escudos of Nuevo Reino de Granada, 1630; eight escudos of the Viceroyalty of Perú, Lima mint, 1724. Coins reproduced courtesy the American Numismatic Society. The relative sizes of the smaller and larger coins are, of course, inaccurate as shown here: a two-escudo piece is roughly the size of one's small fingernail, while the eight-escudo denomination measures somewhere between a U.S. fifty-cent piece and a silver dollar.

40. The author confesses that he found an image of Tycho's face (with the gold prosthesis in place) on the Internet but could not locate the site again. There is a surprisingly large number of sites dedicated to Tycho, his biography, and his achievements. Some of the sites have pictures that have become inaccessible—perhaps that is where the original I saw resides. In any event, one of the most interesting examples of this genre is the on-line exhibition *The Noble Dane: Images of Tycho Brahe*, presented by the Museum of the History of Science at Oxford University (*www.mhs.ox.ac.uk/tycho/.htm*).

41. Photograph courtesy the Metropolitan Museum of Art.

42. Photograph from Ulysses Grant Dietz, "Producing What America Wanted: Jewelry from Newark's Workshops," in *The Glitter and The Gold: Fashioning America's Jewelry*, ed. Ulysses Grant Dietz (Newark: The Newark Museum, 1997), plate 134; reproduced here courtesy The Newark Museum/Art Resource.

43. Photograph from article by Dietz cited above, plate 7; reproduced here courtesy The Newark Museum/Art Resource. The creation of such elaborate gold pieces in Newark was an early result of the nineteenth-century gold discoveries, the rapid growth of railroad transportation to bring the metal back east, and the use of steam-powered machinery for the manufacture of jewelry. The beehive, the brawny arm with hammer, and the plowshares on the shield supported by the two allegorical females all speak the language of thrift, industry, and hard work. (The firm of Durand and Company, which produced the piece, remained prominent in Newark's jewelry industry well into the twentieth century.)

44. Photograph courtesy the Metropolitan Museum of Art.

45. See the photograph of the robe in *Gold: A Book and a Kit*, 44. The volume also depicts the use of gold thread or sequins in fans, shoes, and silk taffeta dresses, the latter reminiscent of attire worn by women in the film *Gone with the Wind*.

46. Jonathan Fuerbringer, "An Icon's Fading Glory. Now, the Gold Rush Is to the Exits," *New York Times*, June 15, 1999.

47. "Going for the Gold—II" at *www.goldinstitute.org/story2.html*, the institute's World Wide Web site. Following the links from *www.goldinstitute.org/about.html* provides a quick overview of the role of gold in today's science, industry, and medicine.

48. "Going for the Gold—II" at *www.goldinstitute.org/story2.html.*

49. "It Could Save Your Life," at *www.goldinstitute.org/story6.html.*

Index

A

B

C

D

E

F

G

L

M

N

O

P

Q

R

S

T

U

V

W

Y

Z

Designed by Sharon Dean, Raleigh, N.C.